ALS

Advances in Life Sciences

Polysialic Acid

From Microbes to Man

Edited by
J. Roth
U. Rutishauser
F.A. Troy II

Birkhäuser Verlag
Basel · Boston · Berlin

Editors' addresses:

Prof. Dr. Dr. Jürgen Roth
Dept. Pathologie der Univ. Zürich
Abteilg. für Zell- und Molekularpathologie
Schmelzbergstr. 12
CH-8091 Zürich
Switzerland

Prof. Dr. Urs Rutishauser
Case Western Reserve University
Department of Genetics
2119 Abington Road
Cleveland, OH 44106
USA

Prof. Dr. Frederick A. Troy II
Dept. of Biological Chemistry
School of Medicine
University of California
Davis, California 95616
USA

A CIP catalogue record for this book is available from the Library of Congress, Washington D.C., USA

Deutsche Bibliothek Cataloging-in-Publication Data

Polysialic acid: from microbes to man / ed. by J. Roth...
– Basel ; Boston ; Berlin : Birkhäuser, 1993
(Advances in life sciences)
ISBN 3-7643-2803-7 (Basel...) Gb.
ISBN 0-8176-2803-7 (Boston) Gb.
NE: Roth, Jürgen [Hrsg.]

Printed from the authors' camera-ready manuscripts on acid-free paper in Germany
ISBN 3-7643-2803-7
ISBN 0-8176-2803-7

TABLE OF CONTENTS

Part 2. Developmental and Cell Biology. Diseased States and Tumors

PREFACE

What do the surface coat of certain neurotrophic bacteria, the vitelline envelope of fish eggs, the glycocalyx of vertebrate neurons, and the membranes of some human cancer cells have in common? One answer, and the subject of this book, is a linear homopolymer of α 2,8-linked *N*-acetylneuraminic acid often referred to as poly-sialic acid. Polysialic acids are a structurally unique group of carbohydrate residues that covalently modify surface glycoconjugates on both prokaryotic and eukaryotic cells, and are used in a remarkably diverse range of important biological contexts. As a result, studies to understand polysialylation of glycoconjugates in sources as distinct as neuroinvasive bacteria and human brain have emerged as an exiting new area of microbiology, developmental biology, neuroscience, and oncology.

The wide distribution of polysialic acid has led to its independent discovery and analysis in fields that only rarely communicate with each other. Thus there has existed the need to bring these parties together in a format that encourages open and substantive exchange of information, ideas and experimental approaches. This need has been met recently in the form of the first International Symposium on Polysialic Acid at Rigi Kaltbad, Switzerland in August, 1992. In addition, it was appreciated that the ability of the conferees to use this information subsequently would be aided by the production of a book including a chapter from each participating laboratory. Being the first comprehensive and detailed review of the subject, the book also serves as a valuable reference for researchers in a number of fields.

The study of polysialic acid can be divided into several categories which also provide a general framework for the chapters of this book. These are the structure of the polymer and its physical/chemical/antigenic properties, a comparison with other sialic acid-containing structures, the availability of specific probes for specific study of polysialic acid on cells and in tissues, the biosynthesis of polysialosyl chains by both prokaryotic and eukaryotic cells, and the elucidation of mechanisms by which polysialic acid influences complex biological systems. The following chapters represent our current knowledge in each of these areas, as well as perspectives for future work. A representative list of the advances described includes:

- Structural studies suggesting that this polyanion can form metastable helices with distinct antigenic properties and a high degree of hydration.
- Evidence that the shared antigenicity between bacterial coats and embryonic cells could result in immunologically-related pathologies.
- The demonstration that a bacteriophage-derived endoneuraminidase can serve as a powerful and specific probe in the analysis of polysialic acid function both in vitro and in vivo.
- The combined use of biochemical and genetic analyses to elucidate the detailed

biosynthetic and transport machinery involved in regulating expression of bacterial polysialic acid capsules.

- A description of the events that link fertilization with the release of polysialic acid from granules to form the vitelline envelope.
- Studies suggesting that differential expression of polysialic acid can regulate cell-cell interactions by altering contact between apposing membranes.
- Correlations between the presence of polysialic acid and a number of morphogenic events during development, and of cell migration in tumor metastasis.
- Evidence that polysialic acid expression on axons is a key component in the specific innervation of targets during formation of the nervous system, and possibly in neural plasticity and regeneration as well.

Even from this brief description, it is clear that a new major player has been added to the list of bioactive polymers. Furthermore it is likely that research over the next few years will uncover additional modes of action that influence an even broader range of biological phenomena. Important advances will need to follow, including further analysis of conformation, identification of the polysialyl transferase(s) and their mode of regulation, and description of the mechanisms by which this polymer is used by cells for such diverse purposes. This state-of-the-art book represents the basis for the future work and thus an invaluable reference in both understanding and designing studies in this exciting new area of glycobiology.

Jürgen Roth Urs Rutishauser Frederick A. Troy II

ACKNOWLEDGEMENTS

The organizers and participants of the

International Symposium on Polysialic Acid

August 2-7, 1992, Rigi-Kaltbad, Switzerland

gratefully acknowledge the generous financial support provided by

Cancer League of the Kanton Zürich

Ciba-Geigy AG, Basel

Department of Pathology, University of Zürich

F. Hoffmann-La Roche AG, Basel

Hochschulstiftung of the Kanton Zürich

Sandoz AG, Basel

Edoardo R.-, Giovanni-, Guiseppe and Chiarina Sassella- Foundation, Zürich

Schweizerische Bankgesellschaft im Auftrage eines Kunden

Swiss Cancer League

Swiss National Science Foundation

LIST OF CONTRIBUTORS

P. Annunziato
Department of Microbiology
School of Medicine and Dentistry
University of Rochester
601 Elmwood Ave., Box 672
Rochester, New York 14642
USA

A. A. Bergwerff
Bijvoet Center
Department of Bio-Organic Chemistry
Utrecht University
P.O. Box 80.075
NL-3508 TB Utrecht
THE NETHERLANDS

D. Bitter-Suermann
Medizinische Hochschule Hannover
Institut für Medizinische Mikrobiologie
Konstanty-Gutschow-Str. 8
W-3000 Hannover 61
GERMANY

E. Bloch-Gallego
URA 1414
Ecole Normale Supérieure
46, rue d'Ulm
F-75005 Paris
FRANCE

G. Boulnois
Department of Microbiology
Medical Sciences Building
University of Leicester
University Road
Leicester LE1 9HN
ENGLAND

J.-R. Brisson
National Research Council Canada
Institute for Biological Sciences
Ottawa, Ontario K1A OR6
CANADA

D. Bronner
Max-Planck-Institut für Immunbiologie
Postfach 1169
D-7800 Freiburg-Zähringen
GERMANY

J.W. Cho
Department of Biological Chemistry
School of Medicine
University of California
Davis, California 95616
USA

J. Diaz-Romero
Instituto de Salud Carlos III
U. Respuesta Inmune/CNBCR
28220 Majadahonda
Madrid
SPAIN

P. Doherty
Department Experimental Pathology UMDS
Guy's Hospital
Medical School
London Bridge
London SE1 9RT
ENGLAND

U. Edwards
Medizinische Hochschule Hannover
Institut für Medizinische Mikrobiologie
Konstanty-Gutschow-Str. 8
W-3000 Hannover 61
GERMANY

D. Figarella-Branger
Université de Luminy
Biologie de la Différenciation Cellulaire
URA CNRS 179
Case 901
F-13288 Marseille, Cedex 9
FRANCE

M. Frosch
Medizinische Hochschule Hannover
Institut für Medizinische Mikrobiologie
Konstanty-Gutschow-Str. 8
W-3000 Hannover 61
GERMANY

H. Higa
Glycomed Inc.
860 Atlantic Avenue
Alameda California 94501
USA

S.H.D. Hulleman
Bijvoet Center
Department of Bio-Organic Chemistry
Utrecht University
P.O. Box 80.075
NL-3508 TB Utrecht
THE NETHERLANDS

Y. Inoue
Department of Biophysics and Biochemistry
Faculty of Science
University of Tokyo
Hongo-7, Bunkyo-ku
Tokyo 113
JAPAN

S. Inoue
Department of Biophysics and Biochemistry
Faculty of Science
University of Tokyo
Hongo-7, Bunkyo-ku
Tokyo 113
JAPAN

F. Ito
Department of Biophysics and Biochemistry
Faculty of Science
University of Tokyo
Hongo-7, Bunkyo-ku
Tokyo 113
JAPAN

M. Iwasaki
Department of Biophysics and Biochemistry
Faculty of Science
University of Tokyo
Hongo-7, Bunkyo-ku
Tokyo 113
JAPAN

K. Jann
Max-Planck-Institut für Immunbiologie
Postfach 1169
W-7800 Freiburg-Zähringen
GERMANY

H. J. Jennings
National Research Council Canada
Institute for Biological Sciences
Ottawa, Ontario K1A OR6
CANADA

A. Joliot
URA 1414
Ecole Normale Supérieure
46, rue d'Ulm
F-75005 Paris
FRANCE

J.P. Kamerling
Bijvoet Center
Department of Bio-Organic Chemistry
Utrecht University
P.O.Box 80.075
NL-3508 TB Utrecht
THE NETHERLANDS

A. Kanamori

Department of Biophysics and Biochemistry
Faculty of Science
University of Tokyo
Hongo-7, Bunkyo-ku
Tokyo 113
JAPAN

K. Kitajima

Department of Biophysics and Biochemistry
Faculty of Science
University of Tokyo
Hongo-7, Bunkyo-ku
Tokyo 113
JAPAN

S. Kitazume

Department of Biophysics and Biochemistry
Faculty of Science
University of Tokyo
Hongo-7, Bunkyo-ku
Tokyo 113
JAPAN

M. Kulakowska

National Research Council Canada
Institute for Biological Sciences
Ottawa, Ontario K1A OR6
CANADA

P.M. Lackie

Abteilung für Zell- und Molekularpathologie
Departement Pathologie
Universität Zürich
Schmelzbergstr. 12
CH-8091 Zürich
SWITZERLAND

L. Landmesser

Department of Physiology and Neurobiology
University of Connecticut
Box U-42, Room TLS 416
75 North Eagleville Road
Storrs, CT 06269-3042
USA

I. Le Roux

URA 1414
Ecole Normale Supérieure
46, rue d'Ulm
F-75005 Paris
FRANCE

G. S. Long

University of Cambridge
Departement of Clinical Biochemistry
Addenbrookes Hospital
Hills Road
Cambridge, CB2 2QR
ENGLAND

J.P. Luzio	University of Cambridge Department of Clinical Biochemistry Addenbrookes Hospital Hills Road Cambridge, CB2 2QR ENGLAND
F. Michon	National Research Council Canada Institute for Biological Sciences Ottawa, Ontario K1A OR6 CANADA
S. Olive	Université de Luminy Biologie de la Différenciation Cellulaire URA CNRS 179 Case 901 F-13288 Marseille, Cedex 9 FRANCE
I.M. Outschoorn	Instituto de Salud Carlos III U. Respuesta Inmune/CNBCR 28220 Majadahonda Madrid SPAIN
M.S. Pavelka	Department of Microbiology School of Medicine and Dentistry University of Rochester 601 Elmwood Ave., Box 672 Rochester, New York 14642 USA
C. Pazzani	Department of Microbiology Medical Sciences Building University of Leicester University Road Leicester LE1 9HN ENGLAND
R.P. Pigeon	Department of Microbiology School of Medicine and Dentistry University of Rochester 601 Elmwood Ave., Box 672 Rochester, New York 14642 USA
A. Prochiantz	URA 1414 Ecole normale Superieure 46 rue d'Ulm F-75005 Paris FRANCE

C. M. Regan

University College
Department of Pharmacology
Belfield, Dublin 4
IRELAND

G. Reuter

Biochemisches Institut
Christian-Albrechts-Universität
Olshausenstr. 40
W-2300 Kiel
GERMANY

I. Roberts

Department of Microbiology
Medical Sciences Building
University of Leicester
University Road
Leicester LE1 9HN
ENGLAND

H.K. Rösner

Institut für Zoologie
Universität Hohenheim
Institut 220
Postfach 700562
W-7000 Stuttgart 70
GERMANY

J. Roth

Abteilung für Zell- und Molekularpathologie
Departement Pathologie
Universität Zürich
Schmelzbergstr. 12
CH-8091 Zürich
SWITZERLAND

G. Rougon

Université de Luminy
Biologie de la Différenciation Cellulaire
URA CNRS 179
Case 901
F-13288 Marseille, Cedex 9
FRANCE

U. Rutishauser

Case Western Reserve University
School of Medicine
Department of Genetics
2119 Abington Road
Cleveland, OH 44106
USA

R. Schauer

Biochemisches Institut
Christian-Albrechts-Universität
Olshausenstr. 40
W-2300 Kiel
GERMANY

P. Scheidegger	Abteilung für Zell- und Molekularpathologie Departement Pathologie Universität Zürich Schmelzbergstr. 12 CH-8091 Zürich SWITZERLAND
L. Shaw	Biochemisches Institut Christian-Albrechts-Universität Olshausenstr. 40 W-2300 Kiel GERMANY
R. P. Silver	Department of Microbiology School of Medicine and Dentistry University of Rochester 601 Elmwood Ave., Box 672 Rochester, New York 14642 USA
A. Smith	Department of Microbiology Medical Sciences Building University of Leicester University Road Leicester LE1 9HN ENGLAND
S.M. Steenbergen	Department of Veterinary Pathobiology University of Illinois College of Veterinary Medicine 2001 So. Lincoln Ave. Urbana, IL 61801 USA
P.W. Taylor	University of Cambridge Department of Clinical Biochemistry Addenbrookes Hospital Hills Road Cambridge, CB2 2QR ENGLAND
T. Terada	Department of Biophysics and Biochemistry Faculty of Science University of Tokyo Hongo-7, Bunkyo-ku Tokyo 113 JAPAN

F. A. Troy
Department of Biological Chemistry
School of Medicine
University of California
Davis, California 95616
USA

W. F. Vann
Laboratory of Bacterial Polysaccharides
Center for Biologies Evaluation and Research
8800 Rockville Pike
Bethesda, MD 20892
USA

A. Varki
University of California
UCSD Cancer Center
Department of Medicine, 0063
9500 Gilman Drive
La Jolla, California 92093-0063
USA

E. R. Vimr
Department of Veterinary Pathobiology
University of Illinois
College of Veterinary Medicine
2001 So. Lincoln Ave.
Urbana, IL 61801
USA

J.F.G. Vliegenhart
Bijvoet Center
Department of Bio-Organic Chemistry
Utrecht University
P.O. Box 80.075
NL-3508 TB Utrecht
THE NETHERLANDS

M. Volovitch
URA 1414
Ecole Normale Supérieure
46, rue d'Ulm
F-75005 Paris
FRANCE

F.S. Walsh
Department Experimental Pathology UMDS
Guy's Hospital
Medical School
London Bridge
London SE1 9RT
ENGLAND

T. Warner
Genentech, Inc.
460 Point San Bruno Boulevard
South San Francisco, CA 94080
USA

L.F. Wright
Department of Microbiology
School of Medicine and Dentistry
University of Rochester
601 Elmwood Ave., Box 672
Rochester, New York 14642
USA

D.E. Wunder
Department of Microbiology
School of Medicine and Dentistry
University of Rochester
601 Elmwood Ave., Box 672
Rochester, New York 14642
USA

R. Yamasaki
Department of Laboratory Medicine
University of California
VA Medical Center 113A
4140 Clement Street
San Francisco, CA 94121
USA

J. Ye
Department of Biological Chemistry
School of Medicine
University of California
Davis, California 95616
USA

G. Zapata
Laboratory of Bacterial Polysaccharides
Center for Biologies Evaluation and Research
8800 Rockville Pike
Bethesda, MD 20892
USA

C. Zuber
Abteilung für Zell- und Molekularpathologie
Departement Pathologie
Universität Zürich
Schmelzbergstr. 12
CH-8091 Zürich
SWITZERLAND

PART 1.

STRUCTURE, IMMUNOLOGY, METABOLISM AND GENETICS OF POLYSIALIC ACID

Polysialic Acid
J. Roth, U. Rutishauser and F. A. Troy II (eds.)

CONFORMATIONS OF GROUP B AND C POLYSACCHARIDES OF NEISSERIA MENINGITIDIS AND THEIR EPITOPE EXPRESSION

Ryohei Yamasaki

Center for Immunochemistry and Department of Laboratory Medicine, University of California, San Francisco, VAMC 113A, San Francisco, CA 94121

SUMMARY: Group B and C capsular polysaccharides of Neisseria meningitidis are both homopolymers of N-acetylneuraminic acid (sialic acid). Sialic acid is α2,8-linked in the B polymer whereas it is α2,9-linked in the C polymer. The α2,8-linked polymer is a poor immunogen, and children under two years of age do not respond to the α2,9-linked polymer although it induces high titers of bactericidal antibodies in adults. The poor immunogenicity of the α2,8-linked polymer and the tolerance to the α2,9-linked polymer has been postulated to be due to their structural similarity to human tissues. Recent immunochemical data indicate that the α2,8-linked polymer has conformational epitopes in the molecule and that epitope expression of the α2,9-linked polymer is complex. We found that a cross-reactive epitope exists between the α2,8- and α2,9-linked polysialic acids. We also determined that both α2,8- and α2,9-linked polysialic acids of N. meningitidis adopt helical structures in solution. Complex epitope expression of α2,8- and α2,9-linked polysialic acids could be due to complex secondary structures of the two polymers.

INTRODUCTION

Meningococcal disease is a major cause of morbidity and mortality among children and adults throughout the world (Peltola, 1983). Of the nine distinct serogroups of meningococci, groups B and C account for approximately 70 to 80 % of the endemic disease in the United States. However, effective polysaccharide vaccines against group B meningococci have not been available, and infants and children under two years of age do not respond to group C polysaccharide (Gold, 1975). The development of better polysaccharide vaccines against meningococcal meningitis has been a challenge to us. This paper describes the recent immunochemical and structural studies of group B and C polysaccharides and discuss how epitope expression of the polymers is associated with their conformations. These studies may also provide insight into experimental approaches on how to overcome the cross-reactivity between the bacterial antigens and human tissues.

THE PRIMARY STRUCTURES OF GROUP B AND C POLYSACCHARIDES

Group B and C polysaccharides of Neisseria meningitidis are both homopolymers of sialic acid (N-acetylneuraminic acid) (Figure 1). Group B polysaccharide is an α2,8-linked polysialic acid (Bhattacharjee, 1975). A polymer of the same primary structure is produced by Escherichia coli and is known as the K1 antigen (Kasper, 1973). E. coli also produces the polymer of which hydroxyl groups are O-acetylated, whereas N. meningitidis do not. Group C polysaccharide is an α2,9-linked polysialic acid (Bhattacharjee, 1975). Some of the hydroxyl groups of the C-polymer are O-acetylated, and it is termed as O-acetyl positive (OAc+). The polymer without O-acetyl groups is designated O-acetyl negative (OAc-).

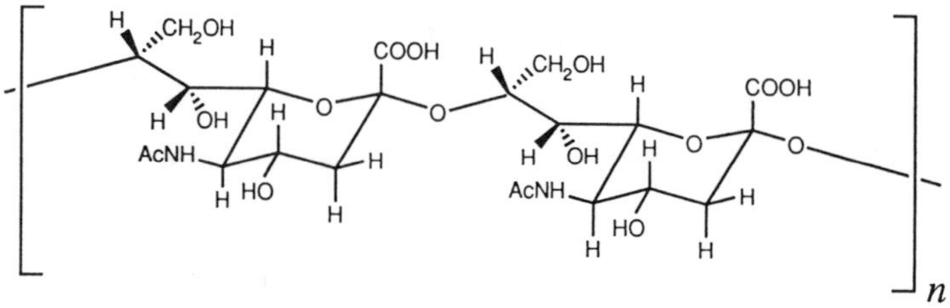

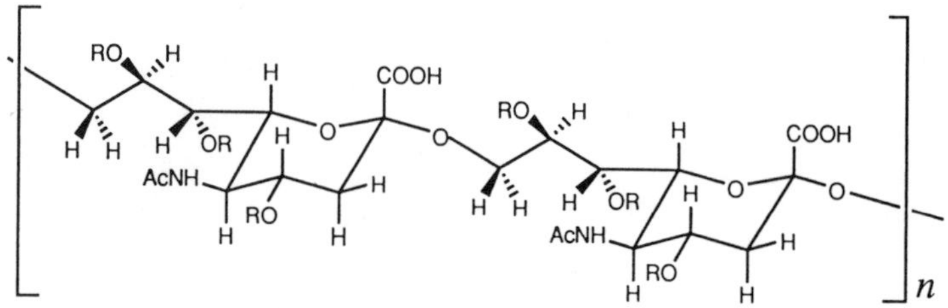

R= H or Ac

Figure 1. A: the primary structure of group B polysaccharide, the repeating disaccharide, α2,8-linked N-acetylneuraminic acid, is shown; B: the primary structure of group C polysaccharide, the repeating disaccharide, α2,9-linked N-acetylneuraminic acid, is shown.

IMMUNOCHEMICAL PROPERTIES OF THE POLYMERS

Although the structural difference between the B and C polysaccharides is primarily due to their difference in glycosidic linkage, the immunological properties of theses polysaccharides are different. The C polysaccharide induces high titers of bactericidal antibodies in adults (Beuvery, 1982; Vodopija, 1983), and the OAc- polymer produces protective antibodies against OAc+ group C meningococci. In contrast, the B polysaccharide is not immunogenic by itself (Wyle, 1972).

The poor immunogenicity of α2,8-linked polysialic acid of N. meningitidis and E. coli K1 has been postulated to be due to its structural similarity to human glycolipids and glycoproteins present in neural and extra neural tissues (Finne et al., 1983, 1987; Bitter-Suermann and Roth, 1987; Roth et al., 1987). The immunologic cross-reactivity between the polysaccharide and neonatal brain tissues from humans and rats could be confirmed by using IgG monoclonal antibody MAb 735. A somewhat similar situation exists for the α2,9-linked polysialic acid. An α2,9-linked sialic acid dimer is expressed on human glycoconjugates during infancy but is lost by 2 years of age (Fukuda, 1985). This might explain why infants less than two years do not respond to the C polysaccharide (Gold 1975).

COMPLEX EPITOPE EXPRESSION OF THE POLYSACCHARIDES

Recent immunochemical analysis indicates that epitope expression of group B and C polysaccharides is not simple. The conformations of α2,8-linked polysialic acid are critical for their epitope expression (Jennings, 1985; Kabat, 1986; Häyrinen, 1989) (Table 1). The binding of anti-B horse polyclonal serum (horse 46) is not inhibited by small oligomers such as a dimer and trimer (Jennings, 1985). A mouse monoclonal antibody specific for the B polymer also binds only larger oligosaccharides, and the critical chain length for the binding has been reported to be 10 (Häyrinen, 1989). These data indicated that the α2,8-linked polysialic acid has a conformational epitope.

Table 1. The supporting data for the presence of conformational epitope in an α2,8-linked polysialic acid

The inhibition of the binding of antibodies requires larger oligosaccharides
Horse polyclonal serum (horse 46) (Jennings et al., 1978)
Human IgM monoclonal antibody (Kabat et al., 1986)
Large oligosaccharides are necessary for the antibody binding
Mouse IgG 2a monoclonal antibody mAb 735 (Häyrinnen et al., 1989)

In addition to the requirement of larger size for the antibody binding and the inhibition of the binding, other immunochemical data support that the conformations of the α2,8-linked polysialic acid is related to its epitope expression (Table 2). Kabat et al. found that a human MAb specific for meningococcal group B and the K1 polymer binds to denatured DNA and polynucleotides such as poly A and G (Kabat et al., 1986). However, the horse 46 serum, raised against group B meningococci, bound to neither denatured DNA nor the polynucleotides. Similarly, the mAb 735 has been shown to have an exclusive reactivity with homopolymers of α2,8-linked polysialic acid since it binds not to other forms of polysialic acids nor to denatured DNA or polynucleotides (Husmann et al., 1990). These data indicate that epitope expression of the α2,8-linked polysialic acid is not simple and that the polymer has more than one epitope within the molecule.

Table 2. Specificities of anti-α2,8 polysialic acid antibodies (Kabat 1986).

	α2,8 polymer	α2,9 polymer	denatured DNA	polynucleotides
Human IgM MAb	+	-	+	+
Horse 46 serum	+	not known	-	-
Mouse IgG 2a MAb 735	+	-	-	-

Epitope expression of group C polysaccharides is not simple either. Rubinstein and Stein obtained 15 different MAbs by immunizing adult mice with OAc+ group C meningococci and found that the specificities of the MAbs are diverse (Rubinstein, 1988). Of the 15 MAbs, seven MAbs bound only the OAc+ polysaccharide. Five MAbs bound better to OAc- polymer than to the OAc one, and these five MAbs also bound to the K92 polysaccharide in which α2,8- and α2,9-linkages alternate. Their results emphasized that epitope expression of the α2,9-polysialic acid is complex.

CROSS-REACTIVE EPITOPE BETWEEN GROUP B AND C POLYSACCHARIDES

We investigated epitope expression of the B (6275) and C [35E (OAc+) and MC19 (OAc-)] polysaccharides by using the anti-B horse (horse 46) and rabbit polyclonal sera as well as their absorbed sera with the above polysaccharides. The horse IgM binds to the homologous (6275) and heterologous C (35E and MC19) polysaccharides, and this heterologous binding was abolished when the horse serum was absorbed with each of the three polysaccharides (Figure 2A). The

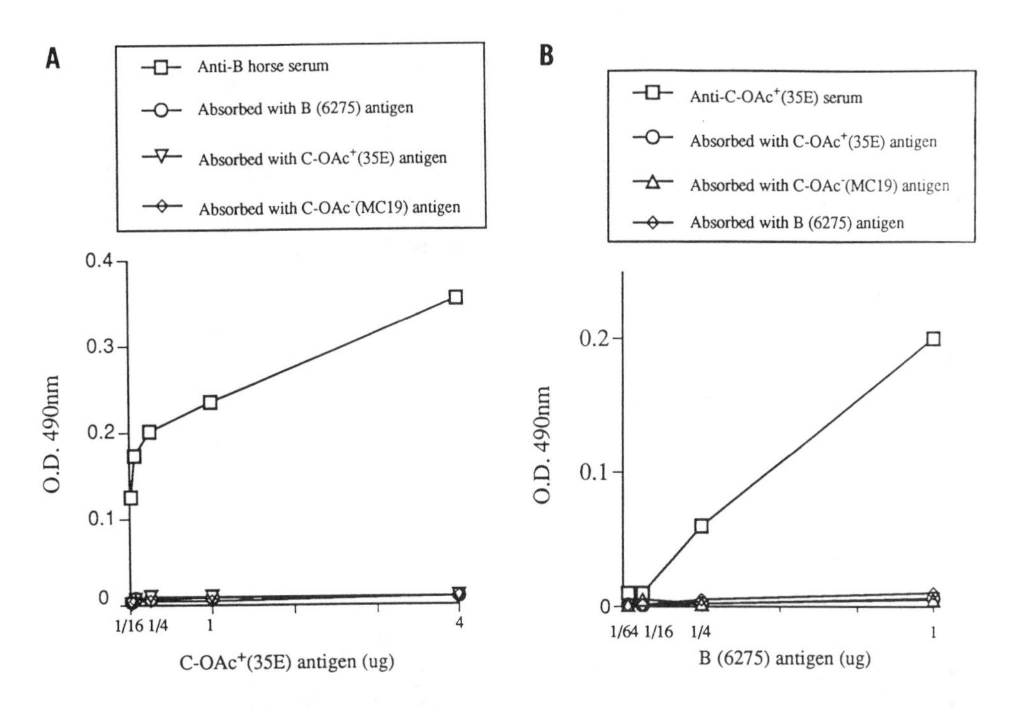

Figure 2: (A) The binding of anti-B horse polyclonal antiserum to group C polysaccharide (OAc^+35E). This binding was abolished when the antiserum was absorbed with group B (6275) or group C (OAc^+35 and OAc^-MC19).The antiserum was absorbed with each antigen at 4°C for 2 h, and the intact antiserum and each absorbed antiserum was diluted to 1/64 for the binding assay. The amounts of the C polymer used were 0.0125, 0.06, 0.25 and 1 μg. The ELISA assay was done as follows: coating the plate with poly-L- lysine; sequential additions of the 35E C (OAc^+) polymer, a blocking solution (0.2% casein in PBS), each antiserum and a secondary antibody. The microtiter plate was incubated for 1 h at room temperature after each addition of the reagent and washed three times with PBS after each step. Goat-anti-horse IgM peroxidase conjugate (Accurate Chemical & Scientific Co., NY) was used as a secondary antibody, and orthophenylenediamine dihydrochloride (Sigma Chemicals, MO) in 100 mM sodium citrate buffer (pH 5.0) was used as substrate (490 nm). ELISA plates were analyzed using a Bio-Rad model 2550 EIA reader (Bio-Rad, Richmond, CA).
(B) The binding of anti-C (OAc^+) polyclonal antiserum to group B polysaccharide (6275). The antiserum was obtained by immunizing rabbits with the 35 E OAc^+ whole organism, and the absorption of the antiserum and the binding assay were done as described above. Each antiserum was diluted to 1/32 for the assay. Goat anti-rabbit IgM (Fc portion) peroxidase conjugate (Accurate Chemical & Scientific Co., NY) was used as secondary antibody.

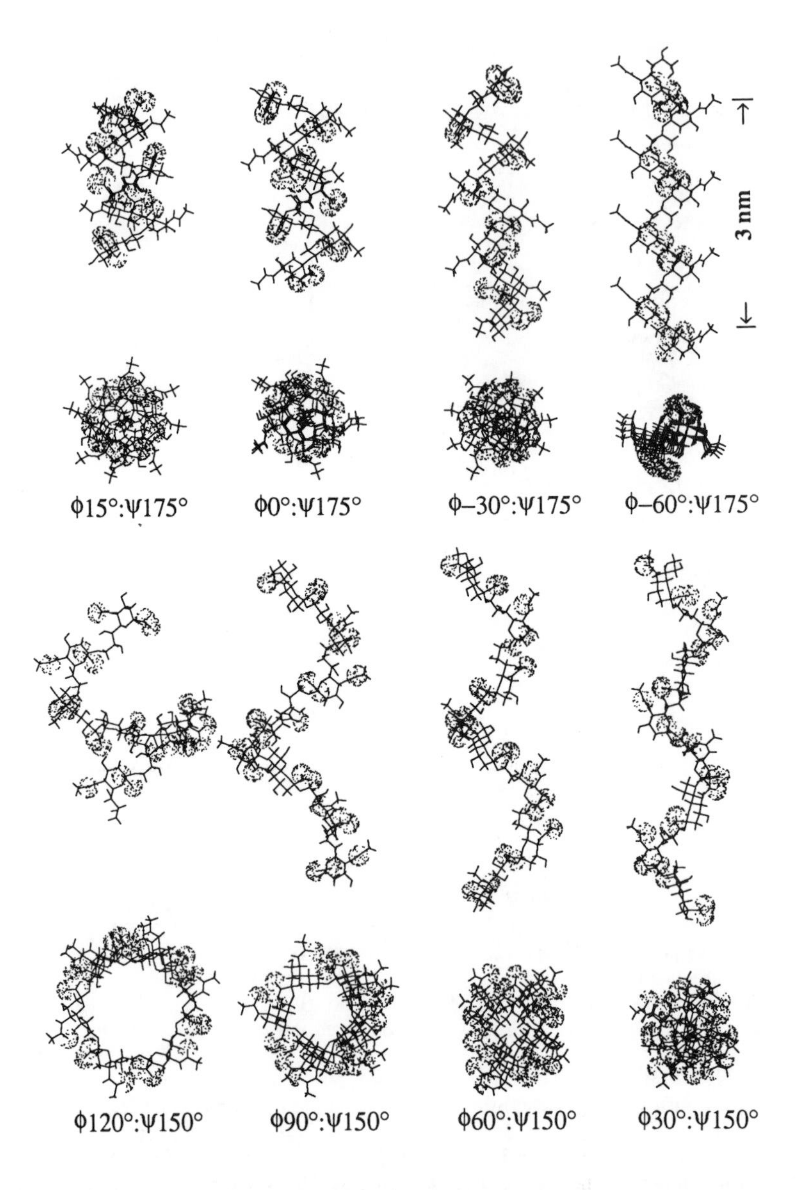

Figure 3: Representatives of possible octamer conformers for 6275 group B and OAc^- group C polysaccharides: A, octamer models for group B polysaccharides (Yamasaki and Bacon, 1991), conformers are expressed by the dihedral angles, ϕ (O6-C2-O8-C8) and ψ (C7-C8-O8-C2); B, octamer models for MC19 OAc^- group C polysaccharide (Yamasaki et al., 1992), conformers are expressed by the dihedral angles, ϕ (O6-C2-O9-C8) and ψ (C8-C9-O9-C2). The glycosidic torsion angle was fixed to 117° (Bock, 1983).

presence of the cross-reactive epitope between the B and C epitope was also confirmed by the anti-B polyclonal (rabbit) antibodies.

As expected from the results with the horse serum, the IgM antibodies of the anti-C [35E (OAc+)] serum bound to the 6275 B polymer (Figure 2B) although its titer to the 6275 B polymer was much lower than those to the OAc- (MC19) and OAc+ (35E) C polymers. No IgG antibody binding to the B polysaccharide was detected with the undiluted serum.

We were not able to confirm the presence of an epitope that is specific for the OAc-positive polysaccharide. However, earlier studies by Rubinstein and Stein (1988) support that the OAc+ 35E polymer has an epitope that is not shared with the OAc- MC19 polymer. Their results together with ours indicate that the 35E polymer has at least three different epitopes in the molecule: (a) an epitope that is common to the intact C (OAc+) and OAc- polymers; (b) an epitope that is governed by the presence of the OAc groups; this epitope; and (c) a conformational epitope that cross-reacts with the B polysaccharide.

The group B polysaccharide of strain 6275 has at least two different epitopes: (a) a conformational epitope present in the B polymer; (b) another conformational epitope that cross-reacts with the C (OAc+ and OAc-) polysaccharides.

CONFORMATIONAL ANALYSIS OF GROUP B AND C POLYSACCHARIDES

The conformations of the two polysaccharides, in particular, the B polymer, have been previously studied. Lindon et al. analyzed the conformations of the group B and C polysaccharides by 13C-NMR and theoretical analysis and indicated the difference in the segmental motion of the C7-C9 exocyclic chain of the two polymers (Lindon et al., 1984). Later on, Michon et al. showed the conformational difference between α2,8-linked oligomers (di- and trimers) and polymers and suggested that the polymer exist in the conformations such that the carboxyl groups of the sialic acid residue within the polymer lined up one side of polymer (Michon et al., 1987). However, the overall conformations of the polymer was not clear.

We have analyzed the conformations of the B polymer by two dimensional nuclear Overhauser effect NMR spectroscopy in conjunction with molecular modeling. Our studies showed that group B polysaccharide adopt helical structures (Figure 3) in solution (Yamasaki 1991). The polymer adopts tighter helical coils than double stranded DNA coils. Small oligomers such as a dimer and a trimer, do not adopt such secondary structures, which explains why the anti-B antibody requires larger oligomers for the binding. We have also analyzed the OAc- group C polysaccharide produced by strain MC19 by two-dimensional NMR and molecular modeling (Yamasaki, 1992). The C polymer can exist in helical conformations in solution. As shown in Figure 3, the helical coils of the OAc- C polysaccharide are more stretched and loose than the B polysaccharide since the C polymer has an extra carbon between the pyranose rings of the repeating unit compared with B polysaccharide (see Figure 1).

Very recently, Brisson et al. (1992) refined their previous conformational work on colominic acid (Michon, 1987) and confirmed our finding that the α2,8-linked polymer adopts helical structures. Some of our findings are different from theirs, which could be partly due to the different experimental conditions. However, the chemical shift difference between the B polymer and colominic acid could be a manifestation of the conformational difference between the two polymers. Although the two polymers have the same primary structure, it does not necessarily mean that their secondary structures are identical. Their molecular weights could be different, and the change in the molecular weight could affect the conformations of the α2,8-linked polymer.

The three-dimensional shapes of the helical structures of the B and C polysaccharide change depending on the glycosidic dihedral angles (see Figure 3) and the polymers may adopt complex tertiary structures through intermolecular hydrogen bonding. This indicates that many epitopes could exist in those polysaccharides, which may explain the heterogeneous epitope expression of the two polymers. The cross-reactive epitope found between the B and C polysaccharides could be due to the presence of an identical structural surface within the two polymers. The presence of the cross-reactive epitope between the two polymers is intriguing in terms of development of effective polysaccharide vaccines against the group B meningococci or E. coli K1. If this cross-reactive epitope is not shared with human tissues and if it is immunogenic, we may be able to use such an epitope as an effective vaccine to overcome the obstacle, the structural similarity between the polymer and human glycoconjugates.

ACKNOWLEDGMENTS

Special thanks go to Dr. J. McL. Griffiss for his encouragement and support. We thank Anne Lucas, Kevin P. Quinn and Aileen Chen for their technical assistance. We also thank Dr. Janice. Kim for reading this manuscript and for her editorial comments. Earlier part of this work was supported by a grant from World Health Organization and partially by a NIH grant (AI 22998). This work was also supported by the Veterans Administration.

REFERENCES

Beuvery, E. C., Leussink, B. Delft, R. W. V. Tiesjema, R. H. and Nagel, J. (1982) Infect. Immun. 37; 579-585.

Bitter-Suermann, D. and Roth, J. (1987) Immunol. Res. 6; 225-237.

Bock, K., (1983) Pure Appl. Chem. 55; 605-622.

Brisson, J., Baumann, H., Imberty A., Pérez, S. and Jennings, H. J. (1992) Biochemistry 31; 4996-5004.

Finne, J., Bitter-Suermann, D., Goridis, C. and Finne, U. (1987) J. Immunol. 138; 4402-4407.

Finne, J., Leinonen, M. and Mäkelä, P. H., (1983) Lancet 2; 235.

Finne, J. and Mäkelä, P. H., (1985) J. Biol. Chem. 1985; 1265-1270.

Fukuda, M. N., Dell, A., Oates, J. E. and Fukuda, M., (1985) J. Biol. Chem. 260; 6623-6631.

Glode, M. P., Lewin, E. B. Sutton, A. Le, C. T. Gotschlich, E. C. and Robbins, J. B. (1979) J. Infect. Dis. 139; 52-59.

Gold, R. M., Lepow, M. L. Goldschneider, I. Draper, T. L. and Gotschlich, E. C. (1975) J. Clin. Invest. 56; 1536-1547.

Häyrinen, J., Bitter-Suermann, D. and Finne, J. (1989) Mol. Immunol. 26; 523-529.

Husmann, M., Roth, J., Kabat, E.A., Weisgerber, C., Frosch, M. and Bitter-Suermann, D. (1990) J. Histochem. Cytochem. 38; 209-215.

Jennings, H. J., Roy, R. and Michon, F. (1985) J. Immunol. 134; 2651-2657.

Kabat, E. A., Nickerson, K. G. Liao, J., Grossbard, L. Osserman, E. F. Glickman, E. Chess, L. Robbins, J. B. Schneerson, R. and Yang, Y. (1986) J. Exp. Med. 164; 642-654.

Kasper, D. L., Winkelhake, J., Zollinger, W. D., Brandt, B. L. and Artenstein, M. S. (1973) J. Immunol. 110;262-268

Lindon, J. C., Vinter, J. G., Lifley M. R. and Mreno, C. (1984) Carbohydr. Res. 133; 59-74.

Michon, F., Brisson, J. and Jennings, H. J. (1987) Biochemistry 26; 8399-8405.

Peltola, H., (1983) Rev. Infect. Dis. 5; 71-91.

Roth, J., Taatjes, D. J., Bitter-Suermann, D. and Finne, F. (1987) Proc. Natl. Acad. Sci. USA 84; 1969-1973.

Vodopija, I., Baklaic, Z. Hauser, P. Roelants, P. Andre, F. E. and Safary, A. (1983) Infect. Immun. 42; 599-604.

Wyle, F. A., Artenstein, M. S. Brandt, B. L. Tramont, D. L. Kasper, D. L. Altieri, P. L. Berman, S. L. and Lowenthal, J. P. (1972) J. Infect. Dis. 126; 514-522.

Yamasaki, R. and Bacon, B. E., (1991) Biochemistry 30; 851-857.

Yamasaki, R., Bacon, B. E. and Kerwood, D. E. (1992) 203rd American Chemical Society Meeting Carbo 0010.

Polysialic Acid
J. Roth, U. Rutishauser and F. A. Troy II (eds.)

INFLUENCE OF BACTERIAL POLYSIALIC CAPSULES ON HOST DEFENSE: MASQUERADE AND MIMIKRY

Dieter Bitter-Suermann

Institut für Medizinische Mikrobiologie, Medizinische Hochschule Hannover
Konstanty-Gutschow-Str. 8, 3000 Hannover 61, Germany

SUMMARY: Pathogenic bacteria, especially the most invasive meningococci, group B streptococci and K1 E.coli which cause sepsis and meningitis have adopted effective strategies to mislead the hosts defense mechanisms, namely the innate immune system (unspecific immune response) and the adaptive immune system (specific immune response). Physiologically sialic acid plays a pivotal role in the balance of activating and inactivating signals for the regulation of complement component C3. By means of a high affinity for the complement regulatory protein factor H, a necessary co-factor for the enzymatic degradation of C3b to C3bi by factor I (C3b inactivator) sialic acid rich surfaces favour the down-regulation of C3b, avoid the amplifying role of C3 in the cytolytic, opsonic and proinflammatory activities of the alternative and classical complement pathway. In addition, sialic acid is a major carbohydrate constituent of glycopeptides and -lipids of host cell surfaces and therefore as a self-determinant guarantees immunotolerance. Therefore these sialic and polysialic acid rich capsules of the bacteria mentioned above are examples of the most intriguing aspects of genetically stable bacterial pathogenicity maintained by masquerade and mimikry. These principles will be discussed with regard to interference with the complement system and the humoral immune response.

Confronted with the immune system of the vertebrates pathogenic microorganisms and parasites have adopted an amazing number of strategies for survival in or on the host, the so-called evasion and escape mechanisms which separately or combined are used for successful multiplication (Hall, 1991; Borst, 1991; Cooper, 1991).

Pathogenic bacteria of medical importance with regard to sepsis and meningitis (see Table I) are especially equipped with polysaccharide capsules and the most virulent bacteria are equipped with sialic acid containing capsules.

Table I

Pathogenic Bacteria with Capsules of Medical Importance

Strept. agalactiae	(B-Streptococci, 4 capsule types, neonatal sepsis and meningitis
Strept. pneumoniae	(Pneumococci, ~ 85 capsule types), pneumonia, meningitis, sepsis
Neisseria meningitidis	(Meningococci, group A, B, C, W135, Y etc.), sepsis, meningitis
E.coli	(Coli-Bacteria, K1, K5, K92 etc.), sepsis, meningitis, UTI
Haemophilus influenzae	(Haemophilus, capsule types a-f, 80% Hib), meningitis, epiglottitis

This extremely successful common principle coupled with increased invasive properties enables these bacteria to mislead or neutralize the defense system of their hosts to overcome innate immunity as well as the adaptive immune response (Fig. 1).

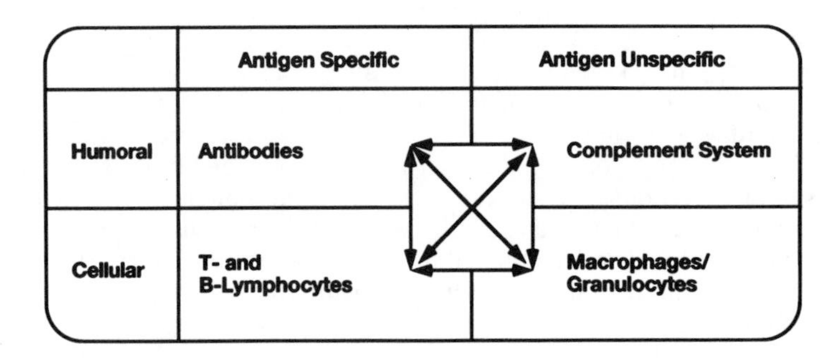

Figure 1. The antigen-nonspecific part of the immune system (complement and macrophages) is innate, inflexible, and cannot be boostered by subsequent antigen contacts. The antigen-specific route comprises antibodies and B- and T-cells, is adaptive, flexible and susceptible to booster effects. The cellular and humoral compartments are highly interconnected and interdependent.

The basic antibacterial defense in the preimmune phase of host-parasite interaction is composed of the phagocytes and the alternative pathway of complement activation and both act in concert. The early immune phase includes IgM antibodies. But, because of missing Fc receptors for IgM again complement (now the classical pathway) is essential for opsonization/phagocytosis and bacteriolysis (Fig. 2).

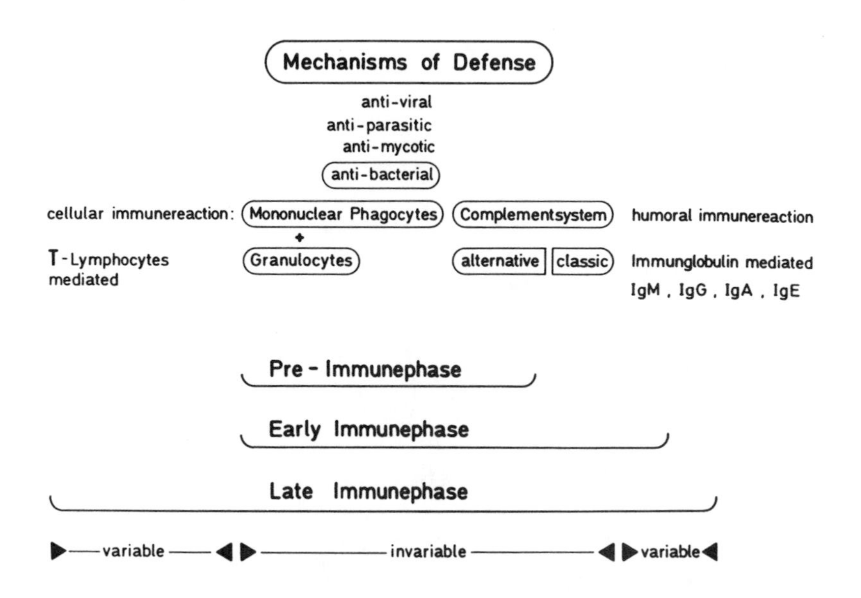

Within this group of virulent encapsulated gramnegative and grampositive bacteria three species, Neisseria meningitidis, Streptococcus agalactiae and E.coli use sialic acid as a common structural element for their capsules. Sialic acid exists either as heteropolymer combined with other carbohydrates (streptococci B and some of the rare meningococcal capsule types) or as mixed α-2,8, α-2,9 linked sialic acid in case of E.coli K92 or as homopolymers of α-2,9-polysialic acid (PSA) in case of meningococci C and finally as α-2,8 PSA with E.coli K1 and meningococci B.

All these bacteria are perfect in disguise: By occupying a principle which the host physiologically is using to downregulate the autoaggressive potential of activation products of the unspecific part of the immune system, the professional phagocytes and the humoral complement system. This host principle of avoiding autoaggression is represented in part by the anionic sialic acid which prevents the central event of alternative pathway activation of the complement system, the C3b mediated amplification and coating of cell surfaces for subsequent phagocytosis (opsonophagocytosis) (Fig. 3).

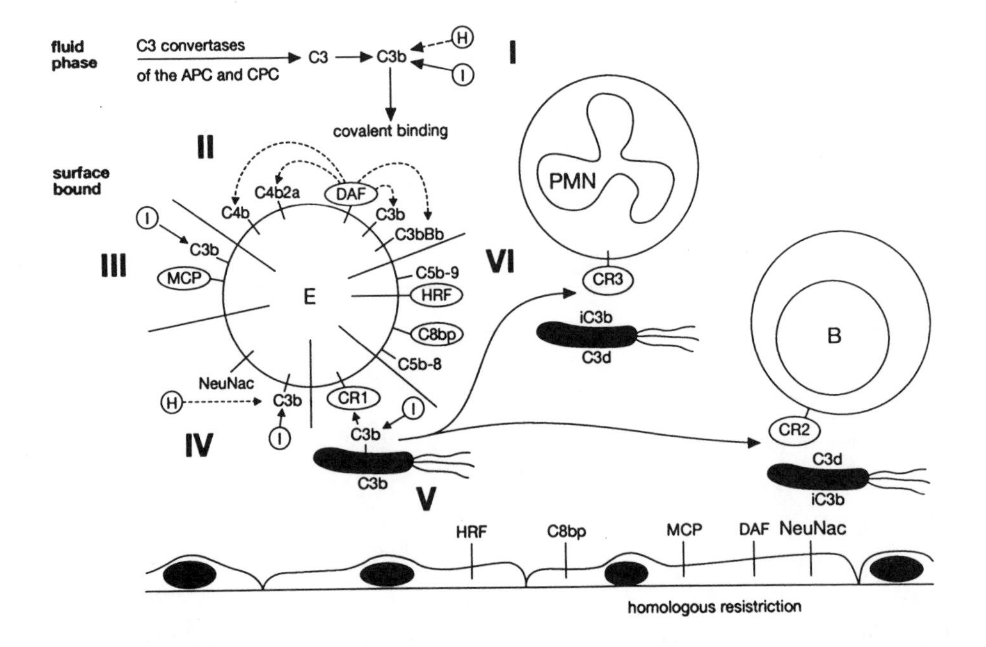

Figure 3. Mechanims of regulation of complement mediated opsonization and cytolysis on autologous cells. E = erythrocytes, PMN = polymorphonuclear leucocytes, B = B-lymphocytes

The central event of a permanent and spontaneous or induced activation of the complement cascade in circulation and at tissue sites is the generation of nascent C3b endowed with an internal reactive thioester bond for covalent binding to hydroxyl- and amino-groups (Levine, 1989). In the fluid phase (Fig. 3, I) this C3b is rapidly attacked by factor H, a member of the RCA gene cluster (regulators of complement activation) (Vik, 1989) and subsequently degraded by the enzyme factor I to C3bi and further to C3d. But, once covalently bound to cell surfaces (Fig. 3, II) C3b is protected from degradation by a rougly tenfold decrease in affinity of factor H for C3b. In such a protected environment the C3b-mediated amplification loop of the alternative pathway of complement activation consisting of C3b, factor B, factor D and properdin should start immediately, leading to a rapid coating of the surrounding surface with additional C3b (opsonization), to the generation of inflammatory split products of C3 and C5 (the anaphylatoxins), and to the initiation of the cytolytic terminal membrane attack complex of complement (cytolysis). That this "horror autotoxicus" does not exist, that autoaggression by complement is counteracted due to membrane anchored members of the RCA family with a broad tissue distribution, namely DAF (decay accelerating factor), MCP (membrane cofactor protein) (Lublin, 1989), and CR1 (complement receptor type 1) is shown in figure 3, II III, V. DAF prevents the assembly of the C3 and C5 convertases of complement pathways by accelerating the decay of C2a and Bb from the convertases. MCP also with a broad tissue distribution and like DAF an intrinsically acting protein behaves in an H-like manner and dissociates factor B from C3b, thereby enabling factor I attack whereas CR1 (Fearon, 1989) reacts with bound C3b on other surfaces (immune complexes, homologous cells, bacterial cells) with the same consequences of factor I-mediated C3b cleavage. In figure 3, IV, cell bound sialic acid now enters the stage and favours factor H binding to this polyanionic surface. This affinity of factor H is not restricted to sialic acid (Fearon, 1978) but true for several other polyanions as DNA, heparin, glycosaminoglycans (Kazatchkine, 1982) and dextran sulfate and even used for affinity purification of factor H (Bitter-Suermann, 1981). By that binding the affinity of factor H for a C3b in the neighborhood is increased tenfold and regains the level of affinity as for fluid phase C3b (Meri, 1990). Finally (Fig. 3, VI) a small group of so-called HRF (homologous restriction factors) regulate the formation and insertion of the membrane attack complex C5b-C8 or C5b-C9 into the lipid bilayer of the membrane in the homologous situation, i.e. complement and cells of the same species (Lachmann, 1991).

C3b-opsonized bacteria are subject to phagocytosis and clearance via adherence to complement receptors (Brown, 1992) CR1 (for C3b) and CR3 (for C3bi). Binding to CR2 (for C3d) on B-cells serves as cofactor signal for the membrane Ig antigen receptor. In conclusion, the content of sialic acid in concert with cell bound RCA molecules like DAF, MCP and CR1 and regulatory molecules for the terminal membrane attack complex of complement like the homologous restriction factors (HRF and CD59) for C8 and C9 control the degree of autoaggression of host erythrocytes, endothelial cells and all self-surfaces exposed to the permanent attack of complement products and phagocytes.

The fact that the alternative pathway of complement activation is essential as the first barrier against invasive bacteria is underlined by experiments of nature, the genetic complement deficiencies (Fig. 4).

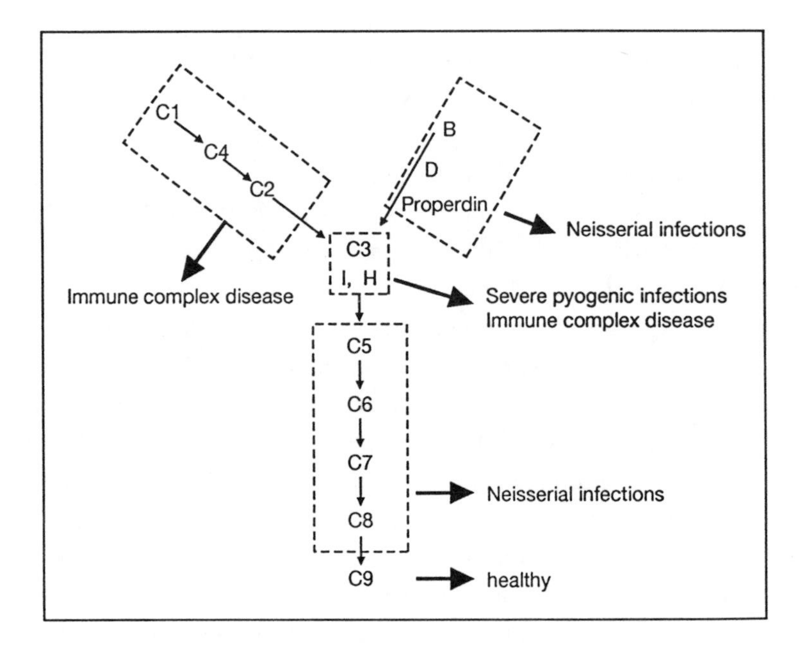

Figure 4. Genetic complement deficiencies and predominant clinical consequences

Properdin deficiencies (properdin is necessary for stabilization of membrane bound C3b and is therefore counteracting factor H) and defects of the terminal complement components in contrast to defiencies within the classical pathway are prone to Neisseria infections especially meningococcal meningitis (Ross, 1984; Morgan, 1991). This means that alternative pathway induced bacteriolytic complement attack governs antibacterial defense (classical pathway defects can be compensated by the alternative pathway but not vice versa) and that bacterial sialic acid induces a shift towards dominance of factor H - dependent down-regulation of C3b and hence evasion of meningococci.
Having in mind the above described role of sialic acid on eucaryotic cell surfaces the mentioned principle of masquerade by sialic acid capsules will be discussed in more detail now from the bacterial point of view (Fig. 5).

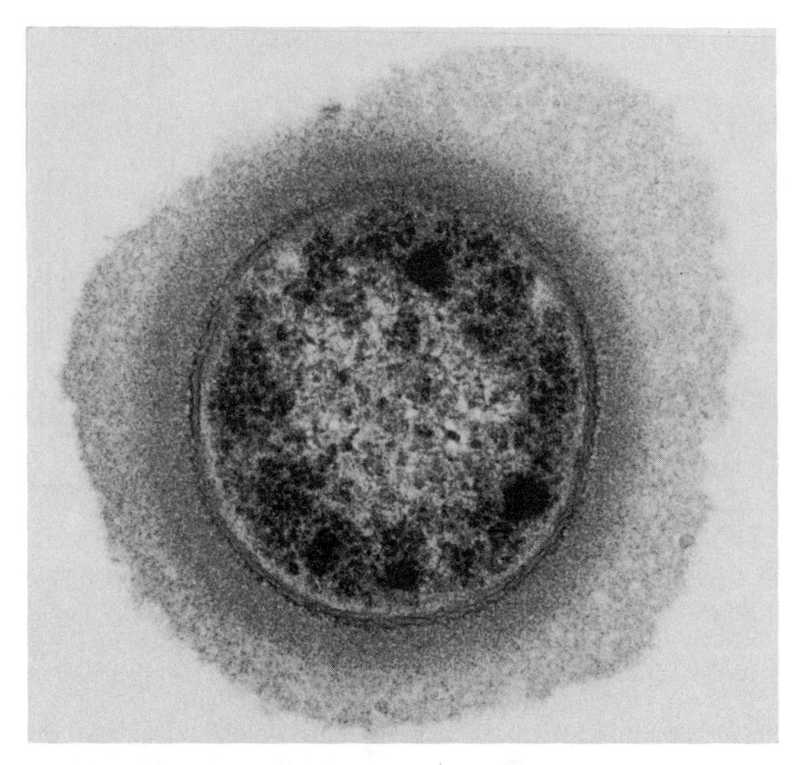

Figure 5. EM-cross section of E.coli K1. The capsule is stabilized with anti-PSA α-2,8 monoclonal antibody 735.

The impressive PSA capsule of E.coli K1 as shown in Fig. 5 interferes with the humoral-cellular tandem of complement and phagocytes in the following ways: In isogenic pairs of K1 E.coli and their capsule deficient mutants (Fig. 6) the alternative pathway activation in C4 deficient guinea pig serum (1:4) and uptake of ^{125}J guinea pig C3 is reduced to 10-50 % by the capsule. Analogous results have been optained by quantitative immunofluorescent assay (Jarvis, 1987).

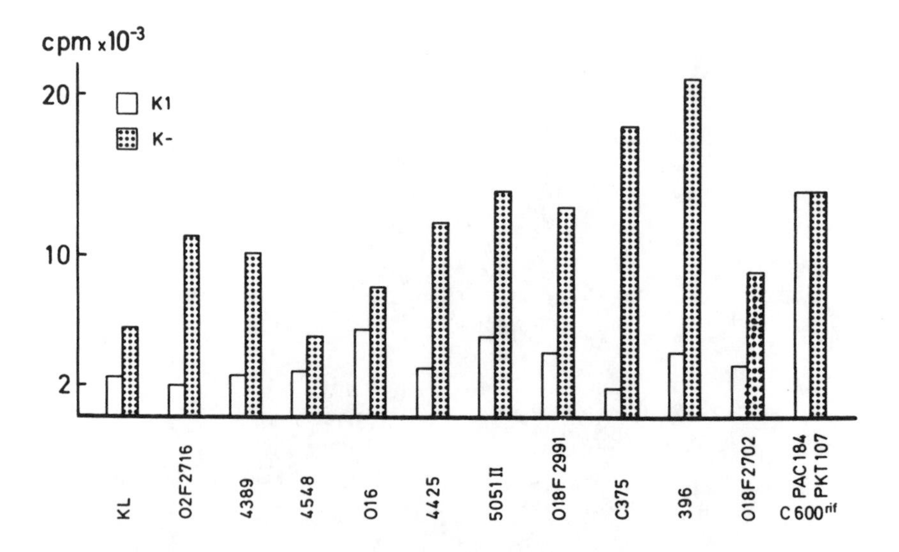

Figure 6. Specific uptake of ^{125}J C3 by 11 pairs of K1/K^{-} strains of E.coli in a concentration of 1 x 10^{8}. C 600^{rif} PAC 184 is a lab. strain of E.coli K12 and pkT 107 a traTp - serum resistance plasmid containing variant.

Therefore it remained to be clarified whether this residual C3b uptake was capsule- or outer membrane localized. Treatment of C3 exposed (C4 deficient guinea pig serum) K1 E.coli C375 (see Fig. 6) with endosialidase did not reduce the cell associated C3b and this is substantiated by immunogold EM with monoclonal antibody against guinea pig C3 prior to endo N treatment and finally labelling with gold - anti mouse immunoglobulin (Fig. 7).

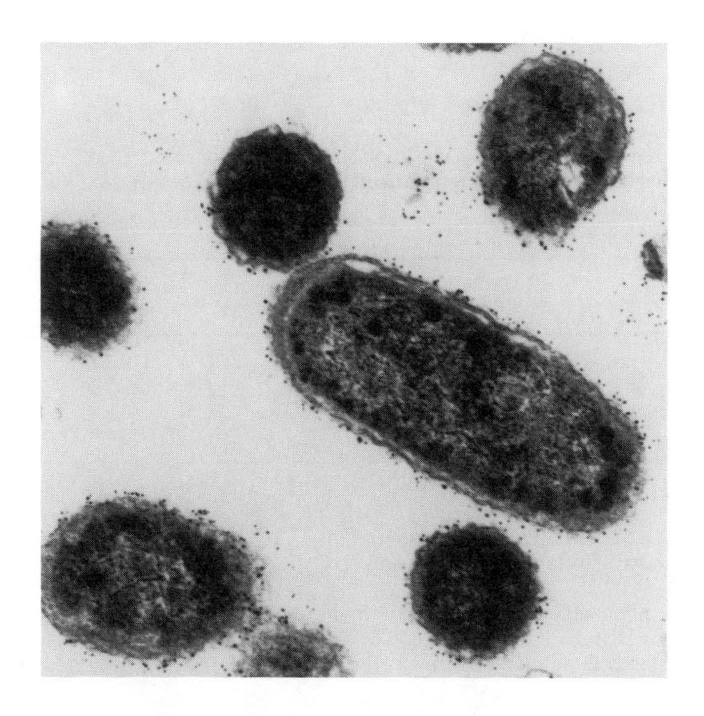

Figure 7. Immunogold EM of desialylated serum treated E.coli K1 (see text).

With regard to phagocytosis such subcapsular C3b is inefficient because of steric hindrance by the capsule for attachment to C3b receptors.

The second important biological effect is bacteriolysis and with PSA capsules both the reduced amount of surface bound C3b and its rapid degradation to C3bi by factor H assisted by sialic acid, limit the lytic effect. Numerous publications have established this sialic acid dependent serum resistance (Pluschke, 1983; Timmis,1985). Not only capsular sialic acid but also LOS-bound sialic acid in Neisseria act by this way (Wetzler, 1992). Table II and Fig. 8 summarize such lytic experiments with pairs of K1 and K- and underline the multifactorial nature of serum resistance. This is also documented (see Table II) by testing phagocytosis and lethality and again this also has been reported by several other groups (Wessels, 1989). Although it is generally accepted that surface bound polyanions, especially sialic acid, regulate and catalyze C3b turnover via factors H and I some unresolved discrepancies still exist with regard to exact molecular mechanisms of PSA interaction with factor H in contrast to terminal sialic acid residues (Meri, 1990; Michalek, 1988; Joiner, 1988).

Table II

Influence of the K1 capsule and the ColV plasmid on virulence of E.coli and on its susceptibility to phagocytosis and killing by serum [a] (taken from Timmis, 1985)

Property [b]	LD_{50} [c]	% Phago-cytosis [d]	% Survival in serum [e]
$K1^{+}$ $ColV^{+}$	10^{5}	26	170
$K1^{+}$ $ColV^{-}$	10^{5}	36	120
$K1^{-}$ $ColV^{+}$	$> 10^{9}$	90	30
$K1^{-}$ $ColV^{-}$	$> 10^{9}$	90	5

a DeLUCA and CABELLO (unpublished data). The Col V codes for traTp serum resistance factor (Bitter-Suermann, 1984).

b The $K1^{+}$ phenotype and carriage of the ColV plasmid of E.coli FC001 (018ab,ac:K1:H7) and its $K1^{-}$ and ColV- derivatives obtained as K1-specific phage-resistant mutants and colicin V nonproducing derivatives obtained by SDS curing, respectively

c The 50% lethal dose obtained by intraperitoneal injection of Swiss-Webster adult mice with bacteria plus hog gastric mucin

d Percentage of mouse peritoneal macrophages having phagocytosed the indicated bacteria

e Percentage of bacteria surviving incubation for 90 min in presence of 10% human serum. Values greater than 100% indicate that bacterial multiplication occurred during the incubation period

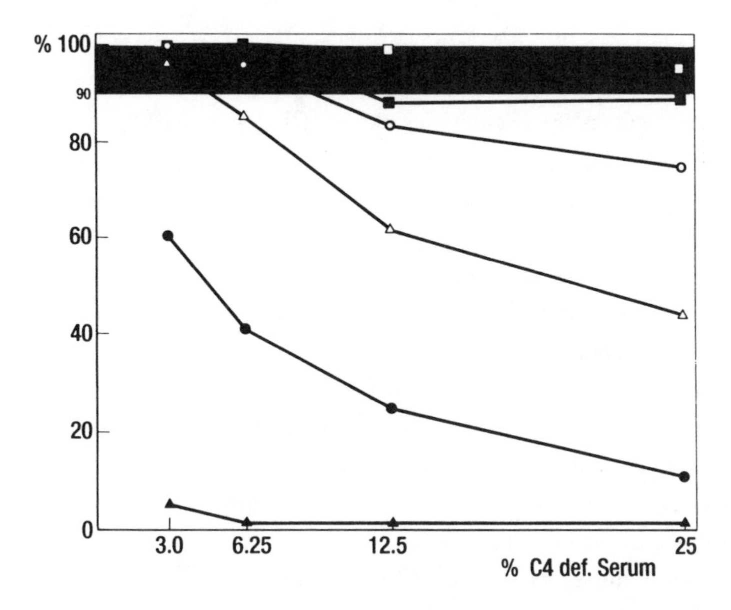

Figure 8. Bactericidal effect of C4 def. guinea pig serum on three different pairs of K1/K⁻ E.coli. The number of surviving bacteria (input of 1 x 10^8 E.coli incubated in 25 % heat inactivated C4 def. guinea pig serum is set as 100 %) is plotted versus different final concentrations of C4 deficient guinea pig serum after 1 h incubation at 37° C. Open symbols are K1 clinical isolates, closed symbols are the respective K⁻ mutants selected as K1-bacteriophage resistant. Note the huge strain dependent variation in serum resistance pointing to the multifactorial nature of serum resistance.

Turning to the second phenomenom, that of mimikry, the interference of PSA capsule with the specific immune response has to be discussed: Delineated from mimesis which is a camouflage painting in animals mimikry is a special state of mimesis meaning that an animal which is protected by toxin production or is uneatable and characterized by a warning outfit is immitated by a harmless and unprotected species with regard to colouring or shape. But, in case of PSA mediated mimikry, we are not dealing with harmless bacteria but with pathogenic species and therefore it is a type of inverse mimikry, that of a wolf in sheep's clothing.

Interference with immune system by antigenic or molecular mimikry is restricted to only the α-2,8 subtype of sialic acid capsules and completes the failure of the complement and phagocytic system in the preimmune phase of the host parasite interaction by a nearly total unresponsiveness of the specific immune system resulting in an absent (secondary IgG) or only marginal (primary IgM) humoral response. Based on a 100% identity with regard to chemical composition and three-dimensional structure of bacterial α-2,8-PSA with the polysaccharide side chains of the embryonal variant of the mammalian neural cell adhesion molecule (e N-CAM) immunotolerance is the basis for this heavily suppressed humoral immune response.

It is a long story to review all attempts and failures to induce antibodies against α-2,8 PSA and to construct vaccines against meningococcal B and K1 E.coli capsules. When we started to produce monoclonals and were unsuccessful after two years we followed an immunological working hypothesis that α-2,8 PSA might be a self-determinant as could be concluded at the same time by the results of Finne et al. (1982). We therefore used the NZB-autoimmune mouse model for immunization. These mice spontaneously produce autoantibodies against numerous self-antigens. Fig. 9 shows the fruitful approach.

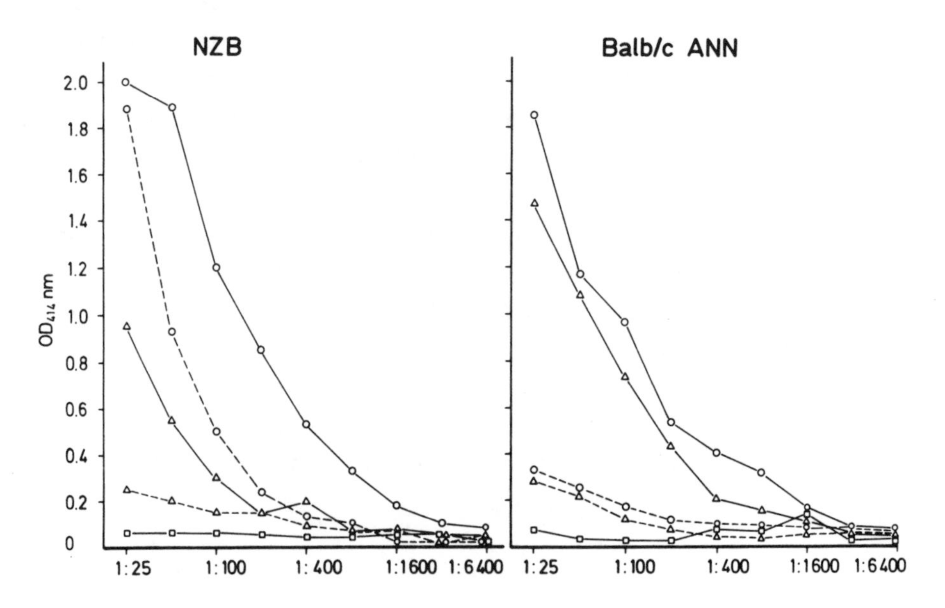

Figure 9. Immune response to meningococal B polysaccharide after immunization of NZB or Balb/c mice with live group B meningococci.

The mean values of antibody titer from 10 mice for each mouse strain are given after 5 days of primary immunization (open triangles) and after five booster injections 5 weeks later (open circles). The preimmune titer (open squares) at respective mouse serum dilutions is negative at baseline. Dotted lines indicate sera treated with 2-mercaptoethanol for reduction of IgM. NZB mice in contrast to Balb/c mice showed a substantial IgG titer after 5 weeks whereas Balb/c mice at both time points only responded with IgM.
In contrast α 2,9 PSA of meningococci C induced an equal and normal immune response in both mice strains. With this approach we successfully generated an IgG 2a monoclonal antibody specific for α 2,8 PSA (Frosch, 1985). This antibody enabled us to underline the validity of the immunotolerance hypothesis. PSA was detected in different embryonic tissues as a developmentally regulated antigen (Roth, 1987; Finne, 1987; Trotter, 1989) and as an oncodevelopmental antigen on different tumor cells (Roth, 1988; Husmann, 1989; Kibbelaar, 1989; Bitter-Suermann, 1987). In addition, we found PSA on embryonic N-CAM in CSF of children without a history of bacterial meningitis (Weisgerber, 1990) and on the surface of NK cells (Husmann, 1989).

Therefore antibodies to α-2,8 PSA should not be induced by vaccination and the naturally occuring IgM antibodies to PSA are either broadly crossreacting with polyanions (Husmann, 1990) or real autoantibodies.

REFERENCES

Bitter-Suermann, D., Burger, R. and Hadding, U. (1981) Eur. J. Immunol. 11: 291-295
Bitter-Suermann, D., Peters, H., Jürs, M., Nehrbass, R., Montenegro, M., Timmis, K.N. (1984) Infect. Immun. 46: 308-313
Bitter-Suermann, D. and Roth, J. (1987) Immunol. Res. 6: 225-237
Borst, P. (1991) Immunoparasitol. 7: 29-33
Brown, E.J. (1992) Infect. Agents Dis. 1: 63-70
Cooper, N.R. (1991) Immunol. Today 12: 327-331
Fearon, D.T. (1978) Proc. Natl. Acad. Sci. USA 75: 1971-1975
Finne, Y. (1982) J. Biol. Chem. 257: 11966-11970
Finne, J., Bitter-Suermann, D., Goridis, C., Finne, U. (1987) J. Immunol. 138: 4402-4407
Frosch, M., Görgen, I., Boulnois, G., Timmis, K. and Bitter-Suermann, D. (1985) Proc. Natl. Acad. Sci. USA 82: 1194-1198
Hall, B.F. and Joiner, K.A. (1991) Immunoparasitol. Today 7: 22-27
Hirsch, R.L., Griffin, D.E. and Winkelstein, J.A. (1983) Proc. Natl. Acad. Sci. USA 80: 548-550
Husmann, M., Görgen, I., Weisgerber, Ch. and Bitter-Suermann, D. (1989) Develop. Biol. 136: 194-200
Husmann, M., Pietsch, T., Fleischer, B., Weisgerber, Ch. and Bitter-Suermann, D. (1989) Eur. J. Immunol. 19: 1761-1763
Husmann, M., Roth, J., Kabat, E.A., Weisgerber, Ch., Frosch, M. and Bitter-Suermann, D. (1990) J. Histochem. Cytochem. 38: 209-215
Jarvis, G.A. and Vedros, N.A. (1987) Infect. Immun. 55: 174-180
Joiner, K.A. (1988) Annu. Rev. Microbiol. 42: 201-230
Kazatchkine, M.D. and Nydegger, U.E. (1982) Prog. Allergy 30: 193-234
Kibbelaar, R.E., Moolenaar, C.E.C., Michalides, R.J.A.M., Bitter-Suermann, D., Addis, B.J. and Mooi, W.J. (1989) J. Pathol. 159: 23-28
Lifely, M.R. and Esdaile, J. (1991) Immunol. 74: 490-496
Meri, S. and Pangburn, M.K. (1990) Proc. Natl. Acad. Sci. USA 87: 3982-3986
Michalek, M.T., Mold, C. and Bremer, E.G. (1988) J. Immunol. 140: 1488-1594
Morgan, B.P. and Walport, M.J. (1991) Immunol. Today 12: 301-306
Pluschke, G., Mayden, J., Achtman, M. and Levine, R.P. (1983) Infect. Immun. 42: 907-913
Ross, S.L. and Densen, P. (1984) Medicine 63: 243-273
Roth, J., Taatjes, D.J., Bitter-Suermann, D., Finne, J. (1987) Proc. Natl. Acad. Sci. USA 84: 1969-1973
Roth, J., Zuber, C., Wagner, P., Taatjes, D.J., Weisgerber, C., Heitz, P., Goridis, C., Bitter-Suermann, D. (1988) Proc. Natl. Acad. Sci. USA 85: 2999-3003
Smiley, M.L. and Friedman, H.M. (1985) J. Virol. 55: 857-861
Timmis, K.N., Boulnois, G.J., Bitter-Suermann, D. and Cabello, F.C. (1985) In: Current Topics in Microbiology and Immunology, Vol. 118, Springer-Verlag, pp. 197-218
Trotter, J., Bitter-Suermann, D., Schachtner, M. (1989) J. Neurosci. Res. 22: 369-383
Vik, D.P., Muñoz-Cánoves, Chaplin, D.D. and Tack, B.F. (1989) In: Current Topics in Microbiology and Immunology, Vol. 153, Springer-Verlag, pp. 147-161
Weisgerber, Ch., Husmann, M., Frosch, M., Rheinheimer, C., Peuckert, W., Görgen, I. and Bitter-Suermann, D. (1990) J. Neurochem. 55: 2063-2071
Wessels, M.R., Rubens, C.E., Benedi, V.-J. and Kasper, D.L. (1989) Proc. Natl. Acad. Sci. USA 86: 8983-8987
Wetzler, L.M., Barry, K., Blake, M.S. and Gotschlich, E.C. (1992) Infect. Immun. 60: 39-43

POLYSIALIC ACID VACCINES AGAINST MENINGITIS CAUSED BY NEISSERIA MENINGITIDIS AND ESCHERICHIA COLI K1

Harold J. Jennings, Jean-Robert Brisson, Malgorzata Kulakowska and Francis Michon

Institute for Biological Sciences
National Research Council of Canada
Ottawa, Canada K1A 0R6

SUMMARY: Although the poor immunogenicity of the GBMP precludes its use as a vaccine agains meningitis caused by group B meningococci and E. coli K1, chemical modification of the polysaccharide remains a viable option to extend the use of the polysaccharide in immunoprophylaxis. This became evident when it was demonstrated that the NPrGBMP-TT conjugate was able to induce in mice high titers of protective bactericidal IgG antibodies. The fact that most of these antibodies were NPrGBMP-specific and not cross-reactive with the GBMP indicated that the NPrGBMP mimics a unique bactericidal epitope on the surface of group B meningococci and E. coli K1. This epitope is GBMP-associated being based on an extended helical epitope similar to that which defines GBMP-specific antibodies. There is evidence to suggest that the epitope is probably intermolecular in nature being composed of the GBMP in association with another more hydrophobic surface molecule. In contrast to the NPrGBMP-TT conjugate, group B meningococci are unable to induce antibodies with exclusive NPrGBMP-specificity. Thus the NPrGBMP is a true synthetic immunogen which has a bacterial antigenic counterpart, but no bacterial immunogenic counterpart.

INTRODUCTION

Meningitis caused by groups B and C Neisseria meningitidis and Escherichia coli K1 remain major world health problems. The capsules of the above bacteria are linear polysialic acids. Both group B meningococci and E.coli K1 have structurally identical capsules which are homopolymers of $\alpha 2 \rightarrow 8$-linked sialic acid while the capsule of group C meningococci is composed of a homopolymer of $\alpha 2 \rightarrow 9$-linked sialic acid (Jennings, 1990).

While the group C meningococcal polysaccharide (GCMP) is immunogenic in humans and constitutes a currently licensed vaccine against meningitis caused by group C meningococci, the poor immunogenicity of the group B meningococcal polysaccharide (GBMP) precludes its use as a vaccine (Jennings, 1990). While the covalent coupling of the GBMP to protein carriers did result in enhanced antibody levels and antibodies of the IgG isotype (Jennings et al., 1986; Devi et al., 1992), these levels were generally low and no bactericidal activity was reported. That molecular mimicry is involved in the poor immunogenicity of the GBMP is supported by the identification of structurally similar $\alpha2\rightarrow8$-linked oligomers of sialic acid in human and animal fetal tissue (Finne et al., 1983). Polysialic chains are developmental antigens and are carried by glycoproteins involved in neural cell adhesion ([E]NCAM) and have also been identified in other tissues including human tumors (Troy, 1992). The extent of human tolerance to the GBMP is explicable in terms of its precise structural homology to the above tissue antigens. These polysialic antigens are long, consisting of up to twelve contiguous sialic acid units (Troy, 1992) and have been identified in lengths in excess of 55 contiguous sialic acid residues in the glycoproteins of human neuroblastoma cells (Troy, 1992).

Despite its poor immunogenicity, antibodies to the GBMP can be produced in special circumstances, and furthermore these antibodies are protective. Examples of the production of polyclonal and many murine monoclonal antibodies of the same specificity have been reported, some of which have been shown to be protective (for references see Jennings, 1989). Recently a human transformed cell line producing protective GBMP-specific antibody has been described (Raff et al., 1988), and a human macroglobulin (IgM^{NOV}) with identical immunological properties has been identified (Kabat et al., 1986). Thus the above experiments indicate that antibody to the GBMP can be produced in animals using an aggressive immunization schedule with whole organisms as the immunogen. However, the best approach to make a vaccine based on the GBMP probably resides in its chemical manipulation.

HELICAL EPITOPE EXHIBITED BY $\alpha2\rightarrow8$-LINKED POLYSIALIC ACID

The presence of a conformational epitope in $\alpha2\rightarrow8$-polysialic acid was initially hypothesized to explain the inability of its oligomer fragments of up to five sialic acid residues to inhibit the binding of the GBMP to an homologous horse antibody (Jennings et al., 1984). Later it was established that a decasaccharide was the minimum sized fragment that was able to bind to GBMP-specific antibodies (Jennings et al., 1985; Finne et al., 1985; Hayrinen et al., 1989). This evidence further strengthened the original hypothesis because only a maximum of six or seven glycose residues are necessary for binding to an antibody in the case of a conventional sequential epitope (Kabat, 1960). In addition, it was determined from NMR studies that the conformation of the linkages of the GBMP and its short oligomer fragments were different (Michon et al., 1987).

The GBMP also shares a common epitope with polynucleotides poly (A) and poly (I) and denatured DNA. The presence of this epitope was revealed by the observation that a human monoclonal macroglobulin (IgM^{NOV}) was precipitated equally well by both the GBMP and poly (A) (Kabat et al., 1986). Because these two biopolymers share no common structural features the cross-reaction was thought to be due to a common epitope composed of a similar spatial arrangement of negative charges. This explanation was supported by the fact that $\alpha 2 \rightarrow 8$-sialooligosaccharides had identical inhibitory properties no matter whether poly (A) or the GBMP were used as binding antigens to IgM^{NOV} (Kabat, 1988). The known propensity of poly (A) to adopt extended helical conformations (Saenger et al., 1975; Yathindra et al., 1976) helped to formulate the hypothesis that the epitope associated with the immune response to the GBMP was in fact helical in nature.

High resolution ^{1}H-NMR studies demonstrated that the GBMP exhibits considerable flexibility in solution because no unique conformer was found to satisfy the NMR constraints and this was also confirmed by potential energy calculations (Brisson et al., 1992). However, analysis of the NMR data indicated that of the four bonds (described by torsion angles, ψ, ϕ, ω_8 and ω_7) linking two contiguous sialic acid residues (Fig. 1), those described by ω_8 and ω_7 had the major preferred conformations shown in Fig. 1.

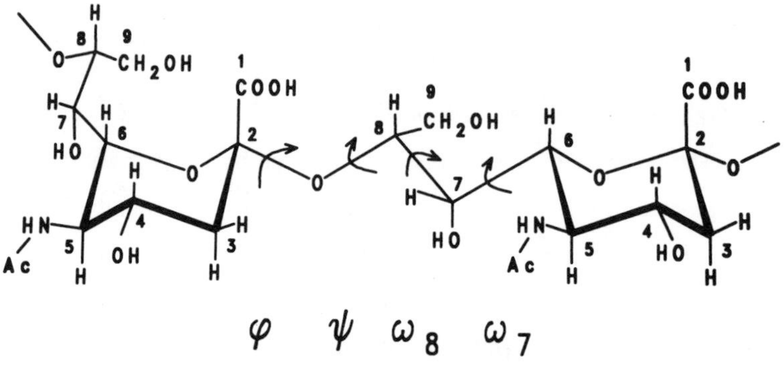

Figure 1. Structure of the disaccharide unit of the GBMP showing the four torsion angles (ψ, ϕ, ω_8, ω_7) which determine the linkage conformation.

Therefore, the origin of the flexibility of the GBMP residues mostly in the bonds described by torsion angles ϕ and ψ. Potential energy calculations carried out by varying these two torsion angles yielded the potential energy contour map shown in Fig. 2. The map is also consistent with the GBMP being a flexible molecule because it reveals the presence of three minimum energy wells A, B and C representing three different energetically favorable conformers.

Thus the topological features of the GBMP can best be described as a random coil. However, the spatial organization of a polymer can be defined by consecutive fragments which have defined helical parameters. These can be referred to as local helices or pseudo helices (Perez and Vergelati, 1985). Thus the three different conformers adopted by the GBMP identified above, propagate three different helices where n=2,4 or 9 (Fig. 2). These helices are shown in Fig. 3 and for comparison the energetically favored n=9 extended helix of poly (A) is shown in the same diagram (Brisson et al., 1992). Thus the immunological properties

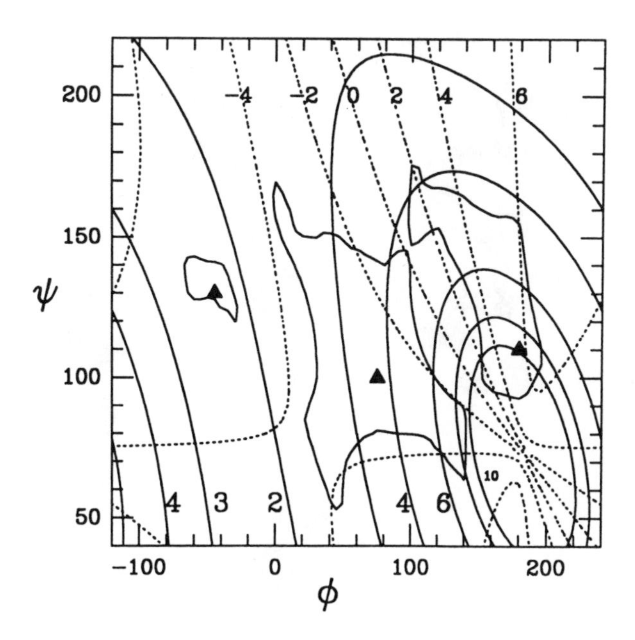

Figure 2. Potential energy contour map for the disaccharide unit of the GBMP with ω_7, ω_8 = (55°, 70°) in which three separate wells are defined (A, B and C from left to right). The helical parameters n (solid lines) and $\underline{h}$ (dotted lines) are also displayed. For poly (A) the energically favoured helices are located in the vicinity of well C.

of $\alpha 2 \rightarrow 8$-polysialic acid can be rationalized in terms of it adopting extended helical forms consisting of eight or more contiguous sialic acid units. Other workers also reported (Yamasaki and Bacon, 1991) that the $\alpha 2 \rightarrow 8$-linked polysialic acid adopts helical structures in solution on the basis of quantitative NMR analyses. However, the results of these analyses did not accommodate an extended helical structure.

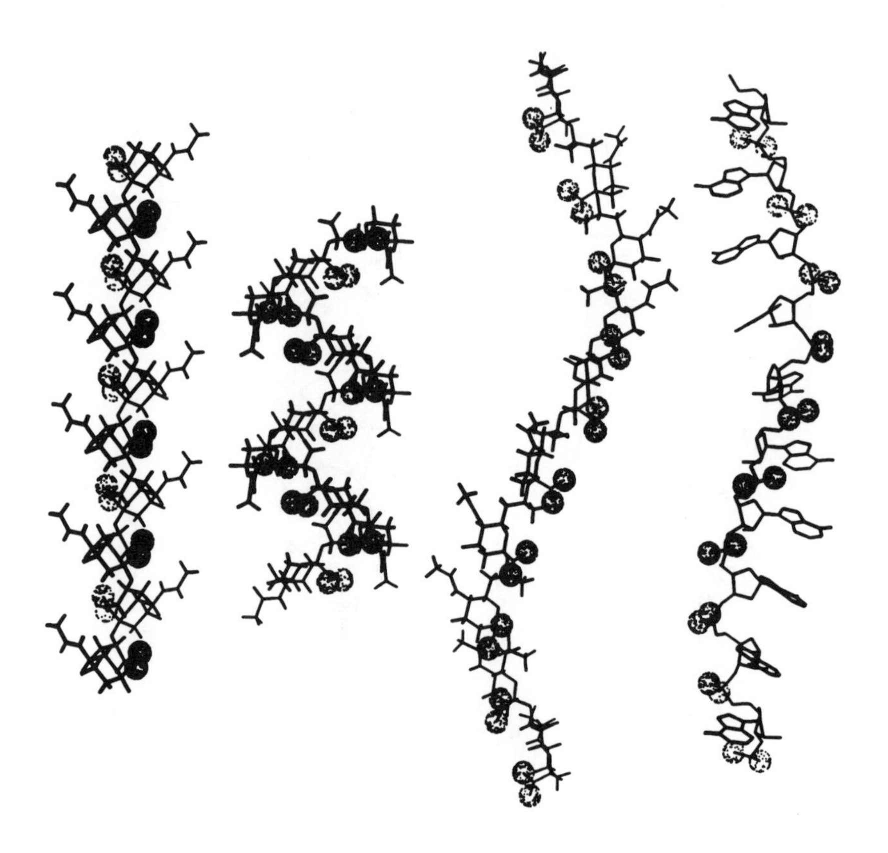

Figure 3. Side view of helices which arise by propagation of conformers represented by wells A, B and C (from left to right) where n=2,4 and 9 respectively. The n=9 helix for poly(A) is shown on the far right.

Does the GBMP exhibit special structural features that stabilize extended helices of $n=8$ to $n=12$? Certainly charges are important because reduction of the carboxyl groups of the GBMP to hydroxymethyl groups changes the immunological properties of the polysaccharide (Jennings, manuscript in preparation). In this case inhibition of the binding of the reduced GBMP to its homologous antibody by its oligomers fragments was more consistent with a conventional carbohydrate epitope as described by Kabat (1960).

In contrast to the GBMP which required a minimum of ten contiguous sialic acid residues for conformational resemblance to the GBMP (Jennings et al., 1985; Finne and Makela, 1985),

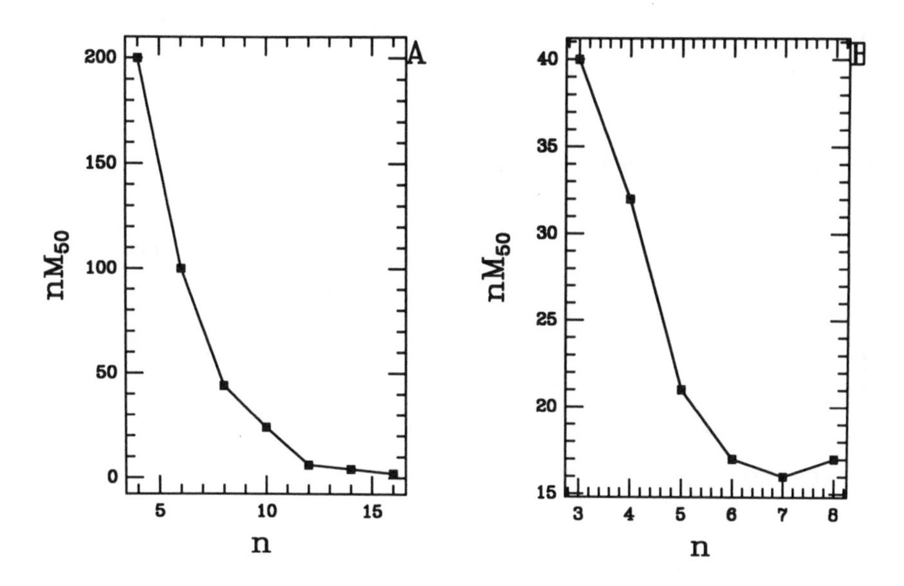

Figure 4. Quantity of linear homologous α2→8-linked oligosaccharides of defined lengths required to cause 50% inhibition of the reaction between the GBMP and reduced GBMP and their homologous antibodies.

the reduced GBMP required only 4 to 5 of its contiguous residues (Fig. 4). Thus, there is no immunological evidence for the presence of an extended helix in the reduced GBMP and this is consistent with recent NMR studies on the reduced GBMP and its constituent oligomers which demonstrate helical conformation is no longer present (Brisson et al., 1992).

However, the charges alone are not sufficient to account for the stabilization of the extended helical conformation and additional structural features are probably involved. Certainly there is no immunological evidence that the GCMP, an isomeric polysialic acid linked $\alpha 2 \rightarrow 9$, can adopt such extended helical structures, because as in the case of the reduced GBMP (Fig. 4), oligosaccharide inhibition studies (Jennings et al., 1985) indicated that only 4 to 5 contiguous $\alpha 2 \rightarrow 9$-linked sialic acid units were required to obtain conformational resemblance to the GCMP (Jennings et al., 1985). This result is entirely consistent with recent NMR studies and potential energy calculations which indicate that only helices of $n=2$ or $n=4$ can be formed by the GCMP (Jennings, manuscript in preparation).

From binding studies that have been reported on different GBMP specific antibodies, all are specific for the extended helical epitope (Brisson et al., 1992). Because our conformational studies indicate that this epitope is only a minor contributor to the total number of epitopes available, we must therefore presume that the dominance of this epitope in the immune response must be the result of immunological selection. The failure of the immune system to produce antibodies specific for the more populous lower order helices present in the polysaccharide random coil, probably occurs because these short helices are conformationally similar to the ones found in shorter sialyl oligomers. These short α(2-8)NeuAc oligosaccharides are also present in human tissue (Finne et al., 1983), and the production of specific antibody to them appears to be even more stringently avoided.

CHEMICALLY MODIFICATION OF α-2→8-LINKED POLYSIALIC ACID

The failure of GBMP-protein conjugates to provide satisfactory levels of protective antibodies against group B meningococci and E. coli K1 (Jennings, 1990) prompted interest in the chemical modification of α-2→8-linked polysialic acid. Initially the goal was to create synthetic antigens capable of modulating the immune system in such a way as to produce enhanced levels of cross-reactive GBMP-specific antibody (Jennings et al., 1986), and in fact this was achieved. More recently it was reported that enhanced levels of cross-reactive GBMP-specific antibodies could also be produced in mice using a conjugate consisting of the E. coli K92 polysaccharide coupled to tetanus toxoid (Devi et al., 1992).

In selecting possible chemical modifications of the GBMP, firstly, the modification had to be accomplished with facility and with the minimum degradation of the polysaccharide, and secondly, in order to produce cross-reactive GBMP-specific antibodies, the conformation of the epitope associated with GBMP-specific antibodies had to be preserved. One successful modification which satisfied the above criteria was that in which the N-acetyl groups of the sialic acid residues of the GBMP were removed by base and replaced by N-propionyl (NPr) groups (Fig. 5). Serological studies on a number of different modifications of the GBMP indicated that retention of both the carboxylate and N-carbonyl groups of the sialic acid residues were essential to the preservation of the

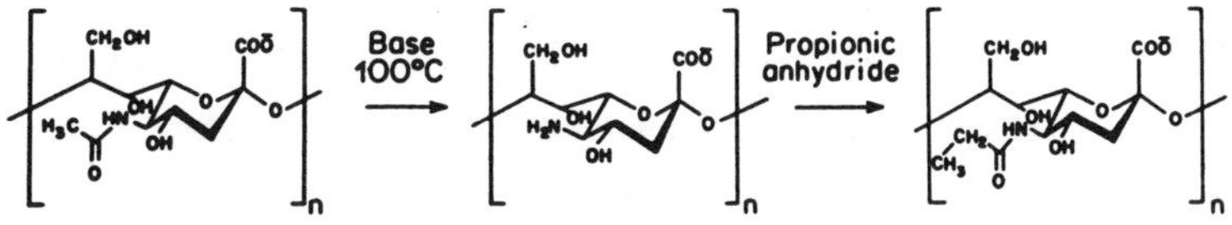

Figure 5. Reaction sequence leading to the formation of the NPrGBMP from the GBMP.

conformation of the epitope responsible for the induction responsible for the induction of GBMP-specific antibodies (Jennings et al., 1986). Due to the poor immunogenicity of the NPr GBMP in mice it was conjugated to tetanus toxoid (TT) thus yielding a virtually synthetic vaccine (Jennings et al., 1986).

The potential of the NPrGBMP to enhance the induction of GBMP-specific antibodies was proven when it was demonstrated that the NPrGBMP-TT conjugate was able to produce much higher levels of GBMP-specific antibodies in mice than the homologous GBMP-TT conjugate. Booster effects were noticeable following second and third injections of both conjugates, indicative of the participation of T-cells. However, the booster effect was more pronounced in the case of the NPrGBMP-TT conjugate and a large proportion of GBMP-specific antibodies of the IgG isotype were detected in the response (Jennings et al., 1986). Because these antibodies were also bactericidal for group B meningococci (Jennings et al., 1987) the NPrGBMP-TT conjugate must be considered as a prototype vaccine against meningitis caused by group B meningococci and E.coli K1. Experience would indicate that an immunogenic form of the GBMP would be an ideal vaccine candidate although its success as a vaccine, would depend firstly on whether this vaccine would in fact break tolerance to $\alpha 2 \rightarrow 8$-polysialic acid, and secondly even if it did, what if any (Devi et al., 1992), the consequence of this would be. Of significance to the controversy surrounding the possible deleterious effects caused by the purposeful induction of GBMP-specific antibodies in humans was an important observation made on the anti-NPrGBMP-TT mouse serum (Jennings et al., 1987). It was deduced from precipitation experiments that the GBMP was unable to precipitate all the NPrGBMP-specific antibodies from this antiserum and more importantly that even following removal of the GBMP-specific antibodies the bactericidal activity of NPrGBMP-specific antibodies remained undiminished. From this evidence it can be inferred that the NPrGBMP mimics a unique bactericidal epitope on the surface of group B meningococci which could be the basis of an efficacious vaccine against disease caused by this organism.

NATURE OF THE BACTERICIDAL EPITOPE

The close structural resemblance of the NPrGBMP with its native precursor would suggest that the bactericidal epitope is still closely associated with the GBMP even though the GBMP does not bind to the bactericidal antibodies (Jennings et al., 1989). This hypothesis is consistent with the fact that two different bacteria (group B, meningococci and E.coli, K1), having the $\alpha 2\rightarrow 8$-linked polysialic acid as their only common antigen, were both able to absorb the bactericidal antibody from the NPrGBMP-TT conjugate mouse antiserum (Jennings et al., 1989). Additionally the bactericidal NPrGBMP epitope has the same conformational requirement as that of the GBMP (Jennings et al., 1989). Inhibition studies using a series of $\alpha 2\rightarrow 8$-NeuNPr oligomers to inhibit the binding of NPrGBMP to affinity purified bacteridical NPrGBMP-specific antibodies revealed the same epitope length dependency as exhibited by the GBMP. Therefore, the bactericidal epitope of the NPrGBMP, although not cross-reactive with the GBMP, expresses the same extended helical conformation as that of the GBMP (Fig. 6). This has recently confirmed in high resolution NMR studies and potential energy calculations (Brisson, unpublished results) which indicate that although random coil in nature the NPrGBMP

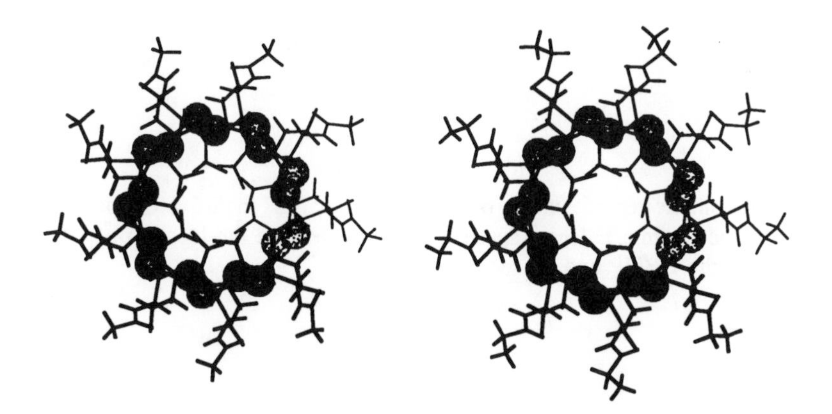

Figure 6. End-on view of the identical helices ($n=9$) which can be formed by both the GBMP (left) and the NPrGBMP (right).

adopts the same extended helical conformation as the GBMP (Fig. 6). Thus the specificity of the immune response to the NPr GBMP is under the same regulatory control as that of the GBMP, no antibodies to the more populous lower order helices associated with shorter oligomers being detected.

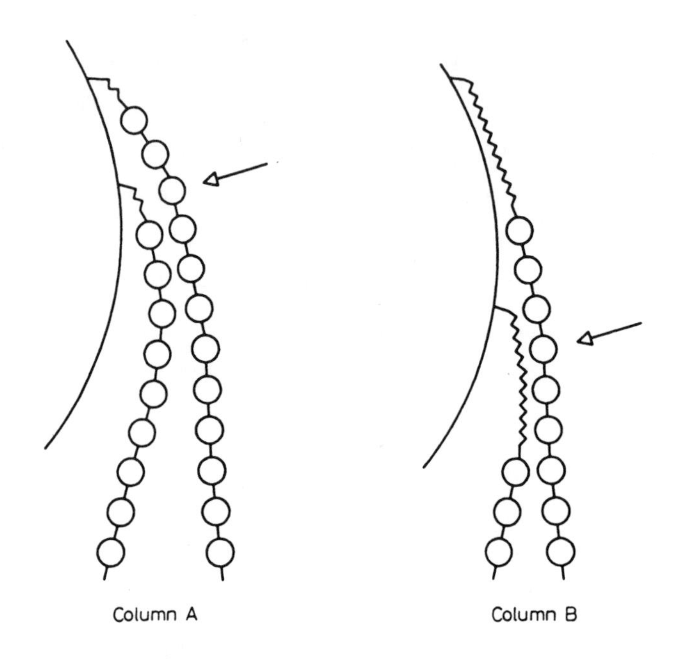

Figure 7. Diagrammatic representation of the GBMP ligands, inclusive of spacers of column A (long spacer) and column B (short spacer). Arrows indicate the location of intermolecular epitopes involving the GBMP and spacer molecules.

This hypothesis is strongly supported by the observation that while the NPrGBMP-specific bactericidal antibodies do not bind to the GBMP, they can be bound by an affinity column in which the latter polysaccharide is linked to the solid support by a long spacer (Jennings et al., 1989). This is shown diagrammatically in Fig. 7. In fact, this binding is dependent on the length of the spacer because no binding occurs when a short spacer is employed. This implies that the long spacer has a function in the process. Its functionality is probably due to the ability of the long spacer and the group B polysaccharide to intermingle on the solid support where, like the NPr-GBMP, it probably also mimics to some extent the intermolecular bactericidal epitope on the surface of group B meningococci.

Is this complex epitope on the surface of group B meningococci and E.coli involved in protection against disease caused by these organisms? On the evidence available this questions remains unanswered. The fact that all the polysaccharide-specific monoclonal antibodies induced in BalbC mice by group B meningococci were GBMP-specific strongly indicates a negative answer (Jennings, manuscript in preparation). However the fact that some of these antibodies were cross-reactive with the NPrGBMP and more importantly were able to bind more strongly to the NPrGBMP than to the GBMP introduces a note of ambiguity. On this evidence (Jennings, manuscript in preparation) one must conclude that unlike the NPrGBMP-TT conjugate, group B meningococci are unable to induce antibodies with exclusive NPrGBMP specificity. Thus it is likely that the NPrGBMP is a truly synthetic immunogen with a natural antigenic but no natural immunogenic counterpart.

ACKNOWLEDGEMENTS

We thank North American Vaccines for financial support. This is N.R.C.C. publication No. 34267.

REFERENCES

Brisson, J.-R., Baumann, H., Imberty, A., Perez, S. and Jennings, H.J. (1992) Biochemistry - 31: 4996-5004.
Devi, S.J.N., Robbins, J.B. and Schneerson, R. (1991) Proc. Nat. Acad. Sci. 88: 7175-7179.
Finne, J., Finne, V., Diagostini-Bazin, H. and Goridis, C. (1983) Biochem. Biophys. Res. Commun. 112: 482-487.
Finne, J. and Makela, P.H. (1985) J. Biol. Chem. 260: 1265-1270.
Hayrinen, J., Bitter-Suermann, D. and Finne, J. (1989) Mol. Immunol. 26: 523-529.
Jennings, H.J., Katzenellenbogen, E., Lugowski, C., Michon, F., Roy, R. and Kasper, D.L. (1984) Pure Appl. Chem. 56: 893-905.
Jennings, H.J., Roy, R. and Michon, F. (1985) J. Immunol. 134: 2651-2657.
Jennings, H.J., Roy, R. and Gamian, A. (1986) J. Immunol. 137: 1708-1713.
Jennings, H.J., Gamian, A. and Ashton, F.E. (1987) J. Exp. Med. 165: 1207-1211.
Jennings, H.J., Gamian, A., Michon, F. and Ashton, F.E. (1989) J. Immunol. 142: 3585-3591.
Jennings, H.J. (1989) Contrib. Microbiol. Immunol. 10: 1-10.
Jennings, H.J. (1990) Curr. Top. Microbiol. Immunol. 150: 97-127.
Kabat, E.A. (1960) J. Immunol. 84: 82-85.
Kabat, E.A., Nickerson, K.G., Liao, J., Grossbard, L., Osserman, E.F., Glickerman, E., Chess, L., Robbins, J.B., Schneerson, R. and Young Y. (1986) J. Exp. Med. 164: 642-654.
Kabat, E.A., Liao, J., Osserman, E.F., Gamian, A., Michon, F. and Jennings, H.J. (1988) J. Exp. Med. 168: 699-711.
Michon, F., Brisson, J.-R. and Jennings, H.J. (1987) Biochemistry 26: 8399-8405.
Perez, S. and Vergelati, C. (1985) Biopolymers 24: 1809-1822.
Raff, H.V., Devereux, D., Shuford, W., Abbott-Brown, D. and Maloney, G. (1988) J. Infect. Dis. 157: 118-126.
Saenger, W., Riecke, J. and Suck, D. (1975) J. Mol. Biol. 93: 529-534.
Troy, F.A. (1992) Glycobiology 2: 5-23.
Yamasaki, R. and Bacon, B. (1991) Biochemistry 30: 851-857.
Yathindra, N. and Sundaralingam, M. (1976) Nucleic Acids Res. 3: 729-747.

Polysialic Acid
J. Roth, U. Rutishauser and F. A. Troy II (eds.)

DETECTION OF ANTIBODIES TO Neisseria meningitidis GROUP B CAPSULAR POLYSACCHARIDE BY A LIQUID-PHASE ELISA.

J.Diaz Romero and I.M.Outschoorn[1]

Inmunologia, Centro Nacional de Microbiologia (C.N.M.V.I.S) and
C.N.Biologia Celular y Retrovirus[1],
Instituto Carlos III, Majadahonda, Madrid 28220, Spain

SUMMARY; A liquid-phase immunoassay for the detection of low affinity antibodies to Neisseria meningitidis group B capsular polysaccharide (PS-B) is described. Incubation of antibody in solution with PS-B in a non-covalent complex with avidin is followed by capture of the immunocomplexes formed by a biotin solid-phase. Different factors affecting the sensitivity and specificity of anti-PS-B ELISA systems were also investigated.

INTRODUCTION

Meningococcal meningitis is a major health problem in many countries. Neisseria meningitidis is classified into serogroups based on immunologically and chemically distinct polysaccharides. Diseases caused by group B meningococci accounts for approximately half the cases of meningococcal meningitis worldwide (Lifely et al.,1987). The capsular polysaccharide of the group B meningococcus (PS-B) is a polymer of $\alpha(2\rightarrow8)$ linked N-acetylneuraminic acid residues and is identical to the capsular polysaccharide of Escherichia coli K1, the major cause of neonatal meningitis. The capsule serves as an essential virulence factor and protective antigen in both pathogens (Lifely et al.,1989), and the same structure is also a surface antigen of Moraxella nonliquefaciens and Pasteurella haemolytica A-2, the latter being a major veterinary pathogen, which suggests that poly $\alpha(2\rightarrow8)$ linked N-acetylneuraminic acid may serve as a virulence factor for yet another bacterial species (Sarvamangala et al.,1991).

Antibodies to the capsular polysaccharide are known to play a major protective role in human immunity to meningococci (Skevakis et al.,1984). Purified PS-B is not immunogenic in humans or experimental animals, despite the fact that anti-B antibodies can be clearly demonstrated in infected individuals (Moreno et al.,1985). However attempts to obtain an effective vaccine against group B meningococci using noncovalent complexes of PS-B and outer membrane proteins have given variable results in volunteers, when sera were examined for a specific antibody increase to the PS-B component (Frasch et al.,1988, Lifely et al.,1991a). No standard immunoassay for the quantitative measurement of total levels of antibody to Neisseria meningitidis capsular polysaccharide in serum exists. The development of polysaccharide-protein conjugate meningococcal vaccines (Moreno et al.,1985, Jennings et al.,1986) has emphasized the need for a standardized and reproducible assay to evaluate the immunogenicity of new vaccines and to correlate antibody levels with functional assays and minimum protective antibody levels.

ELISA: SOME THEORETICAL CONSIDERATIONS

The most commonly used solid phase immunoassays for the measurement of antibodies involve the passive absorption of antigens to a support, in particular to microtitre plates in the enzyme-linked immunosorbent assay (ELISA). ELISA is simple, rapid and versatile and has become widely used for serological diagnosis in microbiology, parasitology and immunology laboratories. It does not involve radiation hazards, nor expensive equipment and the necessary reagents have a long shelf-life. However, controversy has arisen with respect to whether the ELISA measures antibody concentration or affinity. As a discontinuous solid-phase assay usually comprises 2 cycles of 3 washes and a second antibody step during which the first antibody may dissociate, it would not be surprising that low affinity antibodies may go undetected. The possibility that low-affinity antibodies were not detected by ELISA and that data correlation was better with antibody affinity than with concentration remains a handicap (Peterfy et al.,1983).

The strength of antibody binding to multivalent antigens, functional affinity or avidity, is primarily a function of two components: intrinsic affinity and effective valency (Greenspan et al.,1992). The former represents the strength of the interaction for a single epitope-paratope pair, while antibody segmental flexibility and epitope distribution are assumed to determine the number of paratopes that can simultaneously interact with epitopes. The striking dependence of the ELISA on epitope density has been shown in some studies, particularly those using low affinity antibodies (Peterfy et al.,1983, Lew,1984). Hence, one of the most important requirements of the microtitre plates used in ELISA is that they should be able to bind sufficient antigen and that the antigen bound should remain attached throughout the assay. The maximum amount of soluble material binding to the solid surface depends on the characteristics of both of these. Polyvinyl chloride and polystyrene are the surface materials most frequently used in immunosorbent assays due to their inherent 'stickiness' to many serum proteins. Measurements of low affinity antibodies would require special modifications in the conditions of the immunoassay to achieve affinity independence such as high epitope density and high antigen concentration (Griswold et al.,1991). Affinity dependence should be suspected if varying assay conditions, such as increasing antigen concentration or epitope density, alters measurement of antibody concentration.

POLYSACCHARIDE ASSOCIATED COATING PROBLEMS

The measurement of antibody to purified cell surface carbohydrates by ELISA is often hindered by the failure of lipid-free or protein-free polysaccharides to adhere to a solid phase efficiently and reproducibly (Barra et al.,1988). Negatively charged polysaccharides do not attach well to the commonly used polystyrene. In an attempt to address this problem, various methods have been developed to bind polysaccharide to microtitre plates. The use of capture antibodies (Zigterman et al.,1988), are limited in this case by the extreme difficulty in producing immunological reagents with PS-B specificity. Covalent binding to poly-L-lysine (Messina et al.,1985) or tyramine (Anthony et al.,1982), in addition to being complex and tedious, introduce the risk of altering the immunological reactivity of the PS-B. Precoating with poly-L-lysine (Leinonen et al.,1982) generates a low signal to noise ratio (Diaz-Romero et al.,1992), reacting nonspecifically with some sera. Recently, a new ELISA procedure which uses methylated human serum albumin to coat the solid phase with the polysaccharide has been successfully employed with group A and C capsular polysaccharides (Arakere et al.,1991, Carlone et al.,1992), but its' potential use with PS-B remains to be established.

The utilization of the high affinity of avidin for biotin ($K_d = 10^{-15}$) has been previously employed to bind biotinylated Escherichia coli K1 polysaccharide to avidin-coated microtitre wells (Sutton et al.,1985). The procedure introduced multiple biotin groups onto the polysaccharide via adipic hydrazide derivatives of carboxyl groups. This represents an inherent risk of conformational alteration, in cases where conformational or discontinuous determinants on the PS-B have been postulated, and it has been demonstrated that modifications of their carboxyl groups abolished antigenicity (Lifely et al.,1991b). With this in mind, we have developed a selective PS-B biotinylation method (Diaz-Romero et al.,1992) that introduces a single biotin molecule onto the polysaccharide by specific periodate oxidation of terminal neuraminic acid residues and further condensation of the aldehyde produced with biotin-hydrazide (O'Shannessy et al.,1990). This approach is applicable to other polysaccharides susceptible to selective periodate oxidation, such as the capsular polysaccharide from Neisseria meningitidis groups A and C (Jennings et al.,1981), and combined with the use of spin-columns (packed in syringe barrels) in the intermediate purification steps, allows labelling time to be shortened from days to hours.

OPTIMAL ASSAY CONDITIONS: ANALYSIS

In order to detect low concentrations of analyte, optimization of the assay conditions had to be carried out. The choice of an enzyme/substrate chromogen alternative is important to improve enzyme immunoassay sensitivity. Among the different combinations employed as markers in the heterogeneous enzyme immunoassay two are among the most common: horseradish peroxidase/ o-phenylenediamine (HRP/OPD) and alkaline phosphatase/p-nitrophenylphosphate (AP/PNPP). HRP catalyses the reduction of H_2O_2 with the concurrent oxidation of a chromogen producing an optically measurable colour change. OPD seems to be the chromogen giving the highest absorbance readings but also produces one of the highest backgrounds due to low stability resulting in its' spontaneous decomposition within a few hours (Al-Kaissi et al.,1983). In addition, the prolonged incubation of enzyme/substrate does not improve detection limits, since

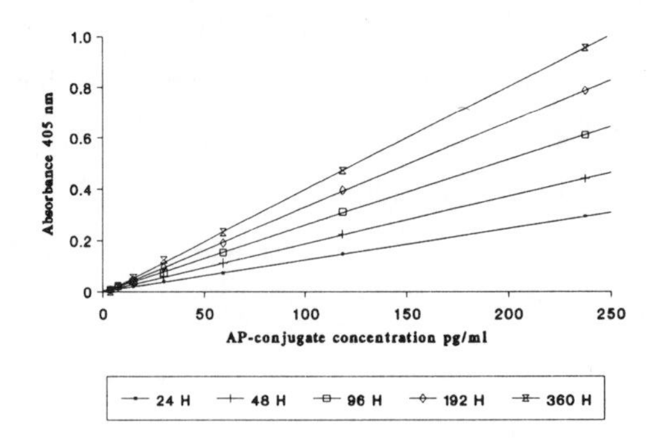

Fig.1. Time-dependent enzymatic kinetics of AP-immunoglobulin conjugate

the enzyme is inactivated over time by its' substrate: H_2O_2 (Ishikawa et al., 1991). One set of reactions catalyzed by AP is the hydrolysis of phosphate esters to give inorganic phosphate and a phenolic leaving group (Thompson et al.,1991). The formation of this phenolic moiety can be followed spectrophotometrically with PNPP as substrate and is the most widely used method to follow AP catalysis. Both PNPP and its' reaction product, p-nitrophenol, are very stable at room temperature, and when the influence of the enzyme/substrate reaction time on sensitivity was analyzed, a time dependent linear increase in detection limit was observed, even up to 15 days incubation (Porstmann et al.,1985, Diaz-Romero et al.,1992) (Fig.1). Thus, we selected the AP/PNPP system as the detection procedure for our experiments, with the added advantage, in contrast to HRP, of making absorbance readings without stopping the reaction.

The effect of assay incubation temperature is especially important to analyse since temperature dependent binding of anti-PS-B antibodies has been described (Mandrell et al., 1982). When this factor was examined together with the influence of antigen density, using a specific mouse monoclonal antibody and biotinylated PS-B with a streptavidin solid-phase, the results obtained were somewhat surprising. While the ELISA titre at 4°C was not influenced by antigen concentration, the assay was antigen density dependent at 37°C (Fig.2), reflecting the lower affinity at this temperature along with the requirement to increase the number of occupied binding sites in order to maintain avidity. Similar behaviour has been observed in certain IgM Waldeström macroglobulins with both cold agglutinin and cryoglobulin activity (Tsai et al.,1977, Weber et al.,1981). Both properties seem to have as common basis the immune binding of sialic

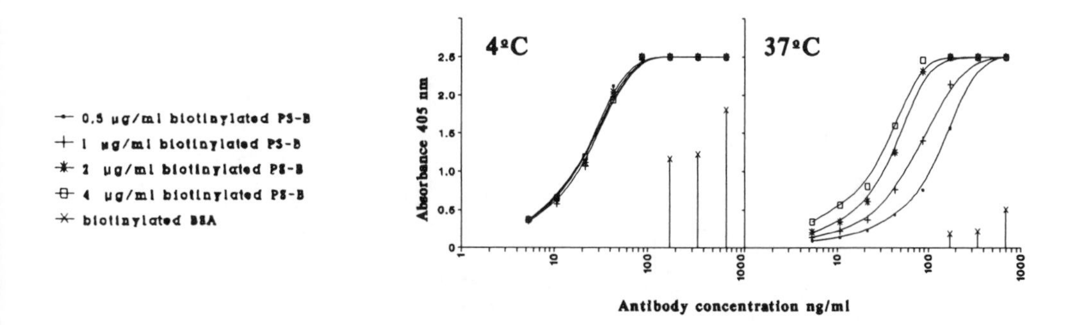

Fig.2. Antigen density and temperature dependent binding of a monoclonal antibody to PS-B

of sialic acid residues: to N-acetylneuraminic acid-containing carbohydrates in erythrocyte membranes or in the macroglobulin itself, respectively. These studies support the suggestion that cold agglutinins in general are simply low affinity antibodies that require multivalent attachment for the stabilization of bonds strong enough to agglutinate erythrocytes, adding a new view regarding the nature of immune response to PS-B.

The selection of blocking agents to inhibit non-specific binding can be critical to the sensitivity and specificity of ELISA-type systems. The non-specific binding is dependent on serum concentration and limits the amount of antibody detectable in whole serum (Zollinger et al.,1976). Bovine serum albumin (BSA), one of the more widely used blocking agents, exhibited a limited capacity to reduce the non-specific binding in our experimental system. When we compared non-fat dry milk (SkiM) and foetal calf serum (FCS) as possible blocking agents, conflicting results were obtained. SkiM seemed to be superior to FCS in ELISA with other capsular polysaccharides (data not shown). However, in the PS-B assay the specific binding with SkiM (measured in wells coated with streptavidin and biotinylated PS-B) was decreased with respect to FCS, and the reduction in the non-specific binding (wells with streptavidin and biotinylated BSA) was influenced by the nature of the specific antibody (Fig.3). Currently we are evaluating the properties of casein: a purified protein very effective as a blocking agent and lacking the problems of interfence and 'masking' associated with milk (Vogt et al.,1987).

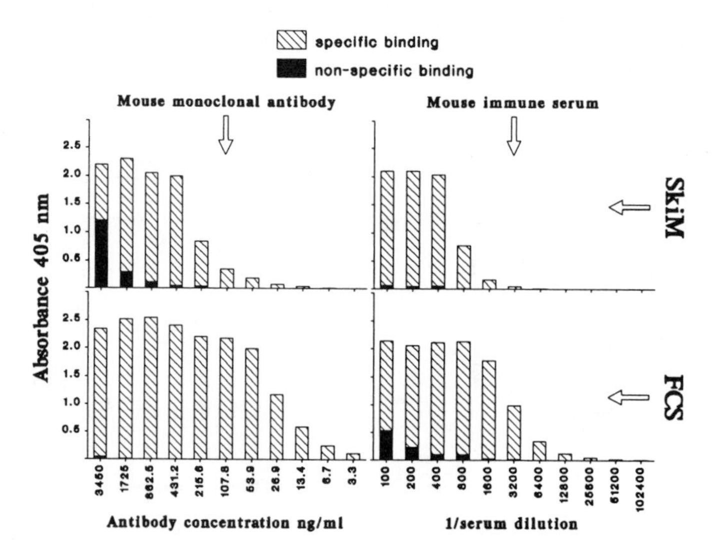

Fig.3. Effects of blocking agent on sensitivity and non-specific binding on anti-B ELISA

THE ANTI-B LIQUID-PHASE ELISA

Our efforts to eliminate nonspecific binding through qualitative and quantitative changes in the blocking buffer were unsuccessful at high serum concentrations, in agreement with previous reports (Zollinger et al.,1976). The main contributing factor to non-specific binding in our system was the incubation time between the solid phase and serum. Longer incubation times are required for enhancing sensitivity when low affinity anti-B antibodies are used, resulting in higher backgrounds. The optimal ELISA format is such that would decrease the incubation time of antibody with solid phase without an inherent reduction in the antigen/antibody interaction time. On this assumption, a liquid-phase ELISA was designed using biotin labelled antigen complexed with avidin (Av).

The preparation of an equimolecular complex of PS-B/Av using a biotin bridge is facilitated by the selectivity of the biotinylation procedure, that introduces exclusively one molecule of biotin/PS-B molecule. By the colorimetric follow-up of the displacement of 4-hydroxyazobenzene-2'-carboxylic acid (HABA) by biotin (Green,1970), it is possible to estimate the number of occupied binding sites in avidin. It is therefore not difficult to obtain avidin with one of its' biotin binding sites occupied by biotinylated PS-B, with 3 free sites remaining for additional interactions. The complex is first incubated at 4°C in solution with the antibody at constant concentration, transferring the resultant immunocomplexes at different dilutions to the wells of a microtitration plate, where they are captured by a biotin solid-phase, obtained by coating with biotinylated BSA. The following steps do not differ from those of a conventional

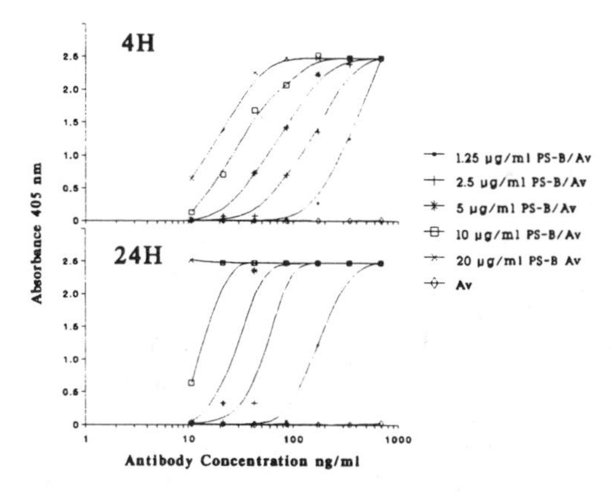

Fig.4. Comparison of different development times in liquid-phase ELISA

assay. The titration of a specific mouse monoclonal with this ELISA at various concentrations of PS-B/avidin complex is shown in Fig.4.

The advantages of this assay with respect to the commonly used ELISA are evident. First, the risk of alteration in presentation and/or conformation of antigen is minimized, avoiding the potential steric hindrance during antigen-antibody association and the unidirectional diffusion restricted to the soluble ligand, since the antigen-antibody reaction is carried out before absorption to the solid-phase. One the other hand, the microtitre well is a limited surface area for antigen binding, resulting in progressive saturation, even when a bridge between antigen and solid-phase is used. Above an optimal level, the antibody binding is not increased by the quantity of antigen used for coating. In a liquid-phase ELISA, a sensitivity enhancement is observed when increasing antigen concentrations are employed (Fig.4).

The main goal of the new assay design, the deletion of nonspecific binding, is currently being examined and experiments with immune sera anti-PS-B obtained with different immunogens will be carried out in the near future. In addition, a series of questions relative to the anti-PS-B immune response will be examined. The almost complete absence of an IgG response to PS-B, even when it is complexed to protein, has been reported (Lifely et al.,1991a). Is this a true isotypic restriction, or merely reflects a defect in measurement of low avidity antibodies lacking the 'bonus' effect of polyvalency?. The ELISA binding data previously obtained with class-switched IgG-IgM monoclonal antibody pairs specific for E.coli K1 capsule (Raff et al., 1991), that share identical V regions, do not rule out the latter possibility. This and other questions could be approached more easily using liquid-phase ELISA.

REFERENCES

Al-Kaissi, E. and Mostratos, A. 1983. Assessment of substrates for horseradish peroxidase in enzyme immunoassay. J.Immunol.Methods 58:127.

Anthony, B.F., Concepcion, N.F., McGeary, S.A., Ward, J.I., Heiner, D.C. et al.1982. Immunospecificity and quantitation of an enzyme-linked immunosorbent assay for group B streptococcal antibody. J.Clin.Microbiol. 2:350.

Arakere, G. and Frasch, C.E. 1991. Specificity of antibodies to O-acetyl-positive and O-acetyl-negative group C meningococcal polysaccharides in sera from vaccinees and carriers. Infect.Immun. 59:4349.

Barra, A., Schulz, D., Aucouturier, P. and Preud'homme, J.L. 1988. Measurement of anti-Haemophilus influenzae type b capsular polysaccharide antibodies by ELISA. J.Immunol.Methods 115:111.

Carlone, G.M., Frasch, C.E., Siber, G.R., Quataert, S., Gheesling, L.L. et al. 1992. Multicenter comparison of levels of antibody to the Neisseria meningitidis group A capsular polysaccharide measured by using an enzyme-linked immunosorbent assay. J.Clin.Microbiol. 30:154.

Diaz-Romero, J. and Outschoorn, I.M. 1992. Selective biotinylation of Neisseria meningitidis group B capsular polysaccharides: application to an improved ELISA for the detection of specific antibodies. (submitted).

Frasch, C.E., Zahradnik, J.M., Wang, L.Y., Mocca, L.F. and Tsai, C.M. 1988. Antibody response of adults to an aluminium hydroxide-adsorbed Neisseria meningitidis serotype 2b protein-group B polysaccharide vaccine. J.Infect.Dis. 158:710.

Green, N.M. Spectrophotometric determination of avidin and biotin. 1970. In: Methods in Enzymology, edited by McCormick, D.B. and Wright, L.D. New York: Academic Press, p. 418.

Greenspan, N.S. and Cooper, L.J.N. 1992. Intermolecular cooperativity: a clue to why mice have IgG3? Immunol.Today 13:164.

Griswold, W.R. and Chalquest, R.R. 1991. Theoretical analysis of the accuracy of calibrated immunoassays for measuring antibody concentration. Mol.Immunol. 28:727.

Ishikawa, E., Hashida, S. and Kohno, T. 1991. Development of ultrasensitive enzyme immunoassay reviewed with emphasis on factors which limit the sensitivity. Molecular and cellular probes 5:81.

Jennings, H.J. and Lugowski, C. 1981. Immunochemistry of groups A, B and C meningococcal polysaccharide-tetanus toxoid conjugates. J.Immunol. 127:1011.

Jennings, H.J., Roy, R. and Gamian, A. 1986. Induction of meningococcal group B polysaccharide-specific IgG antibodies in mice using an N-propionylated B polysaccharide-tetanus toxoid conjugate vaccine. J.Immunol. 137:1708.

Leinonen, M. and Frasch, C.E. 1982. Class-specific antibody response to group B Neisseria meningitidis capsular polysaccharide: use of polylysine precoating in an enzyme-linked immunosorbent assay. Infect.Immun. 38:1203.

Lew, A.M. 1984. The effect of epitope density and antibody affinity on the ELISA as analysed by monoclonal antibodies. J.Immunol.Methods 72:171.

Lifely, M.R., Moreno, C. and Lindon, J.C. 1987. An integrated molecular and immunological approach towards a meningococcal group B vaccine. Vaccine 5:11.

Lifely, M.R., Esdaile, J. and Moreno, C. 1989. Passive transfer of meningococcal group B polysaccharide antibodies to the offspring of pregnant rabbits and their protective role against infection with Escherichia coli K1. Vaccine 7:17.

Lifely, M.R., Roberts, S.C., Shepherd, W.M., Esdaile, J., Wang, Z. et al. 1991a. Immunogenicity in adult males of a Neisseria meningitidis group B vaccine composed of polysaccharide complexed with outer membrane proteins. Vaccine 9:60.

Lifely, M.R. and Esdaile, J. 1991b. Specificity of the immune response to the group B polysaccharide of Neisseria meningitidis. Immunology 74:490.

Mandrell, R.E. and Zollinger, W.D. 1982. Measurement of antibodies to meningococcal group B polysaccharide: low avidity binding and equilibrium binding constants. J.Immunol. 129:2172.

Messina, J.P., Hickox, P.G., Lepow, M.L., Pollara, B. and Venezia, R.A. 1985. Modification of a direct enzyme-linked immunosorbent assay for the detection of immunoglobulin G and M antibodies to pneumococcal polysaccharide. J.Clin.Microbiol. 21:390.

Moreno, C., Lifely, M.R. and Esdaile, J. 1985. Immunity and protection of mice against Neisseria meningitidis group B by vaccination, using polysaccharide complexed with outer membrane proteins: a comparison with purified B polysaccharide. Infect.Immun. 47:527.

O'Shannessy, D.J. and Wilchek, M. 1990. Immobilization of glycoconjugates by their oligosaccharides: use of hydrazido-derivatized matrices. Anal.Biochem. 191:1.

Peterfy, F., Kuusela, P. and Makela, O. 1983. Affinity requirements for antibody assays mapped by monoclonal antibodies. J.Immunol. 130:1809.

Porstmann, B., Porstmann, T., Nugel, E. and Evers, U. 1985. Which of the commonly used marker enzymes gives the best results in colorimetric and fluorimetric enzyme immunoassays: horseradish peroxidase, alkaline phosphatase or β-galactosidase? J.Immunol.Methods 79:27.

Raff, H.V., Bradley, C., Brady, W., Donaldson, K., Lipsich, L. et al. 1991. Comparison functional activities between IgG1 and IgM class-switched human monoclonal antibodies reactive with group B streptococci or Escherichia coli K1. J.Infect.Dis. 163:346.

Sarvamangala, J.N., Devi, J.N., Robbins, J.B. and Schneerson, R. 1991. Antibodies to poly[(2-8)-α-N-acetylneuraminic acid] and poly[(2-9)-α-N-acetylneuraminic acid] are elicited by immunizing mice with Escherichia coli K92 conjugates: potential vaccines for groups B and C meningococci and E.coli K1. Proc.Natl.Acad.Sci.U.S.A. 88:7175.

Skevakis, L., Frasch, C.E., Zahradnik, J.M. and Dolin, R. 1984. Class-specific human bactericidal antibodies to capsular and noncapsular surface antigens of Neisseria meningitidis. J.Infect.Dis. 149:387.

Sutton, A., Van, W.F., Karpas, A.B., Stein, K.E. and Schneerson, R. 1985. An avidin-biotin based ELISA for quantitation of antibody to bacterial polysaccharides. J.Immunol.Methods 82:215.

Thompson, R.Q., Barone, G.C., Halsall, H.B. and Heineman, W.R. 1991. Comparison of methods for following alkaline phosphatase catalysis: spectrophotometric versus amperometric detection. Anal.Biochem. 192:90.

Tsai, C.M., Zopf, D.A., Yu, R.K., Wistar,R.,Jr. and Ginsburg, V. 1977. A Waldeström macroglobulin that is both a cold agglutinin and a cryoglobulin because it binds N-acetylneuraminosyl residues. Proc.Natl.Acad.Sci.USA. 74:4591.

Vogt, R.F.J., Phillips, D.L., Henderson, L.O., Whitfield, W. and Spierto, F.W. 1987. Quantitative differences among various proteins as blocking agents for ELISA microtiter plates. J.Immunol.Methods 101:43.

Weber, R.J. and Clem, L.W. 1981. The molecular mechanism of cryoprecipitation and cold agglutination of an IgM λ Waldeström macroglobulin with anti-Gd specificity: sedimentation analysis and localization of interacting sites. J.Immunol. 127:300.

Zigterman, G.J., Verheul, A.F., Ernste, E.B., Rombouts, R.F., De Reuver, M.J.et al. 1988. Measurement of the humoral immune response against Streptococcus pneumoniae type 3 capsular polysaccharide and oligosaccharide containing antigens by ELISA and ELISPOT techniques. J.Immunol.Methods 106:101.

Zollinger, W.D., Dalrymple, J.M. and Artenstein, M.S. 1976. Analysis of parameters affecting the solid phase radioimmunoassay quantitation of antibody to meningococcal antigens. J.Immunol. 117:1788.

Polysialic Acid
J. Roth, U. Rutishauser and F. A. Troy II (eds.)

MOLECULAR MECHANISMS OF CAPSULE EXPRESSION IN NEISSERIA MENINGITIDIS SEROGROUP B

Matthias Frosch and Ulrike Edwards

Institute of Medical Microbiology, Medical School Hannover, 3000 Hannover 61, Germany

Summary: The enzymes and proteins for biosynthesis and surface translocation of the capsular polysaccharide of N.meningitidis serogroup B, which consists of α-2,8 linked polysialic acid, are expressed by a 24 kb chromosomal gene cluster (cps). Within cps five functional regions have been identified. Region A encodes all enzymes necessary for polysialic acid biosynthesis. The capsular polysaccharide, which averages 200 NeuNAc residues in length, is synthezised completely intracellularly. The gene products of region B substitute the polysaccharide chains with a phospholipid at the reducing end. Phospholipid substitution is crucial for translocation of the polysaccharide to the cell surface, which is directed by the gene products encoded by region C. The region C encoded proteins share strong homologies to members of the the ABC (ATP-binding cassette) superfamily of active transporters. The same ATP-dependent transport mechanism for capsular polysaccharides also seems to direct capsular polysaccharides in H.influenzae and E.coli to the surface, suggesting a common evolutionary origin of capsule expression in these bacterial species.

INTRODUCTION

Neisseria meningitidis, a gram-negative diplococcus, is a major causative agent of sepsis and meningitis in younger children. Based on the chemical composition of the capsular polysaccharide, which is one of the most important pathogenicity factors in meningococci, several serogroups were defined. Among those serogroups, that are associated with meningococcal disease, capsular polysaccharides of sialic acids predominate by far (Devoe, 1982). In serogroup B and serogroup C meningococci the capsular polysaccharide is a homopolymer, composed of α-2,8 (serogroup B) or α-2,9 (serogroup C) linked N-acetylneuraminic acid. Serogroup B meningococci are responsible for about 70% of all cases of meningococcal disease in Central Europe and Northern America. In contrast, serogroup C meningococci are only found in about 15% of sporadic outbreaks (Calain et al. 1988; Band et al., 1983). The predominant occurence of serogroup B meningococci, the important role of the group B capsular polysaccharide in the pathogenesis of meningococcal disease (Jarvis and Vedros, 1987) and its chemical and immunological identity to the carbohydrate moiety of the eukaryotic neural cell adhesion molecule, N-CAM, (Finne et al., 1983) made this structure a research priority for microbiologists, immunologists, cell biologists and molecular biologists. In this review we will focus on the molecular aspects of capsular polysaccharide expression and we will present a model which describes the mechanisms of capsule formation in meningococci.

MOLECULAR CLONING AND CHARACTERIZATION OF THE GENE COMPLEX (cps) ENCODING THE MENINGOCOCCAL B CAPSULAR POLYSACCHARIDE

Previously we described the expression of the meningococcal capsule in E.coli by cosmid cloning (Frosch et al., 1989). Analysis of the cosmid clones revealed that the genes required for capsule biosynthesis and cell surface expression are located on a 24 kb chromosomal DNA fragment (Fig.1). By transposon and deletion mutagenesis five different functional regions could be defined within the meningococcal capsule gene complex (cps). The genes encoding all enzymes necessary for poly-α-2,8 sialic acid biosynthesis are located within region A. Mutants in this region expressed no capsular polysaccharide. In contrast, mutagenesis of region B or region C resulted in intracellular accumulation of the capsular polysaccharide. Due to putative transport defects in region B and region C mutants these clones could not form a capsule on their cell surface. Regulatory functions in capsule expression were assigned to the regions D and E. There is preliminary evidence that these regions participate in biosynthesis and assembly of the meningococcal lipooligosaccharide (LOS). However, these features will not be considered in this review.

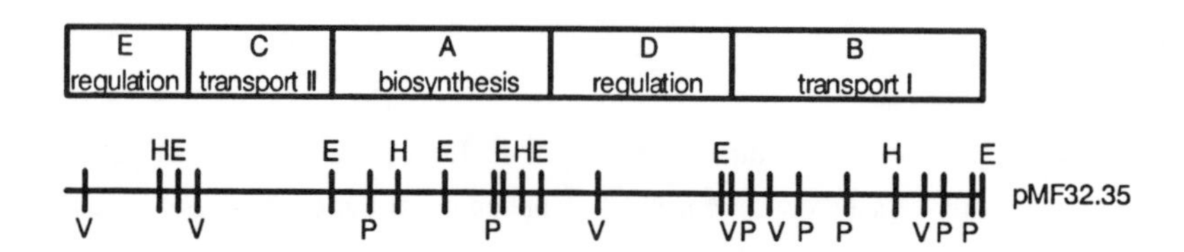

Figure 1. Physical map of clone pMF32.35, which harbours the complete cps gene cluster of Neisseria meningitidis B (strain B1940). Restriction sites of the enzymes EcoRI (E), EcoRV (V), HindIII (H) and PstI (P) are given. The functional regions A-E are indicated above. The complete cps gene cluster has a size of 24 kb.

CHARACTERIZATION OF THE ENZYMES OF THE POLY-α-2,8 SIALIC ACID BIOSYNTHESIS PATHWAY

By biochemical analysis of the poly-α-2,8 sialic acid biosynthesis pathway of Neisseria meningitidis several enzymatic activities were described (Masson and Holbein, 1983). Monomeric N-acetyl neuraminic acid is synthezised by condensation of N-acetylmannosamine and phosphoenol-pyruvate by NeuNAc-condensing enzyme. Prior to polymerization the enzyme CMP-NeuNAc synthetase transfers monomeric NeuNAc molecules into the activated state, CMP-NeuNAc. While biosynthesis and activation of monomeric NeuNAc both occur in the cytoplasm, polymerization of CMP-NeuNAc to form α-2,8 linked polysialic acid occurs at the cytoplasmic site of the inner membrane by the action of the inner membrane associated polysialyltransferase. Undecaprenol was described as an acceptor molecule necessary for the polymerization process (Masson and Holbein, 1985). However it is unknown whether the polysialyltransferase is a bifunctional enzyme which initiates polymerization by formation of an undecaprenol-NeuNAc molecule, as well as catalyzing chain elongation. Alternatively, there may be two different sialyltransferases each specific for one of the two enzymatic functions.

Nucleotide sequence analysis of the biosynthesis region A of the meningococcal cps gene cluster (Fig.1) revealed four open reading frames, which presumably constitute a transcriptional unit. Proteins with molecular weights of 38.000, 18.200, 38.300, and 54.400, respectively, are expressed by the region A genes. All open reading frames were subcloned into expression vector pKK223-3 and analysed for the enzymatic properties described above (Table I). By this strategy we found that the 38.3 kDa protein encoded by the third open reading frame is the NeuNAc-condensing enzyme. The 18.2 kDa protein exhibited CMP-NeuNAc synthetase activity (Edwards and Frosch, 1992) and polysialyltransferase activity was assigned to the 54.4 kDa polypeptide (Frosch et al., 1991). None of the described enzymatic activities was observed for the gene product encoded by the first open reading frame. The meningococcal CMP-NeuNAc synthetase exhibited 59% identity to the same enzyme of the E.coli K1 biosynthesis pathway. However, the E.coli enzyme has about a 2.5 fold larger molecular weight (Zapata et al., 1989). The homologies between both proteins were only observed at the N-terminus, thus suggesting that this region harbours the active site of the E.coli CMP-NeuNAc synthetase.

To analyze whether the polypeptides encoded by the open reading frames 2, 3 and 4 are able to generate α-2,8 linked polysialic acid by de novo synthesis, the three genes were subcloned into expression vector pKK223-3. This construct was transformed into E.coli DH5α and cell lysates of the transformants were analysed by ELISA for their capability to synthesize polysialic acid. The ELISA was performed using monoclonal antibody 735, which is directed

Table I. Enzymatic activities of the gene products encoded by region A.

Clone	Mol.weight of the encoded polypeptide	enzymatic activity of NeuNAc-condensing enzyme	CMP-NeuNAc synthetase	Polysialyl-transferase
ORF1	38.000	0.01	0.002	n.t.
ORF2	18.200	n.t.	0.2	n.t.
ORF3	38.300	0.2	0.002	n.t.
ORF4	54.400	n.t.	n.t.	14
E.coli DH5α	-	0.02	0.006	0.025
pMF32.35	-	0.12	0.15	6.23

Determination of NeuNAc-condensing enzyme and CMP-NeuNAc synthetase activity (given as Units) according to the method of Warren and Blacklow (1962). The enzymatic activities were determined in the crude cytosolic fraction. For determination of the polysialyltransferase activity the membrane preparation was used and enzymatic activity was determined as described by Weisgerber and Troy (1990). Incorporation of ^{14}C labelled CMP-NeuNAc in µg per mg membrane protein was determined using colominic acid as exogenous acceptor.

against the capsular polysaccharide of group B meningococci (Frosch et al., 1985). After induction of the p_{tac} promoter the transformants expressed the capsular polysaccharide in about the same amounts as clone pMF32.35, indicating that other than the polypeptides encoded by the open reading frames 2-4 there are no other enzymes required for polysialic acid synthesis. Furthermore, these data indicate that there is only one sialyltransferase in the capsular polysaccharide biosynthesis pathway of meningococci, which seems to initiate polysialic acid synthesis by formation of an undecaprenol-NeuNAc molecule and that also directs chain elongation. These biosynthetic steps have been considered in the model for capsule expression shown in figure 2.

MECHANISMS OF CELL SURFACE TRANSLOCATION OF THE CAPSULAR POLYSACCHARIDE

The mature capsular polysaccharide is integrated into the outer membrane by a phospholipid moiety, which is linked to the reducing end of the polysaccharide chain (Gotschlich et al., 1981). However, it is unknown at which step in the biosynthesis pathway the phospholipid substitution of the polysaccharide chains occurs and whether phospholipid substitution is linked to the translocation process. As mentioned above, translocation of the capsular polysaccharide to the cell surface is directed by the gene products encoded by regions B and C. In region B mutants the capsular polysaccharide remains soluble in the cytoplasm (Frosch et al., 1989), whereas in region C mutants the capsular polysaccharide is associated with the inner membrane. However, mutants in either or both regions do not express capsular polysaccharide on the cell surface. We have purified the capsular polysaccharide from region B and region C mutants by affinity chromatography and analysed the polysaccharide chains by SDS- and Tris-borate buffered polyacrylamide gel electrophoresis (Pelkonen et al., 1988), and investigated the sensitivity of the polysaccharides to phospholipases. Summarizing the results of these experiments, we observed that region B mutants expressed full length polysaccharide chains, which exhibited no phospholipid moiety. In contrast, the polysaccharide of region C mutants does possess a phospholipid moiety, which presumably anchors the polysaccharide in the inner membrane. From these data we concluded that the gene products of region B direct the phospholipid substitution of the capsular polysaccharide chains. Nucleotide sequence analysis of region B revealed two open reading frames, which encode proteins with molecular masses of 45.1 and 48.7 kDa. These proteins were termed LipA and LipB according to their putative function of phospholipid substitution. Because the lipid-free polysaccharide of region B mutants is not translocated to the cell surface, we concluded that the phospholipid-mediated attachment to the inner membrane is a strong requirement for translocation across both membranes. Consequently, the transport process itself must be directed by the gene products encoded by region C.

The transporter proteins of region C share strong similarities with transporter proteins of the ABC (ATP-binding cassette) protein superfamily of active transporters (Higgins et al., 1986). Nucleotide sequence analysis of region C showed four open reading frames, which encode proteins with molecular masses of 41 kDa, 42 kDa, 30 kDa and 25 kDa, termed CtrA, CtrB, CtrC and CtrD. By use of transposon Tn<u>phoA</u> the 41 kDa CtrA protein was located in the outer membrane. In contrast, CtrB and CtrC were shown to be inner membrane proteins, which both are characterized by strong hydrophobic moments (Frosch et al., 1991). The presence of an outer membrane protein, CtrA, in an ABC transporter system is unusual. Secondary structure analysis of CtrA indicated eight amphipathic membrane spanning beta strands (Frosch et al., 1992), a structure which is known for proteins with porin properties (Jähnig, 1990). Therefore we propose that CtrA may form a pore in the outer membrane through which the mature capsular polysaccharide can pass.

Our findings on phospholipid substitution and surface translocation can be summarized in the model for capsule expression shown in figure 2. Interaction of the capsular polysaccharide with the inner membrane via its phospholipid moiety seems to be an absolute requirement for polysaccharide translocation to the cell surface, since lipid-free polysaccharides from region B

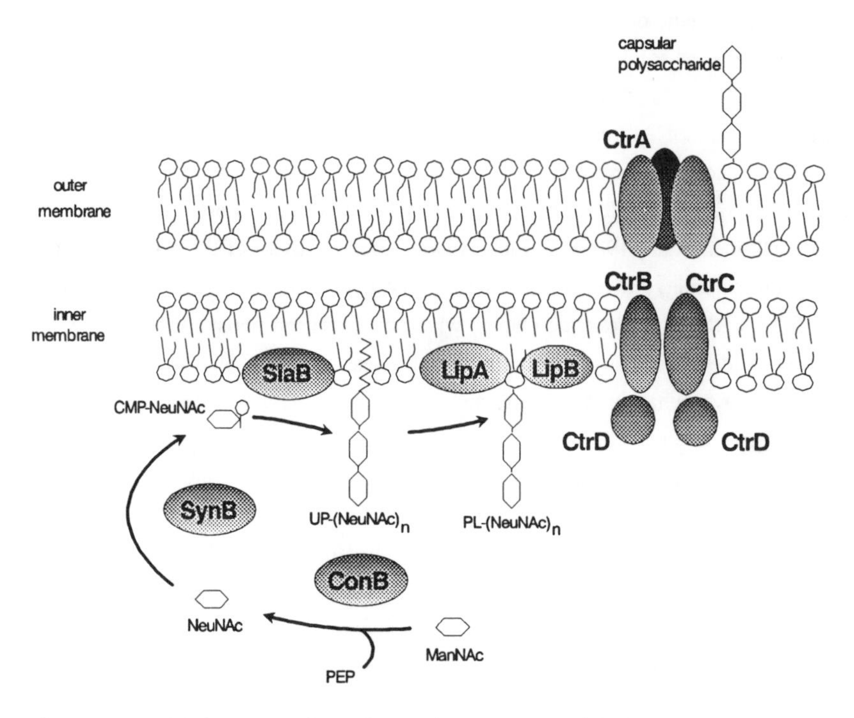

Figure 2. Model for expression of capsular polysaccharide in N.meningitidis. Designation of the gene products: ConB, NeuNAc condensing enzyme; SynB, CMP-NeuNAc synthetase; SiaB, sialyltransferase; LipA and LipB, proteins involved in phospholipid substitution of the capsular polysaccharide; Ctr gene products, capsule transport proteins. UP-NeuNAc, undecaprenol-NeuNAc; PL-NeuNAc, Phospholipid-NeunAc; PEP, phosphoenol pyruvate. We were not able to localize LipA and LipB in the inner membrane. However, the srong hydrophobic profile of these proteins (data not shown) and their putative function in phospholipid substitution of the polysaccharide chains imply the localization of LipA ans LipB in or at the inner membrane.

mutants are not translocated by the ABC transporters. Thus, it seems unlikely, that the ABC transporter proteins of the meningococcal cps gene complex form a hydrophilic pore in the inner membrane through which soluble, hydrophilic polysaccharide molecules can pass directly from the cytoplasm to the periplasm or to the cell surface. Our findings are in good agreement with the "flippase" mechanism for active transport as proposed for the P-glycoprotein (Higgins and Gottesman, 1992). In this model it was suggested that the hydrophobic substrates first interact with the inner lipid leaflet of the cytoplasmic membrane, before interaction with the ABC transporters and subsequent translocation to the outer leaflet of the cytoplasmic membrane or to the external medium. The necessity of phospholipid substitution of the meningococcal capsular polysaccharide and interaction of this lipid with the inner membrane as a condition for cell surface translocation confirms this model. Therefore our understanding of capsular polysaccharide transport provides knowledge of how ABC transporters in general act.

CAPSULE EXPRESSION IN MENINGOCOCCI: A MODEL FOR OTHER GRAM-NEGATIVE ENCAPSULATED BACTERIA?

There is evidence that the model for capsule expression in meningococci is also valid for other gram-negative bacteria which express group II capsular polysaccharides. This capsular polysaccharide type has been found, in addition to Neisseria meningitidis, in Haemophilus influenzae and also in some E.coli serotypes (Jann and Jann,1990). The functional regions within the capsule gene loci of these bacterial species show a similar molecular organization (Boulnois et al., 1987; Frosch et al., 1991; Roberts et al., 1988). Within the capsule gene complexes of different serotypes of N.meningitidis (Frosch et al., 1989; Frosch et al., 1991), H.influenzae (Kroll et al., 1989) and E.coli (Boulnois et al., 1987; Roberts et al., 1988) a central region encoding all enzymes necessary for biosynthesis of the serologically distinct capsular polysaccharides is flanked by two regions that are involved in translocation of the polysaccharide to the cell surface. This common organization of the functional regions of the capsule gene clusters in these bacteria provides evidence for a common evolutionary origin of capsule expression in these species.

This view is supported by the strong nucleotide and amino acid sequence homologies of single genes or gene products of the three bacterial species. This homology includes the proteins of the capsular polysaccharide transport system of N.meningitidis, H.influenzae and E.coli. The capsule transport proteins in H.influenzae, termed BexA, BexB, BexC, and BexD, respectively (Kroll et al., 1990), share identities at the amino acid sequence level between 59% and 82% compared to their meningococcal counterparts CtrA, CtrB, CtrC and CtrD. Furthermore, in E.coli two proteins, KpsM and KpsT, were described (Smith et al., 1990), which show a 24% and a 44% identity to CtrC and CtrD, respectively. The occurrence of proteins with sequence homologies and structural similarities to transporters of the ABC superfamily, indicate a common mechanism of capsular polysaccharide cell surface translocation in N.meningitidis, H.influenzae and E.coli. However, it should be noted that in addition to the ABC transporter proteins KpsM and KpsT, a 60 kDa periplasmic protein was described in E.coli, which seems to be part of the capsular polysaccharide export system. In contrast, no periplasmic proteins were found as part of the capsule expression system of N.meningitidis and H.influenzae. Therefore, there might be differences in the mechanisms of capsular polysaccharide cell surface translocation in E.coli compared to the meningococcal and presumably also the H.influenzae system. Further sequence analysis of the E.coli capsule gene cluster should provide a model for capsule expression in this bacterial species and should demonstrate differences with the model for capsule expression in meningococci.

In addition to the similarities of capsule export in N.meningitidis, H.influenzae and E.coli there are common features of capsular polysaccharide biosynthesis in N.meningitidis B and E.coli K1 (Frosch et al., 1991; Weisgerber et al., 1991), the capsular polysaccharide of which both is composed of α-2,8 linked polysialic acid (Kasper et al., 1973). We have determined the nucleotide sequence of the polysialyltransferase gene of both species. This enzyme catalyzes the specific α-2,8 linkage of the neuraminic acid monomers (Troy, 1979). An amino acid sequence identity of 30% was observed between the sialyltransferases of both bacterial species. Considering conserved amino acid substitutions there is a 42% identity (Frosch et al., 1991). This indicates that beside the capsule transport system the biosynthesis pathways of the E.coli K1 and the meningococcal B capsular polysaccharides both have a common ancestry. The occurrence of identically organized gene clusters with significant nucleotide and amino acid sequence homolgies presumably is due to the horizontal exchange of this virulence marker. Because N.meningitidis and H.influenzae are both characterized by natural competence for DNA uptake (Scocca, 1990; Smith et al., 1981), the horizontal exchange and aquisition of capsule genes can be explained and was recently shown to occur within different strains of meningococci (Frosch and Meyer, 1992). Thus, the stronger nucleotide and amino acid sequence similarities between N.meningitidis and H.influenzae compared to E.coli indicate that the exchange of the capsule genes between E.coli and N.meningitidis or H.influenzae occurred earlier than the exchange between N.meningitidis and H.influenzae.

ACKNOWLEDGEMENTS

This work was supported by grants from the Deutsche Forschungsgemeinschaft to MF. We thank Brian D. Robertson for critical suggestions on the manuscript.

REFERENCES

Band, J.D., Chamberland, M.E., Platt, T., Weaver, R.E., Thornberry, C., Frazer, D.W. (1983) J. Infect. Dis. 148: 754-758.
Boulnois, G.J., Roberts, I.S., Hodge, R., Hardy, K.R., Jann, K.B. and Timmis, K.N. (1987) Mol. Gen. Genet. 208: 242-246.
Calain, P., Poolman, J., Zollinger, W., Sperber, G., Bitter-Suermann, D., Auckenthaler, R., Hirschel, B. (1988) Eur. J. Clin. Microbiol. Infect. Dis. 7: 788-791.
DeVoe, I.W. (1982) Microbiol. Rev. 46: 162-190.
Edwards, U. and Frosch, M. (1992) FEMS Microbiol. Lett., 96: 161-166.
Finne, J., Leinonen, M., and Makela, P.H. (1983) Lancet 2: 355-357.

Frosch, M., Edwards, U., Bousset, K., Krausse, B. and Weisgerber, C. (1991) Mol. Microbiol. 5: 1251-1263.
Frosch, M., Görgen, I., Boulnois, G.J., Timmis, K.N. and Bitter-Suermann, D. (1985) Proc. Natl. Acad. Sci. USA 82: 1194-1198.
Frosch, M. and Meyer, T.F. (1992) FEMS Microbiol. Lett. in press.
Frosch, M., Müller, D., Bousset, K. and Müller, A. (1992) Infect. Immun. 60: 798-803.
Frosch, M., Weisgerber, C. and Meyer, T.F. (1989) Proc. Natl. Acad. Sci. USA 86: 1699-1673.
Higgins, C.F. and Gottesman, M.M. (1992) TIBS 17: 18-21.
Higgins, C.F., Hiles, I.D., Salmond, G.P. Gill, Downie, J.A., Evans, I.J., Holland, I.B., Gray, L., Buckel, S.D., Bell, A.W. and Hermodson, M.A. (1986) Nature (London) 323: 448- 450.
Gotschlich, E.C., Fraser, B.A., Nishimura, O., Robbins, J.B. and Liu, T.Y. (1981) J. Biol. Chem. 256: 8915-8921.
Jähnig, F. (1989) In: Prediction of protein structure and principles of protein conformation (G.D. Fasman Ed.) Plenum Publishing corporation, pp. 707-717.
Jann, B. and Jann, K. (1990) Curr. Top. Microbiol. Immunol. 150: 19-42.
Jarvis, G.A. and Vedros, N.A. (1987) Infect. Immun. 55: 174-180.
Kasper, D.L., Winkelhake, J.L., Zollinger, W.D., Brandt, B.L. and Artenstein, M.S. (1973) J. Immunol. 110: 262-268.
Kroll, J.S., Loynds, B., Brophy, L.N. and Moxon, E.R. (1990) Mol. Microbiol. 4: 1853-1862.
Kroll, J.S., Zamze, S., Loynds, B. and Moxon, E.R. (1989) J. Bacteriol. 171: 3343-3347.
Masson, L. and Holbein, B.E. (1983) J. Bacteriol. 154: 728-736.
Masson, L. and Holbein, B.E. (1985) J. Bacteriol. 161: 861-867.
Pelkonen, S. and Finne, J. (1988) J. Bacteriol. 170: 2646-2653.
Roberts, I.S., Mountford, R., Hodge, R., Jann, K.B. and Boulnois, G.J. (1988) J Bacteriol 170: 1305-1310.
Scocca, J.J. (1990) Mol. Microbiol. 4: 321-327.
Smith, A.N., Boulnois, G.J. and Roberts, I.S. (1990) Mol. Microbiol. 4: 1863-1869.
Smith, H.O., Danner, D.B. and Deich, R.A. (1981) Annu. Rev. Biochem. 50: 41-68.
Troy, F.A. (1979) Annu. Rev. Microbiol. 33: 519-560.
Warren, L. and Blacklow, R.S. (1962) Biochem. Biophys. Res. Commun. 7: 433-438
Weisgerber, C. and Troy, F.A. (1990) J. Biol. Chem. 265: 1578-1587.
Weisgerber, C., Hansen, A., Frosch, M. (1991) Glycobiol. 1: 357-365.
Zapata, G., Vann, W.F., Aaronson, W., Lewis, M.S. and Moos, M. (1989) J Biol. Chem. 264: 14769-14774.

GENETIC AND MOLECULAR ANALYSES OF THE POLYSIALIC ACID GENE CLUSTER OF Escherichia coli K1

Richard P. Silver, Paula Annunziato, Martin S. Pavelka, Ronald P. Pigeon, Lori F. Wright, and David E. Wunder

Department of Microbiology and Immunology, University of Rochester Medical Center, Rochester, New York 14642, U.S.A.

SUMMARY: The E. coli K1 capsule, an α-2,8-linked homopolymer of sialic acid, is an essential virulence determinant, allowing E. coli to evade host defense mechanisms. The 17-kb kps gene cluster, which is divided into three functionally distinct regions, encodes the information necessary for polymer synthesis and expression at the cell surface. In this communication we describe studies of genes within each region of the cluster. We identified a gene, neuD, at the proximal end of region 2. The neuD gene encodes a 23-kDa inner membrane protein which may be a component of the sialyltransferase complex. NeuD has significant similarity to several E. coli acyltransferases. Region 3 encodes two proteins, KpsM and KpsT, that constitute a system for transporting polysialic acid across the inner membrane. KpsM is an integral membrane protein and genetic and biochemical studies indicate that KpsT binds ATP. Region 1, which functions in transport of polymer to the cell surface, includes KpsD, a 60-kDa periplasmic protein. Heparin agarose chromatography was used to purify KpsD to near homogeneity. We presume that kps gene products function as a multicomponent complex and present evidence that translational coupling may provide a mechanism for balanced production of these components within the cell.

INTRODUCTION

Polysialic acid (PSA), an α -2,8 linked homopolymer of sialic acid (NeuNAc), is found on the surface of both prokaryotic and eukaryotic cells, and is associated with diverse biological phenomenon (Schauer, 1982, Troy, 1992). In bacteria, capsular polysaccharides composed of PSA are essential virulence determinants associated with invasive disease (Robbins et al., 1980).

The focus of our work is the polysialic acid capsule of Escherichia coli K1 (Silver and Vimr, 1990). E. coli that synthesize the K1 capsule are important pathogens of the neonate and account for more than 80% of E. coli neonatal meningitis (Robbins et al.,1974). Capsules are the outermost structure on the bacterial cell surface and interact directly with the host immune system. The PSA capsule provides the bacterium with an antiphagocytic barrier that is characterized by its ability to inhibit complement activation by the alternative pathway (Gemski et al., 1980, Pluschke et al., 1983) As a consequence of the structural identity between the K1 polysaccharide and the polysialosyl chains found on host tissue (Finne et al., 1987, Sodestrom et al., 1984), the K1 capsule is also a poor immunogen in humans and other mammals (Wyle et al., 1972).

Although PSA is a relatively simple polymer, the biosynthetic pathway leading to the synthesis of this structure in E. coli is a complex process (Troy, 1992), and involves (a) synthesis of NeuNAc, (b) activation of NeuNAc to CMP-NeuNAc, (c) polymerization of NeuNAc from CMP-NeuNAc into the homopolymer by a membranous sialyltransferase complex and (d) translocation of the sialyl polymer to the bacterial cell surface. These reactions involve a lipid intermediate, NeuNAc-p-undecaprenol, and synthesis requires endogenous sialyl acceptors within the membrane (Troy, 1992). A 20-kDa sialylated membrane protein has recently been identified in E. coli K1 and postulated to be endogenous acceptor (Weisgerber and Troy, 1990). The polymer is composed of about 200 sialic acid residues that terminate in a phosphodiester linkage to 1,2-diacylglycerol (Gotschlich et al., 1981; Schmidt and Jann, 1982).

Our laboratory has been interested in elucidating the molecular and genetic events controlling the synthesis, assembly, and transport of polysialic acid in E. coli K1. The genes encoding proteins necessary for synthesis and expression of the PSA capsule of E. coli K1 have been cloned and characterized (Boulnois and Roberts, 1990; Silver et al., 1984; Vimr et al., 1989). The 17-kb kps gene cluster is divided into three functional regions (Boulnois et al., 1987). The central region 2 contains information for synthesis, activation, and polymerization of sialic acid and is unique for a given polysaccharide antigen (Boulnois and Roberts, 1990; Vimr et al., 1989). In contrast, genes in region 1 and 3 are conserved among E. coli synthesizing serologically distinct capsules (Roberts et al., 1986, 1988; Silver et al., 1987). Region 3 genes are postulated to be involved in transport of polysialic acid across the cytoplasmic membrane (Pavelka et al.,1991; Smith et al., 1990), while region 1 genes appear to function in the transport of polymer to the bacterial cell surface (Boulnois and Roberts, 1990; Silver et al., 1987). Cells harboring mutations in region 2 do not synthesize polymer, while cells harboring mutations in either region 1 or 3 accumulate intracellular polymer (Boulnois and Jann, 1989; Boulnois and Roberts, 1990, Silver and Vimr, 1990; Vimr et al., 1989). This review will highlight our work within each region of the kps gene cluster.

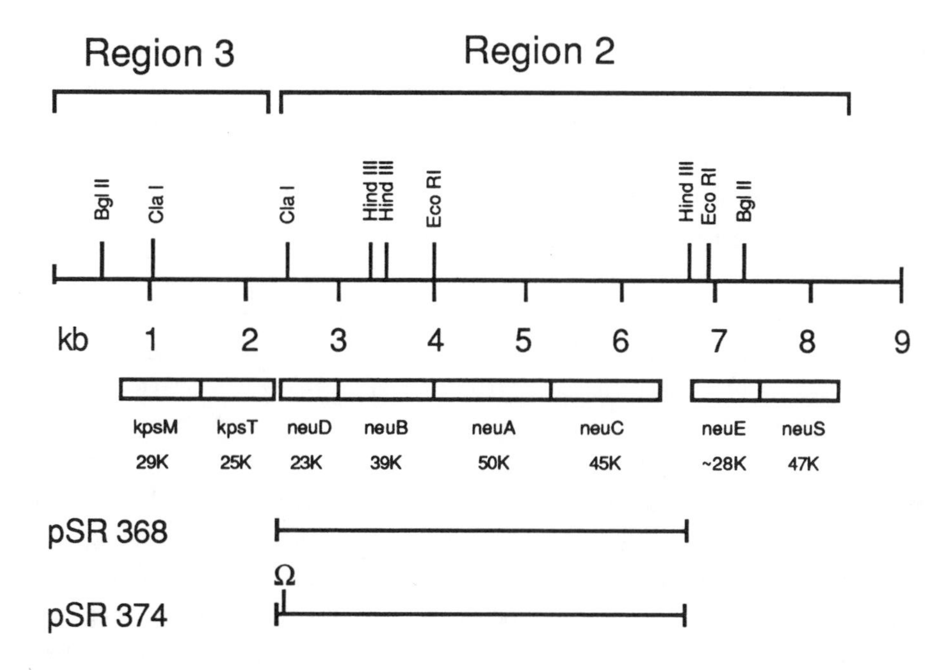

Fig. 1. Genetic organization of regions 2 and 3 of the kps gene cluster of Escherichia coli K1. The location of the kps genes are indicated by the open boxes. The numbers below the boxes refer to the molecular mass of the proteins. The plasmids pSR368 and pSR374 contain the neuD, neuB, neuA, and neuC gene cloned in a Bluescript SK+ vector.

REGION 2: THE neuD GENE

The central region 2 of the kps gene cluster is 5.8-kb and contains information for sialic acid synthesis, activation, and polymerization (Boulnois and Roberts, 1990; Vimr et al., 1989). The genetic organization of the region has been established and most of the nucleotide sequence determined (Fig. 1). Two genes, neuB and neuC appear to be involved in sialic acid synthesis. Cells harboring mutations in neuB or neuC are defective in capsule synthesis and synthesize capsular polysaccharide when exogenous sialic acid is provided (Silver et al., 1984; Vimr et al., 1989; Zapata et al., 1992). The neuA gene encodes the 48.6-kDa CMP-NeuNAc synthetase which catalyzes the activation of NeuNAc to the sugar nucleotide (Zapata et al., 1989). The neuA gene product has been purified (Vann et al., 1987). The 47-kDa product of the neuS gene has recently been identified as the sialyltransferase required for polymerization of NeuNAc

(Steenbergen and Vimr, 1990; Steenbergen et al., 1992; Weisgeber et al.,1991). The neuE gene, located 5' to neuS, encodes a protein of about 28-kD that contains a 12 amino acid conserved sequence with significant homology to the polyisoprenol (dolichol) recognition sequence (Steenbergen et al., 1992; Troy, 1992). It has been suggesed that NeuE is a sialyltransferase that may initiate polymer synthesis by catalyzing the formation of NeuNAc-p-undecaprenol from CMP-NeuNac and undecaprenyl phosphate (Steenbergen et al., 1992; Troy et al., 1992). We have recently identified an additional gene, designated neuD, at the proximal end of region 2. Our observation that cells harboring mutations in neuD do not synthesize intracellular polymer is consistent with the view that neuD is a region 2 gene.

By translation of the nucleotide sequence, NeuD is predicted to have a molecular mass of 22.9-kDa. The protein has a positively charged amino-terminal region followed by a hydrophobic domain suggestive of a signal sequence. To determine if NeuD is a secreted protein, we constructed a fusion between the first 31 amino acids of NeuD and the periplasmic protein alkaline phosphatase (PhoA), lacking its signal sequence. To be enzymatically active alkaline phoshatase must be exported to the periplasm. Cells harboring the fusion plasmid, pSR414, lacked PhoA activity indicating that the amino-terminal portion of NeuD could not promote transport of alkaline phosphatase across the cytoplasmic membrane to the periplasmic space. To identify the neuD encoded gene product, we expressed the gene using the T7 promoter/polymerase system (Tabor and Richardson, 1985). We observed a protein of about 30-kDa suggesting that NeuD has a slower mobility in SDS-PAGE gels than expected from the calculated molecular mass (Fig. 2). We determined the subcellular location of NeuD by screening for labeled protein in membrane, cytosolic, and periplasmic fractions. As can be seen in Fig. 2, most of the protein was recovered in the membrane fraction. These results are consistent with the interpretation that NeuD is an integral protein of the cytoplasmic membrane and that the amino terminal region may constitute a membrane anchoring region.

The biological role of NeuD is not known. Cells harboring mutations in neuD do not synthesize polymer. Recent experiments suggest, however, that the protein is not involved in sialic acid synthesis. First, the addition of sialic acid to cells carrying a non-polar neuD mutation does not lead to synthesis of polymer. Secondly, cells harboring a mutation in neuD still synthesize sialic acid. For these experiments, the plasmid pSR368, carrying the neuD, neuB, neuA and neuC genes, was used. The plasmid pSR374 was derived from pSR368 by insertion of the Omega fragment into the ClaI site of the neuD gene (Fig. 1). The 2.0-kb Omega fragment encodes streptomycin/spectinomycin resistance and carries both transcription termination and translational stop signals (Prentki and Krisch, 1984). Cells harboring either plasmid, however, accumulate sialic acid and CMP -sialic acid (W. F. Vann, personal communication). These results suggest that the neuD gene product is not required for sialic acid synthesis and may be a component of the sialyltransferase complex. It is tempting to speculate that NeuD is the endogenous acceptor essential for the initiation of polymer synthesis.

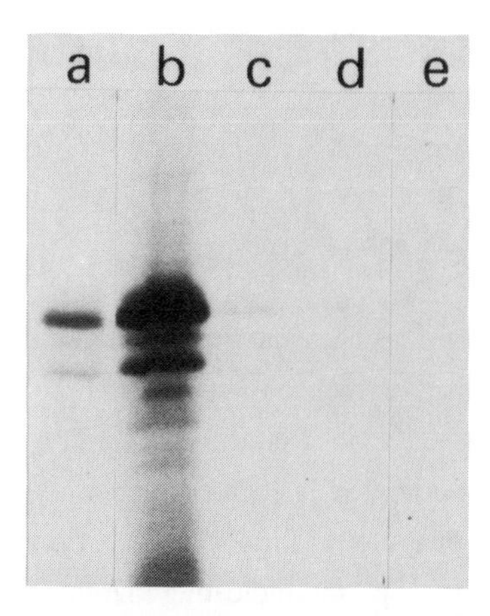

Fig. 2. Autoradiogram of [^{35}S]-methionine labeled NeuD analyzed by elecrophoresis in a 12% SDS-polyacrylamide gel. Lanes: (a) whole cells extract; (b) total membranes, (c) periplasmic fraction, (d) cytoplasmic fraction, (e) vector control.

NeuD shares sequence similarity with several acyltransferases encoded by E. coli, including the lacA and cysE genes. For example, NeuD shows 27% identity with LacA, a thiogalactoside acetyl transferase. When identical and conserved amino acids are considered, the proteins are 57% similar. The strongest homology was found within the carboxy-terminal portion of the proteins (Fig. 3) and contains a common motif associated with a class of acyltransferases from enteric bacteria (Downie, 1989). The homology centers on the sequence IGAGSVV and Downie (1989) suggested that this region may be part of the active site for acetyl transferase. We feel it

```
Neu D      154 SVGEETFVGSVTVVNGQLKLGSKSIIGSGSVVIRNIPSNVVVAGTPTRLI 203
               .:|::.::|| .|:|..:.:|..|:||.||:|.::||.|||.||.|.|:|
Lac A      133 TIGNNVWIGSHVVINPGVTIGDNSVIGAGSIVTKDIPPNVVAAGVPCRVI 182
               .| :.| ||. . | ..:.:| .. |||||:| ..:||:..|||||.|::
Cys E      195 KIREGVMIGAGAKILGNIEVGRGAKIGAGSVVLQPVPPHTTAAGVPARIV 244

Consensus       I   V IG    I   V IG    IGAGSVVT  VP    A G P RV
```

Fig. 3. Amino acid homology between NeuD, LacA (thiogalactoside acetyl transferase), and CysE (serine acetyl tranferase). The sequences were aligned using the GAP program of the University of Wisconsin Genetics Computer Group. Lines indicate identical amino acid residues, while double and single dots represent similar residues with comparison values of $\geq$0.50 and $\geq$0.10, respectively. NeuD and LacA share 43.7% identity and 70.8% similarity, while NeuD and CysE are 31.2% identical and 54.2% similar. The consensus sequence is from Downie, 1989.

unlikely, however, that neuD encodes the O-acetyltransferase responsible for polysialic acid acetylation (Orskov et al., 1979, Higa and Varki, 1988). Cells harboring mutations in neuD have an acapsular phenotype and do not synthesize detectable polymer. Moreover, only OAc^- polymer was detected in E. coli K12 cells harboring the cloned kps gene cluster (Vann and Silver, unpublished observation).

REGION 3: THE kpsMT OPERON:

Capsular polysaccharide synthesis has been shown to occur on the cytoplasmic face of the inner membrane of E. coli, requiring the polymer to traverse two lipid bilayers to be fixed on the cell surface (Boulnois and Jann, 1989, Kroncke et al., 1990). The observation that the functional domain of the sialyltransferase complex is located on the cytoplasmic surface of the inner membrane is consistent with this view (Troy, 1992). Cells with mutations in region 3 accumulate polysaccharide in the cytoplasm, indicating that this region is important for export of the polymer from the cytoplasm.

Region 3 contains an operon of two genes, kpsM and kpsT (Pavelka et al., 1991, Smith et al.,1990). The operon is transcribed from a promoter located 745 base pairs upstream of the kpsM start codon (Silver and Zhao, unpublished observation). The kpsM gene product is a 29-kDa, highly hydrophobic, integral membrane protein. The KpsT protein is a 25-kDa, hydrophilic protein containing a consensus ATP-binding site (Walker et al., 1982). They belong to a large family of prokaryotic and eukaryotic membrane translocators involved in diverse biological processes (Higgins et al., 1988). Most of the family members comprise various membrane transport systems that are believed to move their respective substrates across membranes via energy obtained from ATP hydrolysis (Ames and Joshi, 1991). They have been referred to as the ABC (ATP-binding-cassette) superfamily of transport proteins, or traffic ATPases, and include the periplasmic permeases of enteric bacteria, the Cystic Fibrosis Transductance Regulator, and the P-glycoprotein, a mammalian drug resistance pump (Ames and Lecar, 1992, Blight and Holland, 1992, Higgins et al., 1988). The phenotype of region 3 mutations and the characteristics of KpsM and KpsT have led us and others to propose that these two proteins function together in the transport of capsular polysaccharide across the inner membrane (Pavelka et al., 1991, Smith et al, 1990).

KpsM is a very hydrophobic protein with at least six potential transmembrane helical segments. We have taken a genetic approach, using B-lactamase fusions (Broome-Smith et al., 1990), to analyse the membrane topolgy of KpsM. Thus far, 24 KpsM-BlaM fusions have been generated and characterized yielding a preliminary topology map of KpsM. The results are consistent with the structure predicted from the hydropathy profile of the protein (Pavelka et al., 1991). The second transmembrane domain of KpsM also contains a potential dolichol recognition sequence (Troy, 1992). This conserved sequence may be involved in undecaprenol recognition and provide a mechanism to coordinate synthesis and transmembrane translocation of polysialic acid (Troy, 1992).

Our current model for polysaccharide transport postulates that KpsT is associated with the inner membrane. To examine the location of KpsT within the cell, whole cell lysates, soluble fractions, and inner and outer membranes purified on sucrose gradients, were probed with a polyclonal KpsT antiserum. As expected, KpsT was absent in fractions prepared from a chromosomal kpsT deletion strain. We detected KpsT, however, in all fractions derived from a wild-type strain. It has been documented that membrane purification from E. coli expressing extracellular polysaccharide is difficult, and often leads to cross-contamination (Achtman et al., 1983; Whitfield et al., 1985). Consequently, we examined fractions prepared from an acapsular mutant recently isolated and characterized in our laboratory. This strain has an IS1 element inserted into the kpsM gene and does not transport polymer to the cell surface. KpsT was found in the whole cell lysate and in the soluble fraction, as well as in the inner membrane fraction, but was not present in the outer membrane fraction. These results are consistent with the view that KpsT is likely a peripheral inner membrane protein.

We have also investigated the interaction between KpsT and ATP. We previously reported that a site-directed mutation, changing the conserved lysine at position 44 within the glycine-rich consensus ATP-binding domain of the protein to a glutamic acid residue, results in the production of a non-functional mutant protein (KE44) (Pavelka et al., 1991). We have now performed saturation mutagenesis throughout the ATP-binding domain and have isolated five additional mutants that produce non-functional proteins. Each of the mutations changes a single conserved residue within the consensus ATP-binding domain of the protein and is no longer able to complement a chromosomal kpsT deletion mutation.

The genetic data supports the idea that ATP is important for KpsT function. In addition, we obtained biochemical evidence that KpsT binds ATP with assays using the photo-affinity ATP analog, 8-azido ATP. Binding of the analog to a protein results in a covalent linkage during the irradiation step that can be detected as a shift in the mobility of the protein on SDS-PAGE. The derivatized protein has a slower mobility relative to the underivatized species. As can be seen in Fig. 4, the wild-type KpsT protein was derivatized with good efficiency, while KE44 was derivatized very poorly. Our results indicate that ATP binding, and we presume hydrolysis, is important to KpsT function and capsule expression.

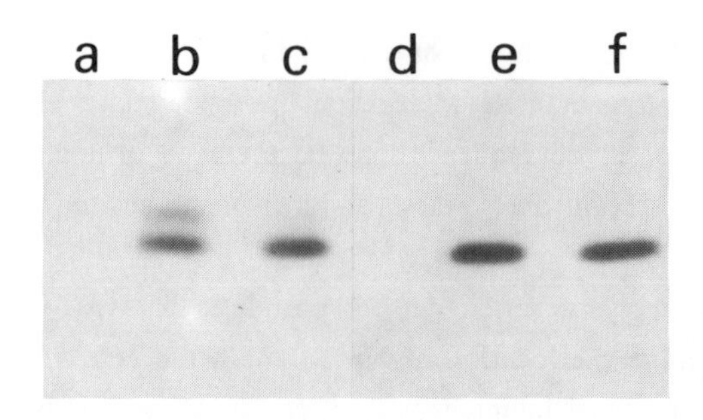

Fig. 4. Binding of 8-azido ATP to KpsT. Total membranes were prepared from strain EV95 (kpsT::Tn10) (Vimr et al., 1989) containing either a Bluescript vector control (a,d); a plasmid overexpressing either wild-type KpsT (b,e); or the KE44 mutant (c,f). Extracts were incubated with 8-azido ATP on ice, and irradiated with short-wave UV light (a, b, c). The reactions were quenched and the products separated on an SDS-PAGE gel, immunoblotted and probed with anti-KpsT sera. Irradiation of the membranes in the absence of 8-azido ATP had no effect on protein mobility (d, e, f). For preparation of polyclonal antisera, inclusion bodies from cells overexpressing KpsT were run on SDS-PAGE and the KpsT band was excised and injected into rabbits.

REGION 1: THE KpsD PROTEIN

We still lack significant insight into the mechanism whereby capsular polysaccharides are translocated from the cytoplasmic membrane, through the periplasmic space, to the bacterial cell surface. It had been postulated that export of polymer occurs at sites of adhesion between the inner and outer membranes (Bayer, 1979). Studies with membranes from K1 E. coli grown at 15°C also suggest a role for zones of adhesion in PSA expression. These membranes are unable to synthesize polymer and undergo an incubation dependent activation of polymer synthesis at 33°C (Troy, 1992). Activation is localized to a low density vesicle fraction and is obligatorily coupled to protein synthesis (Whitfield et al., 1984a, 1984b). These low density vesicles contain proteins found in both the inner and outer membranes and were interpreted as representing zones of adhesion between inner and outer membranes. However, recent evidence based on electron microscopy utilizing cryofixation argues against the existence of adhesion zones (Kellenberger, 1990). How then does a polysaccharide migrate through the periplasmic gel, pass the peptidoglycan barrier, and traverse the outer membrane to the exterior of the cell?

Region 1 of the kps gene cluster is postulated to be involved in this transport process and encodes five proteins with apparent molecular masses of 75, 60, 45, 37 and 27-kDa (Silver et al., 1984; Boulnois and Roberts, 1990). We have shown that one of these genes, kpsD, encodes a 60-kDa periplasmic protein that is necessary for extracellular expression of polysialic acid (Silver et al., 1987). A chromosmal mutation in kpsD results in loss of surface expression of the

K1 capsule (Wunder and Silver, unpublished observation). Immunodiffusion analysis of osmotic shock fractions and culture supernatants of kpsD mutant cells detected polysialic acid in the periplasmic fractions but not in culture supernatants. The kpsD mutant was complemented by a plasmid encoding a wild-type copy of the kpsD gene. These observations are consistent with previously described region 1 mutations in which polysaccharide is not expressed extracellularly, but accumulates in the periplasmic space.

We determined the nucleotide sequence of the kpsD gene and detected an open reading frame encoding a protein of 557 amino acids with a predicted molecular mass of 60.4-kDa (Silver and Aaronson, unpublished observation). A kpsD specific probe hybridized to chromosomal DNA of E. coli that synthesize the K2, K5, K7, K12 and K13 capsular polysaccharides (Silver et al., 1987). KpsD, as well as other region 1 genes, apparently possess the ability to transport a range of acidic polysaccharides to the cell surface. The occurrence of positively charged residues throughout KpsD may be important for interaction with the negatively charged polysaccharide moieties. Interestingly, we have purified KpsD from osmotic shock extracts to near homogeneity by heparin agarose chromatography.

KpsD is a hydrophilic protein with hydrophobic amino and carboxy termini. The amino-terminal hydrophobic region of the protein contains a characteristic 20 amino acid prokaryotic signal sequence for secretion to the periplasmic space. The 19 carboxy-terminal amino acids have the characteristics of a transmembrane α helix. To study this region of the molecule, we constructed a deletion derivative of KpsD which results in the production of a protein lacking the 11 carboxy-terminal amino acids. Surprisingly, the level of expression of the mutant protein was approximately three to five-fold higher than that of wild-type. The protein was, however, non-functional and localized primarily to the cytoplasmic membrane. These experiments indicate that KpsD, like B-lactamase and alkaline phosphatase, requires an intact carboxy-terminus for efficient export to the periplasmic space.

The isolation of polysialic acid from the periplasm of mutant cells need not imply passage of polymer through the periplasm during the translocation process. We cannot rule out the possibility that KpsD, or other region 1 gene products, are needed to connect export machinery of the inner membrane, presumably KpsM and KpsT, directly to a protein in the outer membrane thereby allowing polysaccharide to bypass the periplasmic space. A functional porin in the E. coli outer membrane is required for capsule expression (Foulds and Aaronson, 1984). Although the precise role of porins in the transport process has not been determined, it is an attractive target for specific interactions with region 1 gene products. Interestingly, the presence of a particular porin, protein K, has been correlated with capsule expression in most naturally occurring encapsulated strains of E. coli and there appears to be a temporal correlation between the insertion of protein K into membranes and the expression of the K1 capsule on the cell surface (Paakkanen et al., 1979; Sutcliffe et al., 1983; Whitfield et al., 1985). Although any porin may be able to mediate capsule expression there is no information on the relative efficiency of different porin types in capsule expression and protein K may be the natural component of the transport machinery.

REGULATION OF POLYMER SYNTHESIS: TRANSLATIONAL COUPLING

The notion that the various steps of polymer synthesis, assembly, and transport are coordinated both temporally and spatially is implicit in our current view of polysialic acid synthesis. It seems unlikely that polymer is synthesized in one part of the cell while the transport machinery assembled at another. We presume that all required components for biosynthesis and transport are produced in equimolar amounts and assembled at a defined locus as part of a multicomponent complex. A major challenge is to determine how the synthesis and expression of these components are regulated.

The rate of gene expression in E. coli was, at one time, assumed to be controlled exclusively at the transcriptional level. It is now evident, however, that post-transcriptional events are also of critical importance. The nucleotide sequence of the intercistronic region between kpsM and kpsT shows an interesting feature. The chain termination codon of kpsM overlaps the initiation codon of kpsT by two nucleotides (Pavelka et al., 1991). Such an overlap of a translational initiation signal with the termination codon of an upstream gene was first observed in the tryptophan operon of E. coli and is associated with the phenomenon of translational coupling (Oppenheim and Yanofsky, 1980). In translationally coupled genes the efficient translation of downstream genes is dependent on the translation of the distal portion of the upstream gene.

We have investigated the phenomenon of translational coupling by comparing gene expression distal to the wild-type and a nonsense mutant allele of kpsM. For these experiments an amber mutation in the kpsM gene of pSR356 was constructed. The plasmid pSR356 carries the entire kpsM gene and the first 29 codons of kpsT fused in frame to codon 9 of the lacZ gene of the vector pMLB1034. The fusion of kpsT to lacZ allows us to monitor kpsT translation by synthesis of B-galactosidase. To eliminate the contribution of transcriptional polarity to B-galactoside expression the fusion gene operon was transcribed from a T7 promoter with T7 RNA polymerase. In contrast to E. coli RNA polymerase, interruption of translation does not cause T7 RNA polymerase to terminate transcription prematurely. We produced an amber mutation in codon 129 of kpsM by overlap extension using the Polymerase Chain Reaction.

When compared to cells carrying the wild-type plasmid, cells harboring the mutant plasmid, pSR357, showed a 60% reduction in B-galactosidase activity. We conclude from these experiments that translation of kpsT is coupled, to some degree, to the translation of kpsM. Since it is likely that kps gene products function as a complex within the cell, translational coupling may provide an effective mechanism to insure balanced production of polypeptides that function as a multicomponent complex. Interestingly, a similar overlap of proximal and distal genes is common in region 2 of the E. coli K1 kps gene cluster and has been observed between neuD and neuB, neuB and neuA, neuA and neuC and neuE and neuS (Silver and Wright, unpublished observation; Zapata et al.,1992; Steenbergen et al., 1992). Both transcriptional and post-transcriptional events need to be examined to fully understand regulatory mechanisms involved in the synthesis and expression of polysialic acid in E. coli.

CONCLUSIONS: The biosynthesis of the polysialic acid capsule of E. coli K1 is a complex process involving synthesis, activation, and polymerization of sialic acid subunits into a large polymer which must be transported through two lipid bilayers and anchored to the cell surface. Considerable progress has been made in understanding this process. Knowledge of the organization, function, and control of the genes involved is key to this understanding. Most of the genes of the kps cluster of E. coli have been identified and the genetic organization determined. The major challenge for the future will be to determine the functions of the protein products of the kps cluster. Moreover, while the pathways for the biosynthesis of bacterial cell surface components are becoming well defined, we have little knowledge of the mechanisms that regulate and control their synthesis and transport. We have even less of an understanding of how these pathways are coupled so that membranes retain their normal composition during growth. Synthesis of the K1 polysaccharide, lipopolysaccharide, as well as peptidoglycan, employ the same lipid carrier, polyisoprenol phosphate. Perhaps the availability of this molecule provides a mechanism for the simultaneous regulation of the production of these distinct surface components. In any event, the structures that constitute the bacterial cell envelope cannot be dealt with as unrelated entities and future studies should focus on how a bacterium is able to coordinate synthesis of distinct surface structures to permit orderly growth in vivo.

ACKNOWLEDGMENTS

We wish to thank Willie F. Vann, Eric R. Vimr, and Frederic A. Troy for valuable discussions. We are also grateful to Willie F. Vann for reviewing the manuscript. This work was supported by Public Health Service grants AI-26655 and S07RR05403-29 from the National Institutes of Health (USA).

REFERENCES

Achtman, M., Mercer, A., Kusecek, B., Pohl, A., Heuzenroeder, M., Aaronson, W., Sutton, A., and Silver, R. P. (1983) Infect. Immun. 39, 315-335.
Ames, G. F.-L. and Joshi, A. K. (1990) J. Bacteriol. 172, 4133-4137.
Ames, G. F.-L. and Lecar,.H. (1992) FASEB J., 6, 2660-2666
Bayer, M. E. (1979) In: Bacterial Outer Membranes (M. Inouye, Ed.), John Wiley & Sons, Inc., New York, pp. 167-203
Blight, M. A., and Holland, I.B. (1990) Mol Microbiol. 4, 873-880.
Boulnois, G. J.and Jann, K. (1989) Mol Microbiol. 3, 1819-1823.
Boulnois, G. J.and Roberts, I. R. (1990) Curr. Top. Microbiol. Immunol. 150, 1-18.
Boulnois , G. F., Roberts, I. S., Hodge, R., Jann, K., and Timmis, K. N. (1987) Mol. Gen. Genet. 200, 242-246.
Broome-Smith, J. K., Tadayyon, M. and Zhang, Y. (1990) Mol. Microbiol. 4, 1637-1644.
Downie, J. A. (1989) Mol. Microbiol. 3, 1649-1651.
Finne, J., Leinonen, M., Makela, P.N. (1983) Lancet ii:355-357.
Foulds, J. and Aaronson, W. (1984) Amer. Soc. Microbiol., Abstr. Annu. Meet., D-21.
Gemski, P., Cross, A. S. and Sadoff, J. C. (1980) FEMS Microbiol Lett. 9, 193-197.
Gotschlich, E. C., Fraser, B. A., Nishimura, O., Robbins, J. B. and Liu, T.-Y. (1981) J. Biol. Chem. 256, 8915-8921.
Higgins, C. F., Gallagher, M. P., Mimmack, M. L. and Pearce, S. R. 1988. Bioessays 8, 111-116.
Kellenberger, E. (1990) Mol. Microbiol. 4, 697-705.
Kroncke, K. D., Golecki, J. R. and Jann, K. (1990) J. Bacteriol. 172, 3469-3472.
Oppenheim, D. S. and Yanofsky, C. (1980) Genetics 95, 785-795.
Pavelka, M. S.,Wright, L. F. and Silver, R.P. (1991) J. Bacteriol. 173, 4603-4610.
Prentki, P. and Krisch, H. A. (1984) Gene 29, 303-313.
Roberts, I., Mountford, R., High, N., Bitter-Suermann, D., Jann, K., Timmis,
K. N. and Boulnois, G. (1986) J. Bacteriol. 168, 1228-1233.
Roberts, I., Mountford, R., Hodge, R., Jann, K., Jann,B. and Boulnois, G. (1988) J. Bacteriol. 170, 1305-1310.
Robbins, J. B., McCracken, G. H. Jr., Gotschlich, E. C., Orskov, I., Orskov, F. and Hansson, L. A. (1974) N. Engl. J. Med. 290, 1216-1220.
Robbins, J. B., Schneerson, R., Egan, W. B., Vann, W. F. and Liu, D.-T. (1980) In: The Molecular Basis of Microbial Pathogenicity (H. Smith, J. J. Skebel and M. J. Turner, eds.) Verlag Chemie, pp. 115-132
Schauer, R. (1982) Adv. Carbohydr. Chem. Biochem. 40, 131-194.
Schmidt, A. M. and Jann, K. (1982) FEMS Microbiol. Lett. 14, 69-74.
Silver, R. P., Aaronson, W. and Vann, W. F. (1987) J. Bacteriol. 169, 5489-5495.
Silver, R. P., Vann, W. F. and Aaronson, W. (1984) J. Bacteriol. 157, 568-575.
Silver, R. P., and Vimr, E. R. (1990) In: The Bacteria, Vol 11: Molecular basis of bacterial pathogenesis (B. Iglewski and V. Clark, eds) Academic Press, Inc., New York , pp. 39-60.
Smith, A. N., Boulnois, G. J. and Roberts, I. R. (1990) Mol. Microbiol. 4, 1863-1869.
Soderstrom, J., Hansson, L. and Larson, G. (1984) N. Engl. J. Med. 310, 726.
Steenbergen, S. M. and Vimr, E. R. (1991) Mol. Microbiol. 4, 603-611.
Steenbergen, S. M., Wrona, T. J. and Vimr, E. R. (1992) J. Baacteriol. 174, 1099-1108.
Sutcliffe, J., Blumenthal, R., Walter, A. and Foulds, J. (1983) J Bacteriol. 156, 867-872.
Tabor, S. and Richardson, C. C. (1985) Proc. Natl. Acad. Sci. U S A, 82, 1074-1078.
Troy, F. A. (1992) Glycobiol. 2, 5-23.
Vann, W. F., Silver, R. P., Abeijon, C., Chang, K., Aaronson, W., Sutton, A., Finn, C. W., Lindner, W. and Kotsatos, M. (1987) J. Biol. Chem. 262, 17556-17562.
Vimr, E. R., Aaronson, W. and Silver, R. P. (1989) J. Bacteriol. 171, 106-1117.

Weisgerber, C. and Troy, F.A. (1990) J. Biol. Chem. 265, 1578-1587.
Weisgerber, C., Hansen, A. and Frosch, M. (1991) Glycobiol. 1, 357-363.
Whitfield, C., Adams, D. A. and Troy, F. A. (1984a) J. Biol. Chem. 259, 12769-12775.
Whitfield, C. and Troy, F. A. (1984b) J. Biol. Chem. 259, 12776-12780.
Whiffield, C., Vimr, E. R., Costerton, J. W. and Troy, F. A. (1985) J. Bacteriol. 161, 743-749.
Wyle, F. A., Artenstein, M. S., Brandt, B. L., Tramont, E. C., Kasper, D. L., Altieri, P. L., Berman, S. L. and Lowenthal, J. P. (1972) J. Infect. Dis. 126, 514-522.
Zapata, G., Crowley, J. and Vann, W. F. (1992) J. Bacteriol. 174, 315-319.
Zapata, G., Vann, W. F., Aaronson, W., Lewis, M. S. and Moos, M. (1989) J. Biol. Chem. 264, 14769-14774.

Polysialic Acid
J. Roth, U. Rutishauser and F. A. Troy II (eds.)

MECHANISMS OF POLYSIALIC ACID ASSEMBLY IN ESCHERICHIA COLI K1: A PARADIGM FOR MICROBES AND MAMMALS

Eric R. Vimr and Susan M. Steenbergen

Department of Pathobiology, College of Veterinary Medicine, University of Illinois, Urbana, Illinois, 61801 USA

SUMMARY: We developed a genetic system to investigate the molecular mechanisms of α2,8-linked polysialic acid (PSA) capsule synthesis, translocation, and regulation in the neuroinvasive bacterium Escherichia coli K1. The 12 to 14 genes required for these processes are located in a multigenic *kps* cluster at chromosome unit 64. The cluster is composed of a central group of biosynthetic *neu* genes (region 2) that are flanked on either side by region 1 or 3 *kps* genes encoding general functions for PSA regulation, assembly, and translocation. The polysialyltransferase (polyST) encoded by K1 *neuS* was sequenced and compared to its homolog in K92 E. coli, which synthesizes PSA chains with alternating sialyl α2,8-2,9 linkages. The results indicate that polySTs are processive enzymes which catalyze sequential transsialylations from donor CMP-sialic acid molecules to the nonreducing end of nascent PSA chains. We propose that the polymerase functions in a complex that includes the region 2 gene product of *neuE* and region 1 and 3 gene products of *kpsMTSCDE*. NeuE appears to function in the initiation or termination of PSA synthesis and may interact with polyprenol, as suggested by a dolichol-like binding site located in its predicted C-terminal membrane-spanning domain. These conclusions are supported by phenotypic analysis of mutants with multiple defects in sialic acid synthesis, degradation, and polymerization.

INTRODUCTION

We are investigating the mechanisms of polysialic acid (PSA) synthesis, regulation, and translocation in the neuroinvasive bacterium Escherichia coli K1. In addition to K1 E. coli, PSA chains composed of unbranched sialyl α2,8 linkages are synthesized by a few other pathogenic bacteria, but this unusual biosynthetic capability is more widely evident as the extensive polysialylations of eukaryotic neural cell adhesion molecules (NCAM) and sodium channel polypeptides (reviewed in Silver and Vimr, 1990). In bacteria, the PSA capsule is a virulence factor by virtue of its relatively low immunogenicity and inhibition of complement-mediated killing. PSA chains composed of α2,9 or alternating α2,8-2,9 linkages also appear to serve immune-protective functions in Neisseria meningitidis group C and E. coli K92, respectively (Silver and Vimr, 1990). Structural similarities between microbial and eukaryotic PSA suggest that the molecular machinery for PSA synthesis and regulation may be functionally equivalent in the different species. Our working hypothesis suggests that the mechanism of PSA synthesis in all of these systems is basically similar to the one used by E. coli K1 (Vimr et al., 1992). The variety of techniques available to E. coli geneticists thus makes K1 an attractive system for further investigation of PSA biosynthesis.

In vertebrates, much of our understanding of PSA's role(s) in neural and other tissue morphogenetic pathways comes from perturbation experiments in which a unique phage depolymerase, endo-*N*-acylneuraminidase (endo-*N*), is used to selectively shorten PSA chains on target glycoconjugates. The only known source of this enzyme is the tail component of bacterial viruses that use PSA receptors for infectivity. Endo-*N*'s substrate specificity makes it a powerful tool for investigating processes as distinct as PSA biosynthesis and neural tissue development. Cloning and expression of the endo-*N* structural gene should stimulate potentially novel future approaches to better define PSA's function in animal development and pathology (Petter,1991; Petter and Vimr, 1992).

In this communication, we describe how genetic approaches, combined with the use of endo-*N* and recombinant sialidases as important tools, have led to an initial molecular description of the PSA synthetic apparatus. We suggest that these findings may be directly relevant to the PSA synthetic mechanism in eukaryotes.

MATERIALS AND METHODS

Bacterial strains, media and growth conditions, nucleic acid and protein analysis procedures, endo-*N* and exosialidase purification, and procedures specific to sialometabolism are described in the cited primary references. Inverted inner membrane vesicles were prepared and stored essentially as described in Yamane et al., 1987.

RESULTS AND DISCUSSION

PHYSICAL, GENETIC, AND FUNCTIONAL ORGANIZATION OF THE KPS CLUSTER:

The physical and genetic organization of the E. coli K1 *kps* cluster is shown in Fig. 1.

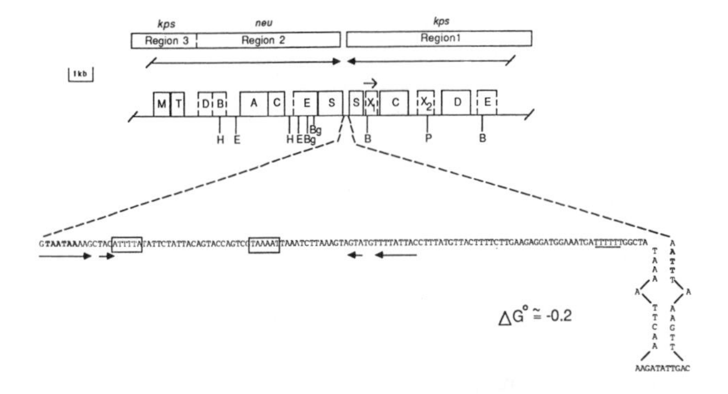

Figure 1. Physical and genetic organization of the E. coli K1 *kps* cluster. The cluster was physically mapped to 64 units by hybridizing non-*kps* flanking sequences to the Kohara mini-collection (Vimr, 1991). *neuABDS* and *kpsTSC* were mapped by complementation and transduction (Vimr and Troy, 1985; Vimr et al., 1989). *kpsM* and *kpsD* were mapped by Silver and his colleagues (Pavelka et al, 1991; Silver et al., 1987). $kpsX_1X_2$ were defined by protein fusions with mini-mudII (Steenbergen, et al., 1992; Steenbergen and Vimr, unpublished), although no functions have been ascribed to these loci. Arrows above gene designations (boxes) indicate transcription direction. Note that $kpsX_1$ is transcribed opposite from other region 1 genes, suggesting a possible regulatory function. *neuS* encodes polyST, whereas *neuE* appears to play a role in the initiation or termination of PSA synthesis (Steenbergen and Vimr, 1990; Steenbergen et al., 1992; Vimr et al., 1992). *neuABC* function in sialic acid synthesis and activation (Vimr et al., 1989; Zapata et al., 1992). Arrows underlining region 1-2 intergenic sequences and the two boxed sequences indicate palindromes that might play a role in the recombination events proposed in the text. An energetically unfavorable *kpsS* transcriptional stop site is drawn as a stem-loop followed by a run of six T's (U's in the message); these structures are reminiscent of Rho-independent terminators. H, *Hind*III; E, *Eco*RI; B, *Bam*HI; P, *Pst*I.

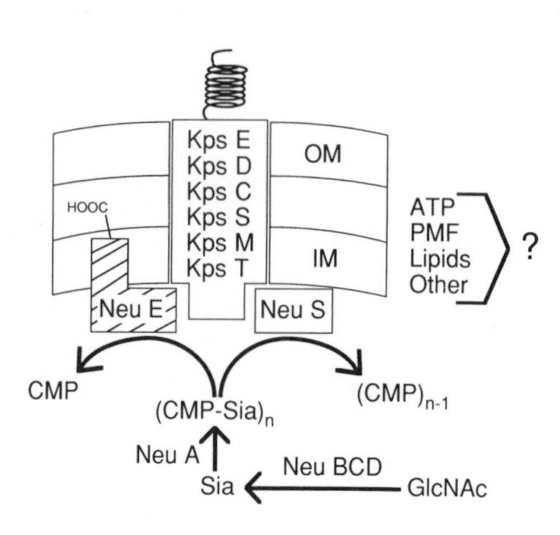

Figure 2. Proposed functional organization of the sialyltransferase complex. Gene products encoded by *kps* regions 1 and 3 are assumed to form a multiprotein complex extending from the inner surface of the inner membrane (IM) to the outer membrane (OM). The schematic implies that there are no freely soluble cytoplasmic or periplasmic intermediates in PSA synthesis or translocation, although it does not exclude them. Polymerization is carried out by the polymerase encoded by *neuS* in association with an unknown number of Kps components. Since NeuS cannot carry out de novo PSA synthesis (Steenbergen et al., 1992), NeuE is assumed to function in the initiation reaction(s), although its role as a distinct sialyltransferase is speculative. Together the envelope components are defined as the sialyltransferase complex. ATP, proton motive force (PMF), outer membrane proteins (OMP) and lipids (e.g., undecaprenol) may also be required for synthesis or translocation. Sialic acid synthesis and activation are shown catalyzed in the cytoplasm by other region 2 gene products.

Evidence for association of *kps* gene products in a multiprotein sialyltransferase complex comes largely from genetic studies, as summarized in Fig. 3. Since the double lipid bilayer structure of the Gram-negative envelope presents special obstacles to macromolecular translocation, some *kps* gene products (Fig. 2) may have no strict correlates in eukaryotic PSA biosynthesis. However, to the extent that some of these gene products function in PSA assembly rather than translocation reactions, these could be functional homologs of as yet undetected eukaryotic gene products. NeuS is assumed to be functionally equivalent to the eukaryotic polyST.

A central group of biosynthetic *neu* genes (region 2) is flanked on either side by accessory *kps* loci of regions 1 and 3. Gene products of the flanking regions are thought to function in polysaccharide assembly reactions (initiation, termination, modification), regulation, and translocation. In contrast, a variety of approaches have shown that the region 2 *neuS* locus encodes polysialyltransferase (polyST); other *neu* genes participate in sialic acid activation (*neuA*), synthesis (*neuBC*), and, possibly, initiation (*neuDE*). The genetic organization of the cluster appears tailor-made for certain kinds of recombination events, potentially explaining evolution of type II capsule diversity in which different region 2 "cassettes" have apparently been shuffled between the shared regions (Boulnois and Jann, 1989). Recombination has almost certainly occurred intragenically between K1 and K92 *neuS* (Vimr et al., 1992), and we presumed also at interregional junctions to join, for example, the common region 1 of *kps* clusters with different region 2 cassettes (Fig. 1; Steenbergen et al., 1992).

A variety of approaches (see below) lead us to propose that most of the *kps* cluster gene products function as a complex which couples polymerization with translocation of nascent PSA chains to the outer membrane (Fig. 2).

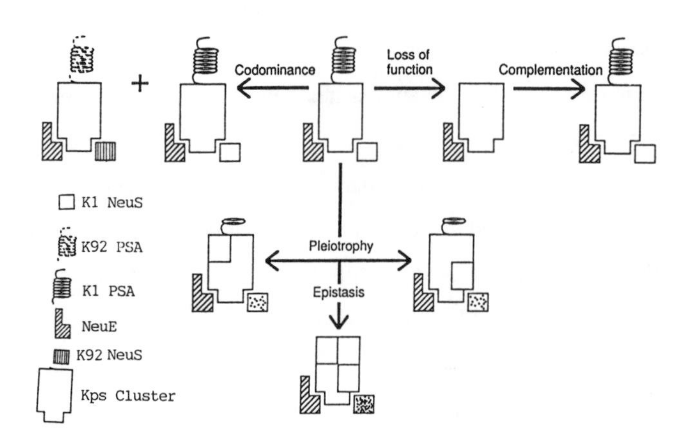

Figure 3. Phenotypes of mutants with defects in *neuS* and *kps* region 1 and 3 genes. Dots indicate relative degree of pleiotropy on polyST function. Shortened PSA chains indicate reduction in levels of polysaccharide synthesis. Rectangles indicate defects in one or more components of the sialyltransferase complex. Other symbols are as indicated.

Strains with null mutations in *neuS* (Fig. 3) are acapsular and lack polyST activity in vitro (Steenbergen and Vimr, 1990) and in vivo (Steenbergen et al., 1992). Complementation of *neuS* strains in *trans* with K1 or K92 *neuS*$^+$ results in α2,8 or α2,8-2,9 PSA synthesis, respectively (Steenbergen et al., 1992); K92 *neuS*$^+$ in *trans* to K1 *neuS*$^+$ is codominant (Fig. 3). These results indicate that NeuS, by "plugging" into the complex, specifies the type of polymer that is synthesized. This suggestion is supported by in vitro membrane-mixing experiments in which NeuS from membranes deficient in region 1 and 3 gene products is functionally reconstituted when added to polymerase-deficient membranes containing these accessory gene products (Steenbergen et al., 1992). Mutants with defects in *kps* region 1 or 3 have pleiotropic phenotypes (Fig. 3), including accumulation of PSA in the cytoplasm or periplasm (Boulnois et al., 1987; Vimr et al., 1989) and reductions in polyST activity (Vimr and Troy, 1985; Vimr et al., 1989). Epistasis in a double mutant with defects in *kpsS* and *kpsT* results in loss of polyST activity (Vimr et al., 1989). In a formal genetic sense, the synergistic-negative effect of the two mutations on polyST activity implies that neither KpsT nor KpsS functions before the other, and that both are required for normal polymerase activity. If these observations are true for other pairs of *kps* mutations, then the pathway for PSA synthesis and translocation is likely to be a concerted one in which synthesis is coupled with membrane transit. Consistent with this view are the results obtained from an in vivo translocation system, which show that PSA synthesis and export to the outer membrane occur in less than two minutes (Vimr, 1992). The observations summarized in Figs. 2 and 3 thus indicate that a thorough investigation of polyST structure and function is necessary for a fundamental understanding of PSA synthesis and translocation.

STRUCTURE AND FUNCTION OF POLYST: Synthesis of PSA chains could occur by a processive mechanism in which nascent chains remain tightly bound to the polymerase between successive transsialylation reactions, nonprocessively (distributively), or by a combination of the two. The view of polyST as a processive enzyme (Steenbergen et al., 1992) was largely intuitive until now and came from our interpretation of Roseman and colleagues measurements of endogenous sialyltransferase activity and this activity's stimulation by exogenous (colominic acid) acceptors (Aminoff et al., 1963; Kundig et al., 1971). Since colominic acid did not compete for the endogenous activity, this result mimicked the classic template challenge experiments conducted with the processive E. coli DNA-dependant RNA polymerase (Kornberg, 1980). The following evidence now strongly supports our initial assumption of processivity: *i*) PSA chain length in vivo is not markedly heterogenous (Pelkonen, 1990; Vimr, 1992); *ii*) the polyST encoded by E. coli K92 *neuS* synthesizes the alternating α2,8-2,9 linkages of the K92 antigen (Steenbergen et al., 1992). K92 *neuS* is homologous with K1 *neuS* and is likely to have evolved from its K1 homolog (Vimr et al., 1992). These observations make it difficult to imagine how any mechanism other than processivity could faithfully synthesize alternating K92 linkages. The results further imply that the K92 polyST has dual linkage specificity (Vimr et al., 1992); *iii*) exogenous oligomeric acceptors smaller than a tetramer are poor substrates (Vimr et al., 1986; Steenbergen and Vimr, 1990). This result is in formal agreement with the solution NMR structure of PSA (Yamasaki and Bacon, 1991); *iv*) K1 and K92 polySTs do not recognize exogenous acceptors from the heterologous system (W. Vann, personal communication); and *v*) polySTs are unable to initiate de novo PSA synthesis (Vimr, et al., 1989). These results lead us to propose the schematic of polyST shown in Fig. 4.

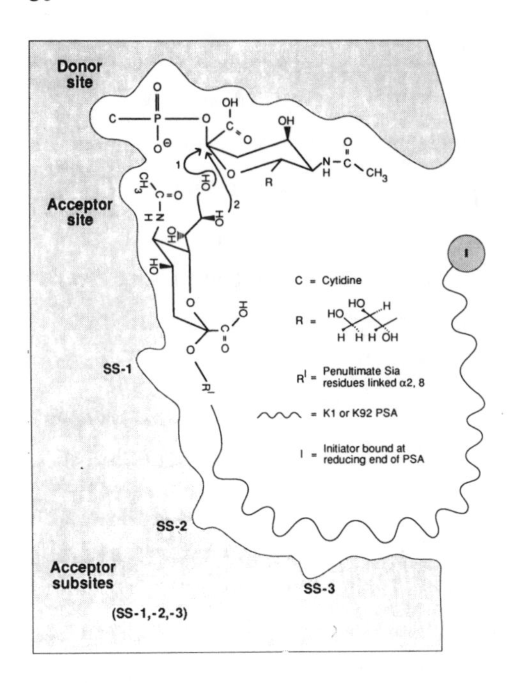

Figure 4. Schematic model of polyST catalytic mechanism.

Although little is known about the initiation and termination of PSA synthesis, the model indicates that at a certain chain length the initiator, which is likely to be membrane-associated, triggers PSA release (termination) by direct interaction with polyST, or through allosteric interaction with accessory *kps* gene products. The latter possibility is attractive, since it potentially explains the epistasis and pleiotropy of mutants with defects in regions 1 and 3 as being the consequence of aberrant protein-protein and protein-polysaccharide interactions (Fig. 3). The model is also consistent with what little is known about mammalian polyST, especially the apparent synthesis of full-length PSA chains prior to their transfer to NCAM acceptors (Breen and Regan, 1988). If an initiator is also required for eukaryotic PSA synthesis, then standard approaches to clone the vertebrate equivalent of *neuS* may not succeed unless the putative initiator is simultaneously expressed. Finally, the model predicts that polyST uses a direct displacement mechanism involving attack of the α face of the donor sialyl residue, which is β-linked to CMP, by the C-8 or C-9 hydroxyl of the terminal sialyl residue at the acceptor site. We hypothesize that the relatively few primary sequence differences between the K1 and K92 polySTs (Fig. 5) will be reflected in functional differences between acceptor subsites (Fig. 4), which we assume help in orienting the growing chain relative to the donor site, and thus potentially explain the dual linkage specificity of K92 polyST (Vimr et al., 1992).

```
Met Ile Phe Asp Ala Ser Leu Lys Lys Leu Arg Lys Leu Phe Val Asn   16
ATG ATA TTT GAT GCT AGT TTA AAG AAG TTG AGG AAA TTA TTT GTA AAT   48
--- --- --- --- --- --- --- --- --- --- --- --- --- --- --- ---
--- --- --- --- --- --- --- --- --- --- --- --- --- --- --- ---

Pro Ile Gly Phe Phe Arg Asp Ser Trp Phe Phe Asn Ser Lys Asn Lys   32
CCA ATT GGG TTT TTC CGT GAC TCA TGG TTT TTT AAT TCT AAA AAC AAG   96
--- --- --- --- --- --- --- --- --- --- --- --- --- --- --- ---
--- --- --- --- --- --- --- --- --- --- --- --- --- --- --- ---

Ala Glu Glu Leu Leu Ser Pro Leu Lys Ile Lys Ser Lys Asn Ile Phe   48
GCT GAA GAG CTA CTA TCA CCG TTA AAA ATA AAA AGT AAA AAT ATT TTT   144
--- --- --- --- --- --- --- --- --- --- --- --- --- --- --- ---
--- --- --- --- --- --- --- --- --- --- --- --- --- --- --- ---

Ile Ile Ser Asn Leu Gly Gln Leu Lys Lys Ala Glu Ser Phe Val Gln   64
ATA ATT AGT AAC CTG GGG CAA TTA AAA AAA GCT GAG TCA TTT GTA CAA   192
--T G-- GC- C-T T-A --- --- --- --G --- --A --- CTT --- A-- ---
--- Val Ala His --- --- --- --- --- --- --- --- Leu --- Ile ---

Lys Phe Ser Lys Arg Ser Asn Tyr Leu Ile Val Leu Ala Thr Glu Lys   80
AAA TTT AGC AAG AGA AGT AAC TAT CTT ATT GTT TTG GCA ACT GAA AAA   240
--- --- --T -G- C-T --- --T -T- --- --C --C --- --- --- A-- ---
--- --- --- Arg --- --- --- Phe --- --- --- --- --- --- Lys ---

Asn Thr Glu Met Pro Lys Lys Ile Val Glu Gln Ile Asn Asn Lys Leu   96
AAT ACT GAG ATG CCA AAA ATT ATT GTT GAA CAA ATA AAT AAT AAA TTA   288
--C --- --A --- --- --G T-A --- C-- --G --- --G --- --A --G --G
--- --- --- --- --- Arg Leu --- Leu --- --- Met --- Lys --- ---

Phe Ser Ser Tyr Lys Val Leu Phe Ile Pro Thr Phe Pro Asn Val Phe   112
TTT TCT TCA TAC AAG GTA CTA TTC ATT CCA ACT TTC CCA AAT GTT TTT   336
--- --- --- --- --A C-- --- --T --A --- --A GAG --- --- ACA ---
--- --- --- --- --- Leu --- --- --- --- --- Glu --- --- Thr ---

Ser Leu Lys Lys Val Ile Trp Phe Tyr Asn Val Tyr Asn Tyr Leu Val   128
TCA CTT AAA AAG GTT ATA TGG TTT TAT AAC GTA TAT AAT TAT TTA GTT   384
--G --- --- --A --- --- --- --- --- --T --- --- --A --- A-- ---
--- --- --- --- --- --- --- --- --- --- --- --- Lys --- Ile ---

Leu Asn Ser Lys Ala Lys Asp Ala Tyr Phe Met Ser Tyr Ala Gln His   144
TTA AAT TCA AAA GCT AAA GAT GCT TAT TTT ATG AGC TAT GCG CAA CAT   432
--- --- --- --- --- --- --- --- --- --- --- --- --- --A --- ---
--- --- --- --- --- --- --- --- --- --- --- --- --- --- --- ---

Tyr Ala Ile Phe Val Tyr Leu Phe Lys Lys Asn Asn Ile Arg Cys Ser   160
TAT GCA ATC TTC GTA TAT TTG TTC AAA AAA AAT AAT ATA AGA TGT TCA   480
--- --- --- --- A-- -GG --- --- --- --- --- --- --- --- --- ---
--- --- --- --- Ile Trp --- --- --- --- --- --- --- --- --- ---

Leu Ile Glu Glu Gly Thr Gly Thr Tyr Lys Thr Glu Lys Glu Asn Pro   176
TTA ATT GAA GAG GGG ACA GGG ACT TAT AAA ACC GAA AAA GAA AAC CCA   528
--- --- --- --- --- --- --- --G --- --- --A --G --- A-- --A ---
--- --- --- --- --- --- --- --- --- --- --- --- --- Lys Lys ---

Val Val Asn Ile Asn Phe Tyr Ser Glu Ile Ile Asn Ser Ile Ile Leu   192
GTA GTA AAT ATT AAT TTT TAT TCA GAG ATT ATT AAT TCA ATT ATC TTG   576
C-- --- --- --- --- --- --- --G TG- --C --- --- --- --- --- ---
Leu --- --- --- --- --- --- --- Trp --- --- --- --- --- --- ---

Phe His Tyr Pro Asp Leu Lys Phe Glu Asn Val Tyr Gly Thr Tyr Pro   208
TTC CAT TAT CCA GAT TTG AAA TTT GAA AAT GTA TAC GGT ACA TAT CCA   624
--- --- --- --- --- --- --- --- --- --- --- --- --C --C -T- ---
--- --- --- --- --- --- --- --- --- --- --- --- --- --- Phe ---
```

```
Ile Leu Leu Lys Lys Lys Phe Asn Ala Gln Lys Phe Val Glu Phe Lys   224
ATT TTG CTT AAG AAA AAA TTT AAT GCG CAA AAA TTT GTT GAG TTT AAA   672
-A- --- T-A --A G-- --- --- G-- --A A-- --- --- T-- --- --- ---
Asn --- --- --- Glu --- --- Asp --- Lys --- --- Phe --- --- ---

Gly Ala Pro Ser Val Lys Ser Ser Thr Arg Ile Asp Asn Val Ile His   240
GGT GCT CCA TCA GTT AAA TCA TCA ACC AGA ATA GAT AAT GTT ATC CAT   720
AC- AT- --- -T- --- --- --G --- --A --- --G --- --- C-C --A ---
Thr Ile --- Leu --- --- --- --- --- --- Met --- --- Leu --- ---

Lys Tyr Ser Ile Thr Arg Asp Asp Ile Ile Tyr Ala Asn Gln Lys Tyr   256
Lys Tyr Ser Ile Thr Arg Asp Asp Ile Ile Tyr Ala Asn Gln Lys Tyr   272
--- --- CG- --C --- --- --- --- --T --- --- -T- -G- --- -GA ---
--- --- Arg --- --- --- --- --- --- --- --- Val Ser --- Arg ---

Leu Ile Glu His Thr Leu Phe Ala Asp Ser Leu Ile Ser Ile Leu Leu   272
TTG ATT GAA CAT ACA TTA TTT GCG GAT TCG TTA ATT TCT ATC TTA CTT   816
-G- --- --C A-C GA- --G -A- --- C-- -TA --- --A --- -C- --G A-G
Trp --- Asp Asn Glu --- Tyr --- His Leu --- --- --- Thr --- Met

Arg Ile Asp Lys Pro Asp Asn Ala Arg Ile Phe Ile Lys Pro His Pro   288
AGA ATA GAT AAG CCT GAT AAT GCA AGA ATA TTT ATA AAA CCT CAC CCT   864
--- --- --- --A T-- --- --C --- --- G-T --- --- --- --- --- ---
--- --- --- --- Ser --- --- --- --- Val --- --- --- --- --- ---

Lys Glu Pro Lys Lys Asn Ile Asn Ala Ile Gln Lys Ala Ile Lys Lys   304
AAA GAG CCT AAA AAA AAT ATT AAT GCA ATT CAA AAG GCA ATA AAA AAG   912
--- --A A-- --- --- T-- --- --- --- --- --- GGT --- --- --T --A
--- --- Thr --- --- Tyr --- --- --- --- --- Gly --- --- Asn ---

Ala Lys Cys Arg Asp Ile Ile Leu Ile Thr Glu Pro Asp Phe Leu Ile   320
GCA AAA TGT CGT GAC ATA ATT CTT ATA ACA GAG CCA GAC TTT TTA ATA   960
--- --G C-- --- --T --- --- A-- --T GT- --A AA- --- --- --- ---
--- --- Arg --- --- --- --- Ile --- Val --- Lys --- --- --- ---

Glu Pro Val Ile Lys Lys Ala Lys Ile Lys His Leu Ile Gly Leu Thr   336
GAG CCG GTA ATA AAA AAA GCA AAA ATA AAA CAC TTA ATT GGA TTA ACA   1008
--- T-A A-- --- --- --- TGC --- --- --- --- --G --- --- --- G--
--- Ser Ile --- --- --- Cys --- --- --- --- --- --- --- --- Ala

Ser Ser Ser Leu Val Tyr Ala Pro Leu Val Ser Lys Arg Cys Gln Ser   352
TCA TCT TCT TTG GTA TAT GCA CCT TTA GTT TCT AAA AGA TGT CAG TCT   1056
--- --- --- --- --- --C --- --- --- --- -A- --- GAG --- A-- A-A
--- --- --- --- --- --- --- --- --- --- Tyr --- Glu --- Lys Thr

Tyr Ser Ile Ala Pro Leu Met Ile Lys Leu Cys Asp Asn Asp Lys Ser   368
TAT TCA ATA GCG CCT CTT ATG ATA AAG TTG TGT GAT AAT GAT AAA TCC   1104
--- --- --- --A --- A-- --T --- --A --- --- A-- --- --A --- ---
--- --- --- --- --- Ile Ile --- --- --- --- Asn --- Glu --- ---

Gln Lys Gly Ile Asn Thr Leu Arg Leu His Phe Asp Ile Leu Lys Asn   384
CAA AAA GGG ATT AAT ACG CTG CGT CTC CAT TTC GAT ATT TTA AAG AAT   1152
--- --- --- --- --- --- --- --- --- --- --- --- --- --- --- ---
--- --- --- --- --- --- --- --- --- --- --- --- --- --- --- ---

Phe Asp Asn Val Lys Ile Leu Ser Asp Asp Ile Thr Ser Pro Ser Leu   400
TTT GAT AAT GTT AAA ATA TTA TCG GAT GAT ATA ACA TCT CCC TCT TTG   1200
--- --- --- --- --- --- --- --- --- --- --- --- --- --- --- ---
--- --- --- --- --- --- --- --- --- --- --- --- --- --- --- ---

His Asp Lys Arg Ile Phe Leu Gly Glu ***   409
CAC GAT AAA AGG ATT TTC TTG GGG GAG TAA   1230
--- --- --- --- --- --- --- --- --- ---
--- --- --- --- --- --- --- --- --- ---
```

Figure 5. Nucleotide sequences and derived primary structures of E. coli K1 and K92 polySTs. The derived primary structure and nucleotide sequences of K1 polyST are shown as the top two lines, respectively. This sequence has GenBank accession number M76370 (Steenbergen et al., 1992). Aligned below the K1 sequences are K92 *neuS* and polyST sequences, which have been assigned GenBank accession number M88479 (Vimr et al., 1992). Nucleotides or amino acid residues identical to the K1 sequences are indicated by dashes.

As shown in Fig. 5, 12.7% of the 1,227 bp composing K1 and K92 *neuS* were different. Fifty-five of these changes were synonymous with the K1 *neuS* sequence and thus did not affect polyST primary structure. Relative to K1 polyST, the 70 altered amino acid residues in the K92 polymerase would be accounted for by an estimated 17 synonymous mutations and 84 missense mutations of the K1 *neuS* sequence. Of the observed amino acid replacements in K92 polyST, 36 were conservative, since the changes did not affect grouping of the altered residue in one of the five amino acid classes defined by Dayhoff (reviewed in Robson and Garnier, 1988). Given the high percent identity between these structures, and that PSA similar to the K1 antigen is found in several other genera, whereas the K92 antigen only has been observed in E. coli, we conclude that K92 *neuS* has evolved from the K1 homolog. Furthermore, the lack of nucleotide changes at the 5' and 3' ends of K1 and K92 *neuS* (Fig. 5) strongly implies that there has been intragenic recombination between *kps* clusters.

On the basis of the results shown in Fig. 5, the 34 nonconservative replacements in K92 polyST were scored by the log-odds matrix of 250 PAMs. Only 17 of these changes had negative scores and were, therefore, considered less likely in related proteins and thus most likely to have an effect on polyST structure. Structural predictions (Fig. 6) indicated that most changes had little or no effects on polyST structure. However, the change from Phe-108 to Glu in K92 polyST (Fig. 5) required 3 nonsynonymous mutations, implying intense selection at this position. Since Phe-108 is bracketed by two Pro residues, it is not surprising that there was a shift from hydrophobic potential in K1 polyST (Fig. 6B) to hydrophilic in the K92 polymerase (Fig. 6A). Although this analysis does not allow any precise conclusions to be made about function, it shows that the evolution of K1 and K92 polySTs did not require any drastic structural alterations (Vimr et al., 1992). How, then, did K92 polyST evolve from its K1 homolog?

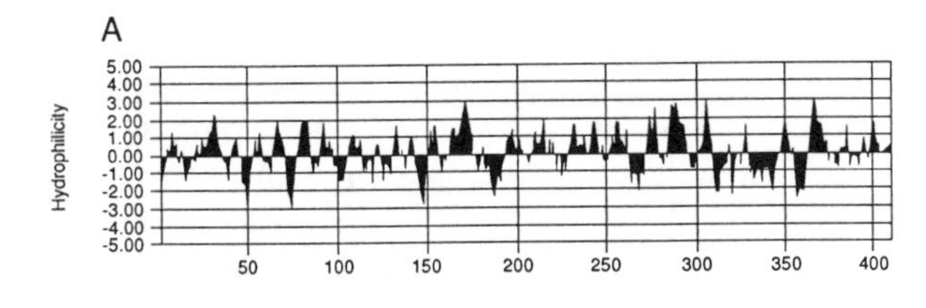

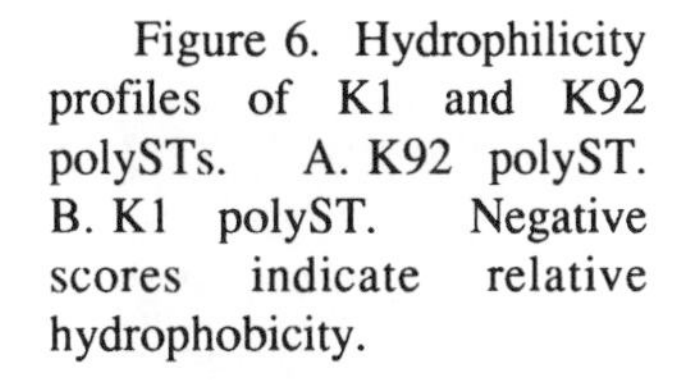
Figure 6. Hydrophilicity profiles of K1 and K92 polySTs. A. K92 polyST. B. K1 polyST. Negative scores indicate relative hydrophobicity.

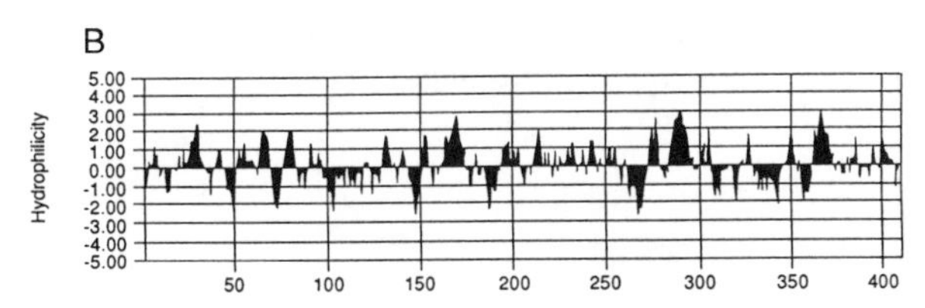

We assume that capsular polysaccharides evolved for reasons that did not initially include pathogenicity. Given this assumption, it is difficult to envision selective pressures that could account for K92 evolution from the K1 system. However, consider the case in which a K1 strain mutates to an acapsular phenotype; the spontaneous rate in *kps* is quite high (0.5 to 1 x 10^{-5}) due to the large number of potential targets. Mutated clusters would continue to accumulate mutations in *kps* genes at the same spontaneous rate as the original mutation, and if these $K1^-$ clones were not selected against in the population, then E. coli strains with derelict (silent), or more poetically, Flying Dutchman *kps* clusters should exist in nature. Thus, spontaneous mutation coupled with promiscuous recombination between derelict and active *kps* clusters could be important for K antigen evolution. Fig. 7 (lane 6) shows that E. coli B DNA probed with a K1 *neuS* fragment gives a strong signal slightly smaller than the 2.9 kb expected for K1 or K92 (Fig. 7, lane 4). Control experiments showed that *E. coli* B is $K1^-$, lacks an active polyST, yet contains region 1 and 3 *kps* DNA (data not shown). We conclude that Flying Dutchman clusters exist and that they may explain evolution of capsule diversity. These original findings are being pursued in our laboratory.

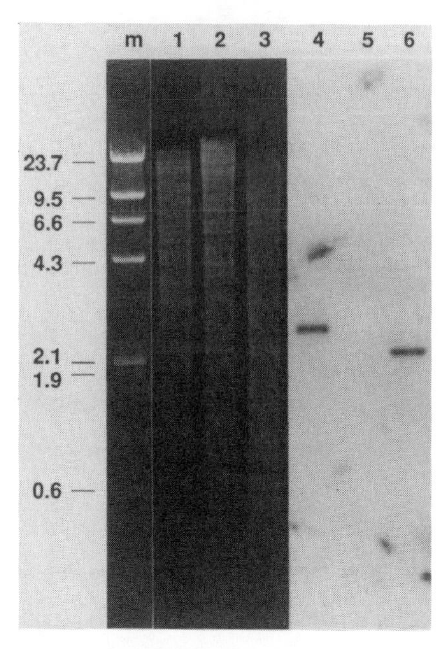

Figure 7. Southern analysis of E. coli and Pasteurella haemolytica DNA probed with a K1 *neuS* fragment. Genomic DNA from E. coli K92 (lane 1), P. haemolytica A2 (lane 2), and E. coli B (lane 3) were digested with *Eco*RI and *Bam*HI and probed with a ^{32}P-labeled *neuS* fragment (Steenbergen et al., 1992), lanes 4-6, respectively. m. λ *Hind*III fragments, stained with ethidium bromide.

INITIATION OF PSA SYNTHESIS: During analysis of K1 polyST structure and function, we noted a potential membrane-spanning domain in an ORF (designated *neuE*) immediately 5' and overlapping *neuS* (Fig. 8; Steenbergen et al., 1992). This domain bore a striking resemblance to the membrane-spanning domains of several yeast glycosyltransferase known to interact with dolichol (Albright et al., 1989), suggesting that NeuE may interact with the bacterial equivalent, undecaprenol (Steenbergen et al., 1992). Since undecaprenyl phosphate is thought to be essential for PSA synthesis (Troy, 1979), we reasoned that NeuE or another *kps* gene product might transfer the initiating sialyl residue(s) to undecaprenol, followed by polyST-catalyzed chain elongation on this endogenous acceptor. Alternatively, the sialyl acceptor could be a protein or another phospholipid, in which case undecaprenol may function in the translocation of PSA to the outer membrane in combination with accessory *kps* products of the sialyltransferase complex (Fig. 2). Together with the inability of polyST to initiate PSA synthesis, these observations imply that there may be a separate sialyltransferase for initiation. Evidence for an "initiase" was obtained by phenotypic analysis of triple mutants with defects in sialic acid degradation, synthesis, and polymerization (Steenbergen et al., 1992). Recombinant sialidases from Salmonella typhimurium LT2 (Hoyer et al, 1991; 1992; Taylor et al., 1992; Warner et al., 1992) and Vibrio cholerae (Vimr et al., 1988; Taylor et al., 1992) were used to establish the chemical identity of sialic acid apparently linked by a quantitatively minor sialyltransferase activity to a lipid or protein membrane intermediate (Steenbergen et al., 1992). The triple mutant provides new approaches to unambiguously characterize the initiation mechanism (Steenbergen et al., 1992; Vimr, 1992).

Direct evidence for a role of NeuE in PSA expression was obtained by marker replacement with a kanamycin cassette engineered into a *neuE* deletion (Fig. 9A). The isogenic chromosomal-replacement strain was acapsular ($K1^-$), demonstrating a requirement for NeuE (Fig. 9A). Similarly, we asked whether *neuS* expressed in an aldolase-deficient (*nanA4*) strain, harboring *neuA*, *neuS*, and region 1 on compatible plasmids, could synthesize PSA after exogenous addition of sialic acid to the medium (Fig. 9B). The lack of in vivo PSA-synthetic activity in this strain further demonstrated a requirement for NeuE, NeuD, or both for PSA synthesis and export. Clearly, the polymerase requires additional gene products besides those encoded by region 1 for endogenous activity.

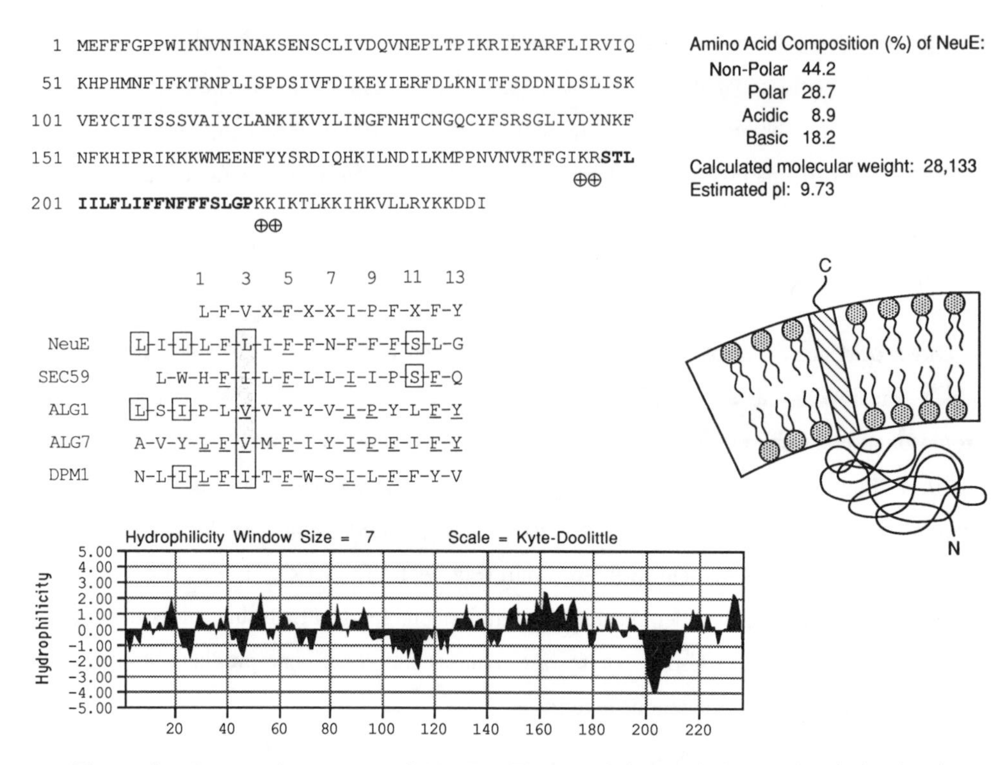

Figure 8. Proposed structure of NeuE. The top left panel shows the derived primary structure of NeuE; the C-terminal membrane-spanning domain is highlighted in bold. This domain is flanked on each side by two positively charged residues that may anchor the domain in the membrane, as shown by the hatched box in the bottom right panel. The membrane-spanning domain is compared in the middle left panel to the consensus sequence of 13-amino acid residues found in several yeast enzymes known to interact with dolichol (Albright et al., 1988); identities are underlined. Additional identities between NeuE and the yeast enzymes are boxed. NeuE hydrophilicity is displayed graphically at bottom left on a scale of -5 to +5, where negative scores indicate relative hydrophobicity. The top right panel shows the amino acid composition of NeuE and its calculated molecular weight and pI.

All region 2 mutants characterized to date lack detectable PSA, a phenotype which is consistent with the function of *neu* gene products in sialic acid precursor synthesis or polymerization. We therefore assumed that the acapsular mutant generated by insertion mutagenesis of *neuE* and marker exchange (Fig. 9A) would likewise lack intracellular PSA. However, whereas this mutant was K1F resistant and thus lacked cell-surface capsular polysaccharide, it had an easily detectable level of immunoreactive, membrane-associated PSA, indicating that *neuE* is not obligatory for polymerization. E. coli RfaL and its homolog in S.

typhimurium were shown recently to contain polyprenyl-binding motifs at their respective C-termini (Klena et al., 1992). RfaL is thought to participate in the transfer of O-antigens from undecaprenol to core oligosaccharides of lipopolysaccharide and thus may function as part of membrane complexes which include specific protein-protein and protein-polysaccharide interactions between *rfb* gene products, polysaccharide, and undecaprenol. The phenotype caused by a *neuE*-null mutation (Fig. 9A) is consistent with NeuE playing a role similar to RfaL's in PSA assembly, although this model does not predict whether NeuE would transfer PSA from or to undecaprenol. We stress that this mechanism is still speculative, since there is little direct evidence for an involvement of undecaprenol in PSA assembly. Although the failure of the *neuE*-null mutation to block PSA synthesis indicates that another *neu* gene product, possibly NeuD, functions as the initiase, it remains a formal possibility that NeuE is the "preferred" initiator, facilitating coupling of PSA synthesis to translocation. This possibility assumes that NeuS can initiate polymerization in the absence of NeuE under physiological conditions, but that the PSA chain synthesized are not recognized by translocation components of the sialyltransferase complex. Whatever the exact function of NeuE, a strain with a null mutation in *neuE* (Fig. 9A) is the first region 2 mutant to have a "translocation" phenotype. Hence, not all region 2 *neu* gene products function as simple synthetic activities, but instead must participate in a variety of protein-protein and protein-polysaccharide interactions, as predicted by the results summarized in Fig. 3.

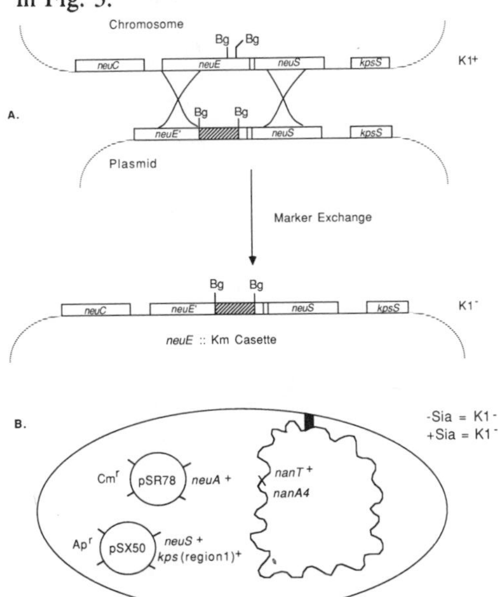

Figure 9. NeuE is required for PSA capsule expression. A. A kanamycin-resistance (Km) cassette (Pharmacia) was ligated into *neuE* deleted of its internal *Bgl*II (Bg) restriction fragment, as indicated by the hatched box. The mutation was crossed into the chromosome of EV36 ($K1^+$) by homologous recombination using a *neuE*::Km suicide plasmid. The resulting mutant was acapsular ($K1^-$). B. EV78 (*nanA4 nanT*$^+$), a derivative of the K-12 strain MC4100, harboring pSR78 (*neuA*$^+$) and pSX50 (*neuS*$^+$ *kps* region 1^+) did not produce PSA in the presence or absence of exogenous sialic acid (Sia).

PSA TRANSLOCATION: NeuS, NeuE, and *kps* regions 1 and 3 gene products are components of a postulated sialyltransferase complex catalyzing a spatiotemporal pathway which includes initiation, elongation (polymerization), termination, and translocation of PSA to the outer membrane (Fig. 2). This hypothesis predicts that loss of any component of this complex would disrupt PSA synthesis or export (Fig. 3), suggesting that inside-out-oriented inner membrane vesicles (IOV) should be defective in one or both of these processes, since the vesicles would lack periplasmic and outer membrane components of the complex. These predictions were tested in an endo-*N* protection experiment, which showed that PSA was synthesized but that the chains were not protected over time from endo-*N* challenge, indicating synthetic competence of IOV but an inability to efficiently transport PSA into the vesicular lumen (Fig. 10). However, 150 to 250 cpm above background remained after endo-*N* challenge, suggesting that a portion of the PSA chains synthesized in the first 5 minutes after CMP-sialic acid addition was sequestered in an endo-*N* inaccessible compartment (Fig. 10). As a control, endo-*N* added to IOV at times zero prevented PSA synthesis (Fig. 10, solid circle). These data imply that nascent chains enter the translocation pathway but that the process aborts in the absence of an intact sialyltransferase complex. That IOV are translocation defective is consistent with PSA synthesis and export being part of a concerted pathway that does not involve soluble periplasmic intermediates. Using different *kpsS*, *kpsT*, and *kpsE* alleles and a variety of fractionation techniques, we were unable to detect periplasmic PSA (E. Vimr, unpublished). Localization results were critically dependent on temperature, suggesting that inconsistencies with previous results (e.g., Kröncke et al., 1990) may be attributable to artifacts of past experimental procedures.

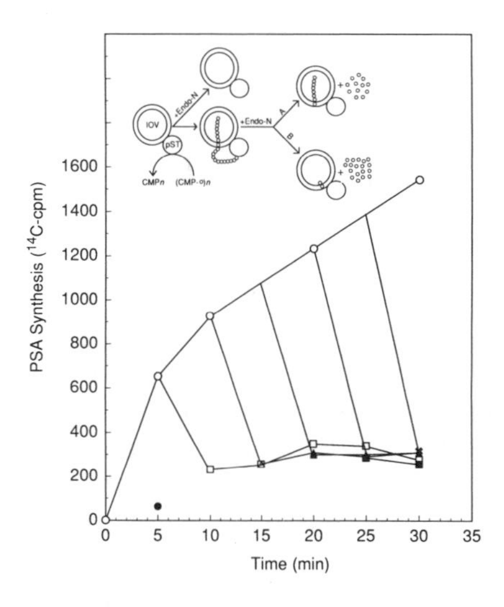

Figure 10. IOV are synthetically competent and translocation defective. IOV were prepared from EV5 (*neuA22*) and exposed to radiolabeled CMP-sialic acid (o) or to label plus endo-*N* (●). At the times indicated, samples from the culture that was not exposed at time zero to endo-*N* were added to tubes containing endo-*N* and assayed for PSA synthesis (□, Δ, ▲, ■, x). As shown in the insert, endo-*N* degrades extra-vesicular PSA. Scheme B indicates the situation in which most PSA chains are not protected from endo-*N* challenge, whereas scheme A indicates extensive transport into the vesicular lumen, and thus protection from endo-*N* challenge. The data tend to support scheme B.

CONCLUSIONS

The activity detected by Roseman and his colleagues almost 30 years ago in extracts of E. coli K1 was the first sialyltransferase to be assayed in a cell-free system (see above). We have now shown that this activity is the polysialic acid polymerase encoded by *neuS*, and that this enzyme is likely to be a member of a new class of processive glycosyltransferases. The apparent ability of K92 polyST to synthesize alternating α2,8-2,9 linkages is consistent with this overall catalytic mechanism. Most of the results described in this communication could not have been obtained except through genetic approaches, and some of the conclusions were unanticipated. Now that many of the "players" in PSA biosynthesis have been identified, we intend to pursue questions pertaining to regulation, translocation, and pathogenesis. We will continue to emphasize genetic approaches, combining these when possible with biochemical studies and animal models. This focus should lead to a fundamental understanding of PSA biosynthesis and improved therapies for K1 meningitis.

ACKNOWLEDGEMENTS

We thank our collaborators and colleagues for advice, strains, and enthusiasm. Our research is supported by National Institute of Allergy and Infectious Diseases grant R01 AI-23039.

REFERENCES

Albright, C.F., Orlean, P., and Robbins, P.W. (1989) Proc. Natl. Acad. Sci. USA 86:7366-7369.
Aminoff, D., Dodyk, F., and Roseman, S. (1963) J. Biol. Chem. 238:1177-1178.
Boulnois, G., and Jann, K. (1989) Mol. Microbiol. 3:1819-1823.
Boulnois, G.J., Roberts, I.S., Hodge, R., Hardy, K., Jann, K., and Timmis, K.N. (1987) Mol. Gen. Genet. 200:242-246.
Breen, K.C., and Regan, C.M. (1988) Development 104:142-154.
Hoyer, L.L., Hamilton, A.C., Steenbergen, S.M., and Vimr, E.R. (1992) Mol. Microbiol. 6:873-884.
Hoyer, L.L., Roggentin P., Schauer, R., and Vimr, E.R. (1991) J. Biochem. 110:462-467.
Klena, J.D., Pradel, E., and Schnaitman, C.A. (1992) J. Bacteriol. 174:4746-4752.
Kornberg, A. (1980) DNA replication, W.H. Freeman and Co., San Francisco, pp. 121-125.
Kröncke, K.-D., Golecki, J.R., and Jann, K. (1990) J. Bacteriol. 172:3469-3472.
Kundig, F.D., Aminoff, D., and Roseman, S. (1963) J. Biol. Chem. 246:2543-2550.
Pavelka, M.S., Wright, L.F., and Silver, R.P. (1991) J. Bacteriol. 173:46-3-4610.
Pelkonen, S. (1990) Curr. Microbiol. 21:23-28.
Petter, J.G. (1991) PhD Dissertation, University of Illinois, Urbana.
Petter, J.G., and Vimr, E.R. (1992) Submitted for publication.
Robson, B., and Garnier, J. (1988) Introduction to proteins and protein engineering, Elsevier/North Holland Publishing Co., Amsterdam, pp. 323-327.
Silver, R.P., Aaronson, W., and Vann, W.F. (1987) J. Bacteriol. 169:5489-5495.
Silver, R.P., and Vimr, E.R. (1990) In: The bacteria, vol. 11. Molecular basis of bacterial pathogenesis (B. Iglewski and V. Miller Eds), Academic Press, Inc., New York, pp. 39-60.
Steenbergen, S.M., and Vimr, E.R. (1990) Mol. Microbiol. 4:603-611.
Steenbergen, S.M., Wrona, T. J., and Vimr, E.R. (1992) J. Bacteriol. 174:1099-1108.
Taylor, G., Vimr, E., Garman, E., and Laver, G. (1992) J. Mol. Biol. 226: in press.
Troy, F.A. (1979) Annu. Rev. Microbiol. 33:519-560.
Vimr, E.R. (1991) J. Bacteriol. 173:1335-1338.
Vimr, E.R. (1992) J. Bacteriol. in press.
Vimr, E.R., Aaronson, W., and Silver, R.P. (1989) J. Bacteriol. 172:1106-1117.
Vimr, E.R., Bassler, B.L., and Troy, F.A. (1986) Abst. Annu. Meet. Am. Soc. Microbiol. K152, p. 154.
Vimr, E.R., Bergstrom, R., Steenbergen, S.M., Boulnois, G., and Roberts, I. (1992) J. Bacteriol. 174:5127-5131.

Vimr, E.R., and Troy, F.A. (1985) J. Bacteriol. 164:854-860.
Warner, T.G., Harris, R., McDowell, R., and Vimr, E.R. (1992) Biochem. J. 285: in press.
Yamane, K., Ichihara, S., and Mizushima, S. (1987) J. Biol. Chem. 262:2358-2362.
Yamasaki, R., and Bacon, B. (1991) Biochemistry 30:851-857.
Zapata, G., Crowley, J.M., and Vann, W.F. (1992) J. Bacteriol. 174:315-319.

Polysialic Acid
J. Roth, U. Rutishauser and F. A. Troy II (eds.)

POLYSIALIC ACID CAPSULE SYNTHESIS AND CHAIN TRANSLOCATION IN NEUROINVASIVE ESCHERICHIA COLI K1: "ACTIVATED" INTERMEDIATES AND A POSSIBLE BIFUNCTIONAL ROLE FOR UNDECAPRENOL

Frederic A. Troy, II, Jin-Won Cho and Jean Ye

Department of Biological Chemistry, University of California School of Medicine, Davis, California 95616, USA

SUMMARY: The polysialic acid (polySia) capsule is a neurovirulent determinant in Escherichia coli K1 and Neisseria meningitidis. Our studies seek to determine how synthesis, translocation and surface expression of these antigens are regulated in E. coli K1. A surprisingly complicated genetic and biochemical pathway is involved in the regulation. The kps gene complex is encoded in ~17 kb of DNA that codes for ca. 14 proteins required for polySia chain synthesis (NeuE, NeuS), translocation (KpsM, KpsT) and export. Synthesis also involves preassembly of Sia residues on undecaprenyl phosphate (P-C_{55}). Two unresolved questions are the focus of this study. First, are Sia residues fully polymerized while linked to P-C_{55}, and if so, is this attachment important in translocation? Second, what is the significance of our unexpected finding that KpsM and NeuE both contain a consensus polyisoprene recognition sequence?

Using [U]-^{14}C-labeled inside-out vesicles (IOV), we have shown that a high energy bond links full length "activated" chains of polySia to the cytoplasmic surface of the inner membrane. The linkage is labile, based on the spontaneous release of [^{14}C]polySia chains from the membrane. This lability is consistent with the chains being attached to P-C_{55}. We hypothesize that full length "activated" polySia chains may be translocated across the IM while linked to P-C_{55}, perhaps catalyzed by a translocase(s). Alternatively, a polysialylated reactive intermediate could be transferred en bloc to a translocator, which then shuttles the polySia through the membrane. The undecaprenol-induced perturbation of lipid bilayer structure to a more non-bilayer (Hex_{II}) conformation may facilitate this translocation step.

The deduced amino acid sequence of NeuE, NeuS, KpsM, and KpsT were subjected to computer analysis for the presence of a consensus "dolichol recognition" sequence. Our analyses showed that NeuE and KpsM each contained a presumed polyisoprenol (PI) recognition sequence. These data further substantiate the model that PIs may interact with transferases through specific consensus sequences. The unexpected finding that KpsM, a protein with no known biosynthetic function, but rather only implicated in chain translocation also contained a PI binding domain, leads us to hypothesize a bifunctional role for the PIs. The central idea is that these "super-lipids" may function as a flexible matrix or scaffolding to organize and tether proteins of multienzyme complexes to coordinate not only biosynthetic reactions, but also translocation processes.

INTRODUCTION AND BACKGROUND

Our studies are focused on the unresolved problem of how synthesis and surface expression of the α-2,8-linked polySia capsule in neuroinvasive E. coli K1 is controlled. Our studies have concentrated on the structural and molecular aspects of the membranous polysialyltransferase (polyST) complex in these neurotropic strains (Troy, 1992). This multienzyme complex catalyzes synthesis and translocation of the capsule. Our aim has been to characterize the specific bacterial proteins and their genes to provide a molecular basis for understanding microbial pathogenicity and biogenesis of the enzyme/translocator complex. The polySia capsule was chosen for study because it is a homopolymer of Sia, and Sia is uniquely localized to the capsule. This feature provides a strategic advantage for biosynthetic studies, and for studying reactions relevant to expression and regulation of polySia biosynthetic genes.

The poor immunogenecity of the polySia capsule can be explained by the finding that polysialosyl units structurally identical to the polySia capsule of E. coli K1 and Neisseria meningitidis serogroup B (Gp B) are evolutionarily conserved structures present on embryonic neural cell adhesion molecules (N-CAM) (Finne et al., 1983; Vimr et al., 1984). Such a highly conserved "self" is an example of molecular mimicry, wherein the adult host fails to recognize as foreign, molecular structures it previewed during embryonic development (Wyle et al., 1972). Thus, basic studies on polysialylation in E. coli should continue to provide new information relevant to how synthesis and surface expression of polySia, an oncodevelopmental antigen in human kidney and brain, is controlled in eucaryotic cells (reviewed in Troy, 1992). A related objective of our studies has been to determine the conformation and motional properties of undecaprenol in membranes, and to investigate the interaction of this polyisoprenol with NeuE and KpsM. It is anticipated that these studies will provide basic and critical information necessary to understand how these superlipids function as carriers of glycosyl units in transmembrane glycoconjugate synthesis.

Organization of the kps Gene Cluster

A complicated genetic and biochemical pathway is involved in regulating surface expression of the polySia capsule. The genetic organization of the kps gene cluster is shown in Figure 1. This complex has been cloned and is encoded in ca. 17-19 kb of DNA that consists of three coordinately regulated regions (Silver et al., 1981; Boulnois et al., 1987; Silver and Vimr, 1990). The multigenic cluster encodes for at least 14 proteins that are required for the synthesis, activation, and polymerization of Sia (region 2), energetics and translocation (region 3) and export of polySia chains to the cell surface (region 1). The gene complex maps at 64 map units on the E.

coli chromosome (Orskov et al., 1976; Vimr, 1991), and is absent from common laboratory isolates of E. coli.

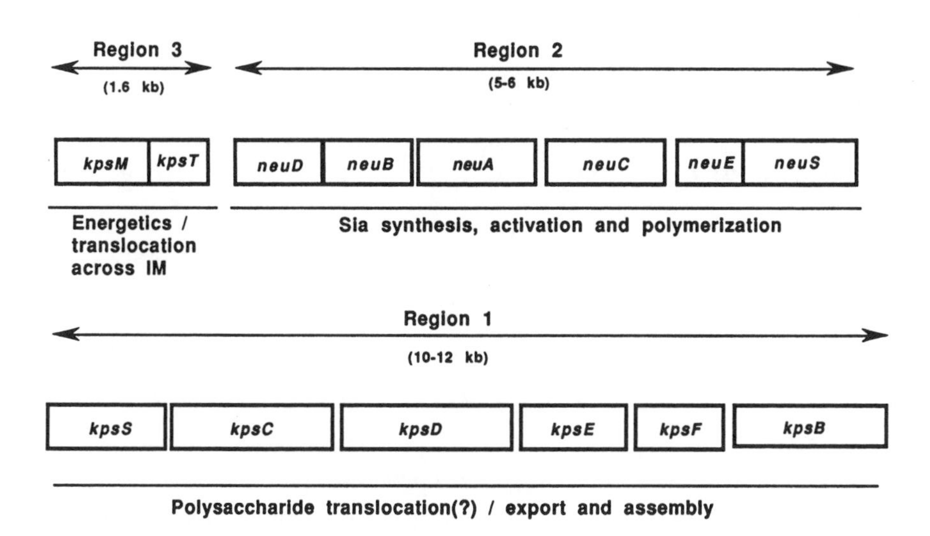

Figure 1. Genetic organization of the *kps* gene cluster in *E. coli* K1.

The molecular organization of regions 1 and 3 gene loci that flank the region 2 cluster is architecturally similar in all E. coli group II capsular polysaccharides. These loci share common components and appear to be functionally interchangeable (Roberts et al., 1988). This gene organization is also common to serogroups of N. meningitidis Gp. B and Haemophilus influenzae, leading to the suggestion that capsule expression in gram negative bacteria evolved from a common molecular origin (Frosch et al., 1991). In contrast, region 2 genes are specific for each type II capsular polysaccharide, and code for those enzymes required for sugar synthesis, activation and polymerization. Thus, the size of the region 2 cluster reflects the structural complexity of the polysaccharide (Boulnois and Jann, 1989).

A major objective of our program is to determine how two kps region 2 gene products, NeuE and NeuS, catalyze polySia chain initiation and polymerization, and how the two proteins encoded by region 3, KpsM and KpsT, catalyze translocation of polySia chains across the inner membrane. None of the enzymatic activities of the polyST have been solubilized, purified and reconstituted.

Biosynthesis of PolySia: Role of Region 2 kps Gene Products

The major steps in synthesis of Sia(Neu5Ac), CMP-Neu5Ac and polySia in *E. coli* K1 are summarized in reaction 1-5.

$$\text{(1) ManNAc+PEP} \xleftrightarrow{\text{Enz. 1}} \text{Neu5Ac+Pi}$$

$$\text{(2) Neu5Ac+CTP} \xleftrightarrow{\text{Enz. 2}} \text{CMP-Neu5Ac+PPi}$$

$$\text{(3) CMP-Neu5Ac+P-}C_{55} \xleftrightarrow{\text{polyST}} \text{Neu5Ac-P-}C_{55}\text{+CMP}$$

$$\text{(4) n(Neu5Ac-P-}C_{55}) \xrightarrow{\text{polyST}} (\text{Neu5Ac})_n\text{-P-}C_{55}\text{+(n-1)(P-}C_{55})$$

$$\text{(5) }(\text{Neu5Ac})_n\text{-P-}C_{55} \xrightarrow[\text{+ endogenous acceptor}]{\text{polyST}} (\text{Neu5Ac})_n\text{-acceptor+P-}C_{55}$$

where: Enz. 1, Neu5Ac synthase (NeuB); Enz. 2, CMP-Neu5Ac synthetase (NeuA); polyST, CMP-Neu5Ac:poly-α-2,8-sialosyl sialyltransferase complex (NeuE/NeuS); PEP, phosphoenolpyruvate; P-C_{55}, undecaprenlyphosphate. Reactions 1 & 2 are catalyzed by soluble enzymes, while reactions 3-5 are catalyzed by enzymes of the membrane-bound polysialyltransferase complex.

Genes for the enzymes catalyzing reactions 1-5 are encoded in region 2 of the kps gene complex. In E. coli K1, Sia synthase (Enz. 1) is a cold sensitive enzyme that is synthesized at 15°C, but is reversibly inactivated at low temperatures (Merker and Troy, 1990). This explains why these cells are acapsular when grown at 15°C (Troy and McCloskey, 1979). CMP-Sia synthetase (Enz. 2), and possibly the polyST activities catalyzing reactions 3-5, may also be cold sensitive (Merker and Troy, 1990). Sia synthase has not been purified from E. coli K1. The enzyme is labile even to dialysis, and is difficult to assay in soluble extracts at protein concentrations of <20 mg/ml^{-1} (Merker and Troy, 1990). CMP-Sia synthetase (Enz. 2) from E. coli has been cloned and sequenced by Vann and collaborators (Vann et al., 1987).

The exact number of enzymes in the polyST complex required to initiate and polymerize Sia (Rx 3-5) is not known. Chain synthesis is initiated (Rx 3) by the transfer of Sia from CMP-Sia to P-C_{55}, which functions as an intermediate carrier of sialyl residues (Troy et al., 1975). The addition of each Sia residue is coupled to the energetically favorable hydrolysis of CMP-Sia, and the subsequent hydrolysis of PPi. The extent to which Sia residues are polymerized on P-C_{55} is not known, but similar oligomeric sialyl-P-C_{55} intermediates have been described for polySia synthesis in N. meningitidis Gp. B (Masson and Holbein, 1985). As will be described in this paper, we have recently obtained evidence that Sia residues may be fully polymerized on P-C_{55} by the sequential addition of monomers (n=ca. 200 in Rx 4). Such a pre-assembled "activated" polysialyl lipid may be translocated across the inner member, while linked to P-C_{55}, presumably catalyzed by a translocase. Alternatively, a polysialylated reactive phosphoryl intermediate could be transferred en bloc to a translocator, designated "endogenous acceptor" in Rx 5, which then shuttles the polySia chains across the membrane. PolySia chain growth occurs by the addition of Sia to the non-reducing terminus of the growing chain (tail growth) in which the activated linkage in CMP-Sia is used for its own addition (Kundig et al., 1971; Rohr and Troy, 1980). Neither the activity responsible for initiating polySia synthesis (an initiase) or for catalyzing chain polymerization (a polymerase) has been isolated.

Steenbergen and Vimr concluded that polySia chain initiation and elongation were catalyzed by a single enzyme, NeuS (Steenbergen and Vimr, 1990). We believe, however, that it is more likely that at least two enzymes may be involved (Troy, 1992). While one transferase could possess two active sites, the identification of an undecaprenol recognition sequence in NeuE, but probably not NeuS (see below), adds support to our hypothesis that separate enzymes are required to initiate (NeuE) and polymerize (NeuS) polySia chains (Troy, 1992). Fortunately, this hypothesis makes several predictions that can now be experimentally tested. On the basis of the evidence available, however, the possibility that NeuE might function at some step other than synthesis, possibly translocation, cannot be excluded.

Topology of the Polysialyltransferase Complex in E. coli K1

Because of the key importance in knowing the topology of the polyST complex to understanding the molecular mechanism of capsule assembly, we sought to determine the transmembrane organization of the complex (Janas and Troy, 1989; Troy et al., 1990a,b). Our experimental strategy was to use membrane vesicles of defined orientation to assay the enzyme activity. Sealed right-side-out vesicles (ROV) have the same topology as intact cells, whereas inside-out-vesicles (IOV), prepared by disrupting cells in a French press, have the opposite orientation (Owen and Kaback, 1978). Thus, sealed IOV have a unique orientation, since enzymes normally located on the inner surface of the cytoplasmic membrane appear on the exterior side of the vesicles. Enzymes showing such a topological orientation can be assessed using impermeable substrates, e.g. CMP-[^{14}C]Sia, and sensitivity to proteases. Therefore, the level of polyST activity was determined in sealed ROV and IOV, using CMP-[^{14}C]Sia. The effect of partial trypsin treatment on inactivation of polyST activity was also assessed both before and after inversion. The results of these studies showed clearly that the functional domain of the polyST complex was located on the cytoplasmic surface of the IM (reviewed in Troy, 1992). This conclusion was confirmed by showing that there was only a slight decrease in polyST activity when ROV were treated with trypsin and then inverted. In contrast, >90% of the polyST activity was lost when ROV were inverted before trypsinolysis, i.e. in IOV. Based on immunoelectron microscopic analysis, a similar topological arrangement was proposed by Kröncke et al. (1990) for the glycosyltransferases catalyzing synthesis of the E. coli K5 capsular heteropolysaccharide.

Full Length PolySia Chains Are Translocated across the Inner Membrane and Translocation Is Not Linked to Synthesis

The new information regarding topology of the polyST complex requires that polySia chains must somehow traverse the IM before being exported to the outer membrane. After polySia chain synthesis is initiated on P-C_{55}, an important question is: how long do the chains become before they are translocated across the membrane, i.e. does polymerization precede or follow translocation? If chain translocation were to precede polymerization, then only short sialyloligomers would be found intracellularly. In contrast, if polymerization was complete before chain translocation, then long chains of polySia would be found inside. The crucial experiment to distinguish between these two possibilities was to carry out an in vivo ^{14}C-Sia labeling experiment in E. coli K1 cells unable to degrade sialic acid (nanA4 mutation: strain RHM18; Troy, et al. 1990c). After 3 hrs. of labeling, a cold osmotic shock was used to release the periplasmic contents. After disruption of the spheroplasts, the cytoplasm and membranes were separated by centrifugation. ^{14}C-Sia labeled components in these fractions were quantitated and analyzed by PAGE in the presence and absence of SDS, and before and after treatment with proteinase K. The results of this experiment showed that only high M_r ^{14}C-polySia chains (DP>85-200 Sia residues) were present in the cytoplasmic fraction from RHM18 and a translocation defective mutant (Troy et al., 1990c). No sialyloligomers were observed, either attached to the membrane or released to the cytoplasm. High M_r polySia chains were also identified in the periplasmic and membrane fractions from RHM18. In contrast, no polySia chains were translocated from the cytoplasm to the periplasm in a translocation-defective mutant. Thus in intact cells, full length ^{14}C-polySia chains can be fully polymerized on the cytoplasmic surface of the IM. These results verify our earlier finding, that full-length polySia chains can be synthesized by IOV (Weisgerber and Troy, 1990), a finding which shows that the polyST in its native topography can synthesize full-length chains. A second important conclusion that emerges from these results is that the polySia chains are fully polymerized before being translocated across the inner membrane. Therefore, chain polymerization is not obligatorily coupled to chain translocation, since full-length polymers can be isolated from inside the cell (Troy et al., 1991). This new finding now requires that any model invoked to explain synthesis and translocation must account for how full length polySia chains, in contrast to sialyloligomers, transit the inner membrane. While earlier studies by Boulnois and Jann (1989) had suggested that synthesis of the E. coli K5 capsule was coupled to its translocation, Jann and colleagues have now confirmed that full length K5 polysaccharide chains do accumulate intracellularly (K. Jann (1992), personal communication). Thus, these results confirm our original hypothesis that in vivo synthesis of full length capsular polysaccharide chains can occur independent of chain translocation. This conclusion is also in accord with the fact that the polySia chains, and the K5 polysaccharide, grow by the addition of new sugar residues to the non-reducing terminus (tail growth; Rohr and Troy, 1980). Given tail elongation, it is difficult to envisage a model in which synthesis is coupled to translocation. The molecular mechanism of the translocation process is unknown, and is the subject of the new experimental results presented in this paper.

Energy-dependence of the Polysialyltransferase as Studied in IOV

In 1984 we showed that the proton electrochemical potential gradient, DmH$^+$, was important for activation of polySia chain synthesis, since collapse of the gradient with CCCP inhibited activation (Whitfield and Troy, 1984). At that time we were unable to define more precisely why the membrane potential was important, but concluded that it may have been required for some step in chain translocation, or for maintaining full activity of the polyST complex. To follow-up the importance of DmH$^+$ in polySia chain synthesis, we re-examined the energy-dependency of the polyST activity in IOV prepared from E. coli K1. The activity of the polyST was modulated by providing additional energy in the form of ATP or electron-chain substrates (NADH, lactate), by selectively decoupling DmH$^+$ and the electrochemical potential (Dy), using CCCP, valinomycin and nigericin, and by using a nonhydrolyzable ATP analog (AMP-PNP) or N_3ATP, as an azido photoaffinity analog of ATP. The results of these studies showed that full activity of the polyST complex required DmH$^+$, Dy and the high energy phosphoryl potential of ATP, although it is still not possible to determine the molecular mechanism that links coupled ATP hydrolysis to polySia chain synthesis, or how the proton motive force is coupled. Translocation of polySia chains across the inner membrane also requires energy provided by DmH$^+$ and Dy (Troy et al., 1991; see below). On the basis of these results it is possible that energized membranes may be required to integrate functions of the polyST complex (NeuE and NeuS) with the superfamily of ABC (ATP-binding cassette) transporters (KpsT; KpsM).

Translocation of PolySia Chain across the Inner Membrane

Given that the functional domain of the polyST is localized on the cytoplasmic surface of the inner membrane, and polySia chains are fully polymerized inside the cell before being translocated across the inner membrane, a pertinent question becomes, "what is the molecular mechanism and energetics of translocation?" To investigate this problem, we developed an *in vivo* labeling procedure that allowed us to study directly the translocation step (Troy et al., 1991). The system uses spheroplasts prepared from K1 cells that are unable to degrade Sia because of a defect in Sia aldolase (nanA mutation). The strategy to study the vectorial translocation of polySia chains was as follows. The Sia permease was first induced by growing cells in the presence of Sia (Vimr and Troy, 1985). After pulse-labeling spheroplasts with ^{14}C-Neu5Ac, synthesis and translocation were followed kinetically, and the theory of compartmental analysis was applied to determine polySia chain distribution among the different compartments. A key feature of the method was the ability to distinguish between polySia chains that were bound to the membrane on the cytoplasmic and periplasmic surfaces. This was accomplished by the fact that polySia chains which were translocated across the inner membrane were sensitive to depolymerization by Endo-N-acetylneuraminidase (Hallenbeck, et al. 1987), whereas those remaining inside the vesicles were not. The effects of CCCP, a modulator of DmH$^+$ and valinomycin, a modulator of Dy, on translocation rates, was also assessed by this procedure. The following results were obtained: 1) polySia chains translocated across the IM remain membrane bound. The chains are then released from the membrane, and become "out-free polymers." It is not known if the release of polySia from the periplasmic surface of the inner membrane is an obligatory step required for their subsequent export to the outer membrane, although periplasmic proteins from kps$^+$, but not kps negative, strains appear to influence the kinetics of release (Janas, T., Janas, T. and Troy, F.A., unpublished results). 2) CCCP, a modulator of DmH$^+$, and valinomycin, a modulator of Dy, inhibited polySia chain translocation 85% and 50%, respectively. The effect of CCCP and valinomycin was on chain translocation and not on inhibition of polySia chain synthesis. Therefore, we conclude that both the DmH$^+$ and Dy are required for the vectorial translocation of polySia chains across the IM. While the molecular mechanism linking DmH$^+$ and Dy to polySia chain translocation is not known, we postulate that energized membranes may be required to integrate conformational changes and interactions between polyST activities (NeuE and NeuS) and the IM translocator complex (KpsM and KpsT) and undecaprenylphosphate (Troy, 1992).

EXPERIMENTAL STRATEGY

The experimental strategy developed to determine if polySia chains were attached to the cytoplasmic surface of the IM by a high energy covalent bond (designated "activated" polySia) was based on the following key findings: 1) the E. coli K1 polyST can polysialylate the sialyloligosaccharide moiety of many glycosphingolipids (GSL), added as exogenous acceptors. GD3 intercalates into the inner membrane of inside-out vesicles (IOVs) and is the best GSL acceptor (Cho and Troy, 1991); 2) in IOV, both the polyST and any "activated" polySia chains (attached to the membrane) appear on the exterior side of the vesicle (Janas and Troy, 1989).

Thus, activated polySia chains can be transferred by the polyST to exogenous GSL acceptors; 3) PolySia chains on the exterior surface of IOVs can be prelabeled by growing cells in ^{14}C-Glc, or by adding CMP-[^{14}C]Sia. The requirements for an en bloc transfer of activated [^{14}C]polySia chains to exogenous GSL acceptors could then be studied directly using non-denaturing gels. ^{14}C-polySia chains that are transferred to G_{D3} will penetrate the gel, only if they have been released from the ceramide moiety by ceramide glycanase. Therefore, activated ^{14}C-polySia chains transferred to G_{D3} were detected by PAGE after release of (^{14}C-polySia)$_n$-Sia-Gal-Glc from polysialylated G_{D3}, by ceramide glycanase (Figure 2). This procedure provides a facile way to measure the *en bloc* transfer of activated polySia chains to membrane-bound GSL acceptors, in the absence of an exogenous energy source.

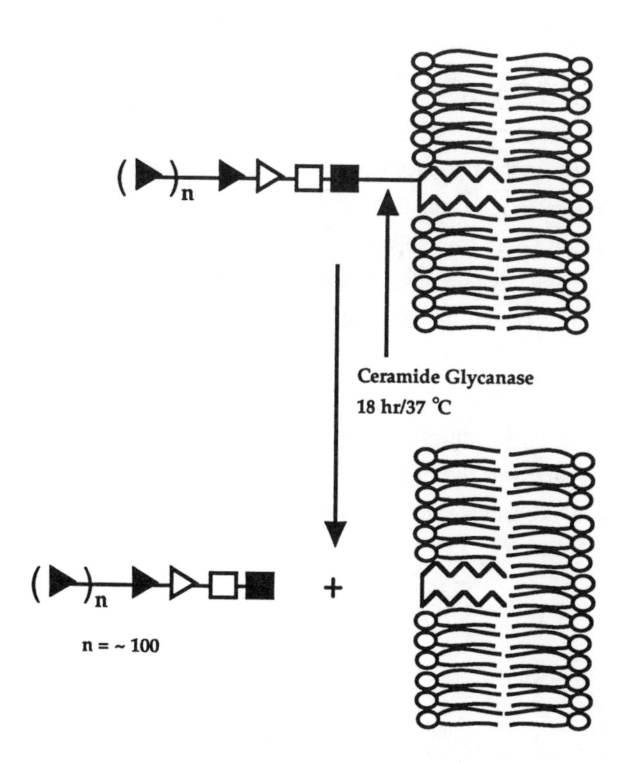

Figure 2. Schematic diagram of ceramide glycanase cleavage site and the release of (polySia)$_n$-Sia-Gal-Glc from polysialylated G_{D3}. Symbols represent α-2,8-linked Sia (▶), α-2,3-linked Sia (▷), Gal (□), Glc (■). The ceramide moiety of G_{D3} is represented by [symbol].

EXPERIMENTAL PROCEDURES

Identification of "Activated" PolySia Chains on the Cytoplasmic Surface of the Inner Membrane

Unlabeled, pre-existing polySia chains located on the cytoplasmic (external) surface of IOV prepared from E. coli K1 (EV11) were labeled with [^{14}C]Sia, by incubation with CMP-[^{14}C]Sia ("polysialylated IOV"), as previously described (Weisgerber and Troy, 1990). Unlabeled, pre-existing chains were also first depolymerized with endo-neuraminidase (Endo-N, Hallenbeck et al., 1987), and the resulting oligoSia chains that remained membrane bound were then labeled with CMP-[^{14}C]Sia in the presence of G_{D3} ("Endo-N pre-treated IOV"). The "activated" ^{14}C-oligoSia or ^{14}C-polySia chains that were transferred from the membrane to G_{D3} were released from the oligo- or polysialylated G_{D3} with ceramide glycanase and detected by PAGE .

Confirmation that PolySia Chains Are Linked to the Cytoplasmic Surface of the IM by a High Energy Bond: En Bloc Transfer of Pre-existing, "Activated" PolySia to G_{D3}

Cells (EV11) were grown in the presence of U-[^{14}C]glucose as the sole carbon source (Rohr and Troy, 1981). Washed, sealed IOV containing pre-existing U-[^{14}C]polySia chains were incubated at 33°C in the presence of G_{D3}. No exogenous energy sources, including CMP-Sia, were added. At the end of 30 min, the incubation mixture was divided into two, and one-half was treated with ceramide glycanase (37°C/18 hr). The control was not treated. The samples were then analyzed by non-denaturing PAGE.

Computer-aided Analysis for Identification of Presumed Polyisoprenol-recognition Sequences

The searching scheme used was the FASTA program (Lipman and Pearson, 1985) in Wisconsin GCG package (Version 7.0; Devereux et al., 1984).

RESULTS AND DISCUSSION

Identification of "Activated" PolySia Chains on the Cytoplasmic Surface of the Inner Membrane

Pre-existing polySia chains on the cytoplasmic surface of the IOV were labeled with [^{14}C]Sia, transferred to exogenously added G_{D3}, and analyzed as described under "Experimental Procedures."

As shown in Figure 3, "activated" ^{14}C-polySia chains pre-exist on the cytoplasmic surface of IOV.

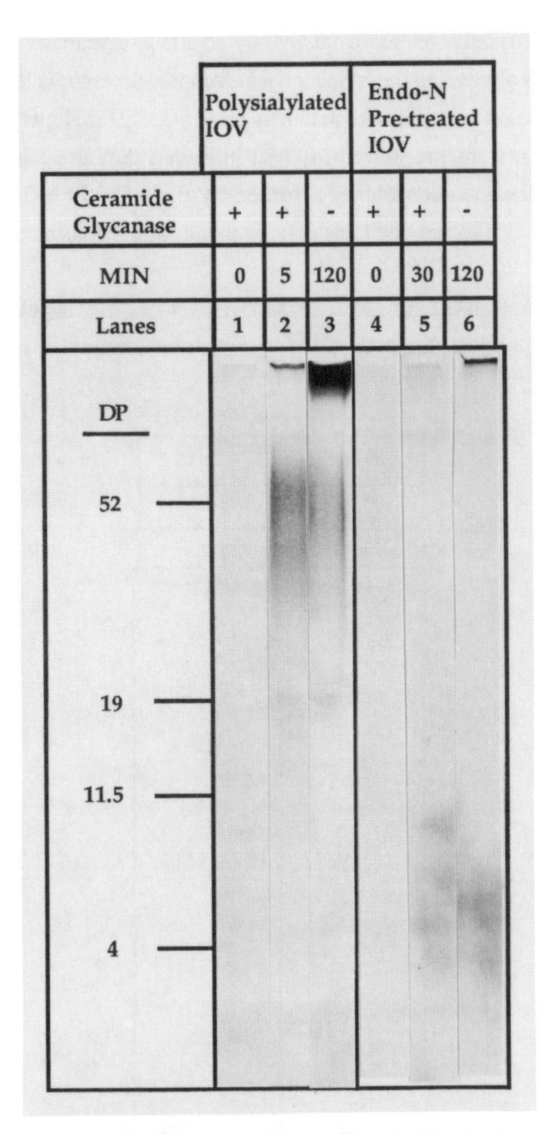

Figure 3. Identification of "activated" polySia chains on the cytoplsmic surface of the inner membrane.

These chains appear to be attached to the membrane by a high-energy bond because they can be transferred to G_{D3} within 5 min, forming polysialylated G_{D3} (lane 2). After depolymerizing the polySia chains with Endo-N to oligoSia, the membrane-bound sialyloligomers could also be transferred to G_{D3}, forming oligosialylated G_{D3}, (lane 5). The controls (lane 3 and 6) show that most of the ^{14}C-polysialylated or ^{14}C-oligosialylated G_{D3} molecules remain membrane-bound, and penetrate the gel only after release from G_{D3} by ceramide glycanase. The high energy bond linking the "activated" polymers to the endogenous membrane acceptor is labile, since some chains are spontaneously released from the membrane at pH 7.8/33°C, as shown in lanes 3 & 6. Once the polysialylated chains are transferred to G_{D3}, however, they are relatively stable. This is expected because of the greater stability afforded by the α-2,3 or α-2,8-ketosidic linkages, compared with "activated" linkage that joins polySia to the endogenous receptor.

Confirmation that PolySia Chains Are Linked to the Cytoplasmic Surface of the Inner Membrane

This confirmation was provided by showing the en bloc transfer of pre-existing, activated polySia to G_{D3} (Figure 4).

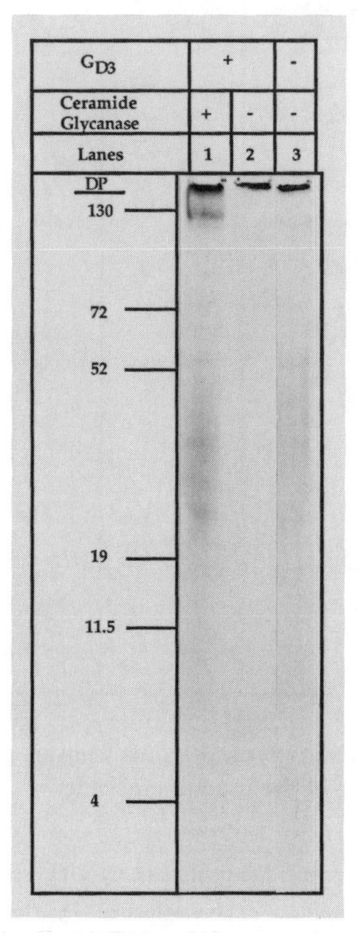

Figure 4. Confirmation that polySia chains are linked to the cytoplasmic surface of the inner membrane by a high energy bond: *en bloc* transfer of pre-existing, "activated" polySia to G_{D3}.

As shown in lane 1, some of the [^{14}C]-labeled polySia chains (DP≈20>70) that were linked to the cytoplasmic surface of the IM were transferred by the polyST to G_{D3}. It was possible to detect this transfer because the [^{14}C]polysialylated oligosaccharide moiety of G_{D3} migrated into the non-denaturing PAG only after release from the membrane by ceramide glycanase. In the absence of ceramide glycanase, the polySia chains remained covalently linked to G_{D3}, and therefore were unable to penetrate the gel (lane 2). In the absence of G_{D3}, some of the polySia chains that were linked to the membrane were released spontaneously (lane 3), thus confirming the labile nature of the "activated" linkage between pre-existing polySia chains and the cytoplasmic membrane. Once the [^{14}C]polySia were transferred to G_{D3}, however, the new glycosidic linkage between polySia and the oligosaccharide moiety of G_{D3} was stable (lane 2). On the basis of these results, we conclude that pre-existing U-[^{14}C]polySia chains on the cytoplasmic surface of the IM are "activated" by virtue of a high energy bond. These chains are transferred en bloc t o G_{D3} in the absence of CMP-Sia, or any other exogenous source of energy. The linkage appears to be labile, based on the spontaneous release of [^{14}C]polySia chains from the membrane (30 min, pH 7.8, 33°C). The nature of the "activated" linkage and to what the polySia chains are attached is under investigation. The lability of the activated chains suggest they may be attached to undecaprenylphosphate (P-C_{55}), a glycosyl carrier lipid known to function in polySia chain synthesis in both E. coli K1 (Troy et al., 1975) and N. meningitidis Gp B (Masson and Holbein, 1985). The extreme lability of the sialylmonophosphorylundecaprenol linkage has been noted previously (Troy et al., 1975).

Polyisoprenol-recognition Sequences in the Polysialyltransferase/Translocator Complex

At least four proteins encoded by the kps gene cluster constitute the polyST/inner membrane translocator complex. NeuE and NeuS are postulated to catalyze polySia chain initiation and polymerization, respectively (Troy, 1992). KpsT and KpsM have been implicated in the energetics and transmembrane translocation (Pavelka et al., 1991). The deduced amino acid sequences of these four proteins were subjected to computer analysis for the presence of a 13 amino acid consensus "dolichol recognition" sequence, as described under "Experimental Procedures." This sequence was previously identified in several yeast glycosyltransferases (Albright et al., 1989). Our analyses has shown that NeuE and KpsM each contain a presumed polyisoprenol (PI) recognition sequence (Table I). These data further verify the proposal that PIs may interact with transferases through specific consensus sequences (Albright et al., 1989).

Table I. Comparison of Proteins from the *kps* Gene Cluster of *E. coli* K1 with the Potential Dolichol-Recognition "Consensus" Sequence

PROTEIN	FRAGMENT	SIMILARITY SCORE (SC)	Z-VALUE
Consensus KpsM	L F V X F X X I P F X F Y V F L L N G L I P F F I F	37	3.85
Consensus NeuE	L F V X F X X I P F X F Y L I I L F L I F - F N F F	34	3.22
Consensus KpsT	L F V X F X X I P F X F Y A F G L S M A F K F D Y Y	31	2.59
Consensus NeuS	L F V X F X X I P F X F Y K F V E F K G A P S V K S	27	1.76

"Consensus" refers to the consensus dolichol-recognition sequence, a 13 amino acid sequence found in three yeast glycosyltransferases (Albright *et al.*, 1989). It was used as a query sequence to compare with the indicated protein sequences by FASTA program of the "Wisconsin Package" (Devereux *et al.*, 1984). The similarity score and the z-value were calculated according to Lipman and Pearson (1985). The average similarity score (18.6) and the standard deviation (4.78) required for calculating the z-value were obtained by comparing the consensus sequence with the Release 18.0 of Swiss Protein Database. The solid vertical lines denote identical amino acids, while the broken lines denote a conserved amino acid replacement. The dash denotes an arbitrary gap to maximize the alignment.

The finding that KpsM, a protein with no known biosynthetic function, but rather implicated only in chain translocation also contained a PI binding motif in one of its six transmembrane spanning domains (Fig. 5) was unexpected. This observation led us to hypothesize a bifunctional role for the PIs. The central idea of this proposal is that the polyisoprenyl "super-lipids" may function as a flexible matrix or scaffolding to organize and tether proteins of multienzyme complexes to coordinate not only biosynthetic reactions, but also translocation processes (Troy, 1992).

E. coli **NeuE**

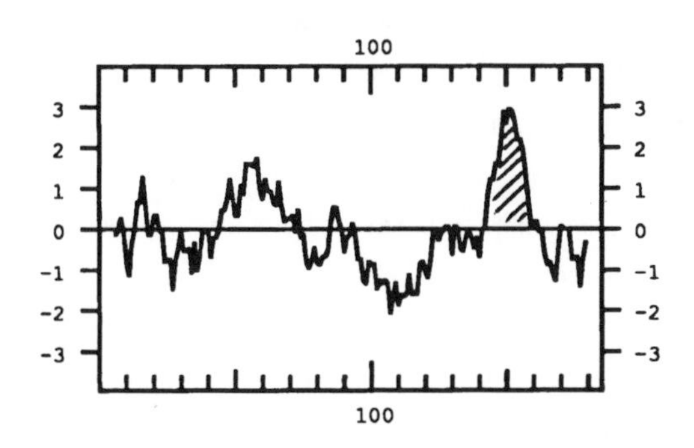

E. coli **KpsM**

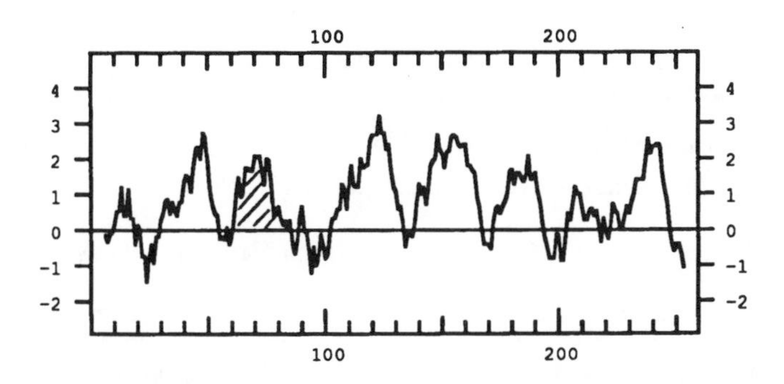

Figure 5. Comparison of hydropathy plots (Kyte and Doolittle, 1982) for NeuE (Weisgerber *et al.*, 1991) and KpsM(Pavelka *et al.*, 1991) from *E. coli* K1. The conserved, 13 amino acid potential membrane-spanning domain in each protein that has homology with polyisoprenol (dolichol)-recognition sequences is indicated by shading.

NMR Studies of Polyisoprenols in Model Membranes and their Implication in PolySia Chain Translocation

The molecular mechanism whereby the extremely hydrophilic, anionic, polySia chains transit the membrane is unknown. This problem remains as much a mystery as how dolichylphosphate-linked oligosaccharides traverse the endoplasmic reticulum (reviewed in Hirschberg, 1992). An interesting physiochemical property of the long chain polyisoprenols (undecaprenol, dolichol and their phosphorylated derivative) that has emerged recently from ^{2}H- and ^{31}P NMR studies is their unusual property to induce non-bilayer structure in model membranes (reviewed in deRopp et al., 1987). This finding could be important for our conceptual understanding of how oligo-polysaccharide chains move across or through membranes. For example, ^{2}H NMR studies of ^{2}H-

labeled undecaprenol, dolichol, and their phosphate esters showed that the free polyisoprenols at 5 mol% significantly disrupted the membrane bilayer structure, and that this effect was more pronounced with dolichylphosphate at 1 mol% (Knudsen and Troy, 1989). Further, this disruption does not appear to be localized to only regions about the PI itself, but rather to extended regions of the membrane. Thus, undecaprenol/undecaprenylphosphate (and the dolichol derivatives) may "modulate" membrane behavior by activating its matrix arrangement to more motional states. In doing so, the PI induced perturbation of lipid bilayer structure to a more non-bilayer or hexagonal (Hex_{II}) conformation may facilitate the translocation process (Knudsen and Troy, 1989). A potential driving force for the actual movement of these polyanionic chains could be the -150 mV potential generated in the E. coli K1 inner membrane by the proton motive force (DmH+, Dy, as discussed earlier). This model makes several predictions that can be experimentally tested, and our future studies will be aimed at testing these predictions.

NeuE and NeuS Have Homology with DNA/RNA Polymerases

A second interesting feature to emerge from our analyses of the deduced protein sequences of NeuE and NeuS is that they appear to be related to DNA and RNA polymerases, and other DNA-binding proteins. A few representative proteins are listed in Table II.

Table II. Protein Sequences Similar to NeuE and NeuS

Databank Sequences Matched		**Homology** *	
Accession#	**Protein**	**Identity(%)**	**Overlap (aa)**
Similar to NeuE			
P10582	**DNA polymerase (Maize)**	**23.5**	**136**
P21422	**DNA-directed RNA polymerase, beta chain(Protozoa)**	**20.0**	**160**
P04386	**Regulatory protein GAL4 (Yeast)**	**22.7**	**75**
Similar to NeuS			
P21402	**DNA polymerase (Fowlpox virus)**	**23.1**	**104**
P06856	**DNA polymerase(Vaccinia virus, strain wr)**	**23.0**	**61**

* The homology data was generated by FASTA program (Lipman and Pearson, 1985) in the "Wiscosin Package" (Devereux et al., 1984): percent identity in the number of overlapped amino acids (aa).

As shown, when the protein sequences were analyzed by the FASTA program (Lipman and Pearson, 1985) in the Wisconsin package (Devereux et al., 1984), each protein showed significant identity and overlap with DNA polymerases. NeuE also showed homology with a protozoal DNA-directed RNA polymerase and the regulatory protein Gal4, from yeast. This intriguing result was also unexpected, and implies that these proteins may be evolutionarily related. Functionally, these proteins share the common property of recognizing/binding polyanions. Whether this is the extent of their commonality remains to be determined. Nevertheless, the similarity has interesting evolutionary implications with respect to origin of kps and how E. coli K1 strains may have acquired the gene cluster, particularly region 2 genes coding for enzymes responsible for synthesis, activation and polymerization of Sia. The fact that most E. coli species, except certain human pathogens, lack these enzymes suggests that these human-derived isolates may have acquired the genes from their host (Vimr et al., 1983).

Acknowledgments: The expert secretarial assistance of Becky Greer is gratefully appreciated. This work was supported by research grant AI09352 from the National Institutes of Health.

REFERENCES

Abeijon, C. and Hirschberg, C. B. (1992) Topography of glycosylation reactions in the endoplasmic reticulum. Trends Biochem .Sci. 17, 32-36.

Albright, C. F., Orlean, P. and Robbins, P. W. (1989) A 13-amino acid peptide in three yeast glycosyltransferases may be involved in dolichol recognition. Proc. Natl. Acad. Sci. U S A 86, 7366-7369.

Boulnois, G. J. and Jann, K. (1989) Bacterial polysaccharide capsule synthesis, export and evolution of structural diversity. Mol .Microbiol. 3, 1819-1823.

Boulnois, G. J., Roberts, I. S., Hodge, R., Hardy, K. R., Jann, K. B. and Timmis, K. N. (1987) Analysis of the K1 capsule biosynthesis genes of Escherichia coli: definition of three functional regions for capsule production. Mol. Gen. Genet. 208, 242-246.

Cho, J.-W. and Troy, F. A. (1989) Gangliosides as Exogenous Acceptors to Map the Acceptor Sugar Requirements of the Poly-a-2,8-sialosyl Sialyltransferase in *Escherichia coli* K1, In Sharon, N., Lis, H., Duksin, D. and Kahane, I. (eds.) Proc. Xth International Symp. on Glycoconjugates 143, Organizing Committee Press, Jerusalem, Israel, pp. 209-210.

de Ropp, J. S., Knudsen, M. J. and Troy, F. A. (1987) ^{2}H NMR Investigation of the dynamics and conformation of polyisoprenols in model membranes. Chemical Scripta 27, 101-108.

Devereux, J., Haeberli, P. and Smithies, O. (1984) A comprehensive set of sequence analysis programs for the VAX. Nucleic Acids Res. 12, 387-395.

Finne, J., Leinonen, M. and Makela, P. H. (1983) Antigenic similarities between brain components and bacteria causing meningitis. Implications for vaccine development and pathogenesis. Lancet **ii**, 355-357.

Frosch, M., Edwards, U., Bousset, K., Kraube, B. and Weisgerber, C. (1991) Evidence for a common molecular origin of the capsule gene loci in gram-negative bacteria expressing group II capsular polysaccarides. Molec. Miocrobiol. 5, 1251-1263.

Hallenbeck, P. C., Vimr, E. R., Yu, F., Bassler, B. and Troy, F. A. (1987) Purification and properties of a bacteriophage-induced endo-N-acetylneuraminidase specific for poly-a-2,8-sialosyl carbohydrate units. J. Biol. Chem.. 262, 3553-61.

Janas, T. and Troy, F. A. (1989) The Escherichia coli K1 poly-a-2,8-sialosyl sialyltransferase is topologically oriented toward the cytoplasmic face of the inner membrane., In Sharon, N.,

Lis, H., Duksin, D. and Kahane, I. (eds.) Proc. Xth International Symp. on Glycoconjugates 142, Organizing Committee Press, Jerusalem, Israel, pp. 207-208.

Knudsen, M. J. and Troy, F. A. (1989) Nuclear magnetic resonance studies of polyisoprenols in model membranes. Chem. Phys. Lipids 51, 205-12.

Kroncke, K. D., Boulnois, G., Roberts, I., Bitter-Suermann, D., Golecki, J. R., Jann, B. and Jann, K. (1990) Expression of the *Escherichia coli* K5 capsular antigen: immunoelectron microscopic and biochemical studies with recombinant E. coli. J. Bacteriol. 172, 1085-1091.

Kundig, F. D., Aminoff, D. and Roseman, S. (1971) The Sialic Acids. XII. Synthesis of colominic acid by a sialyltransferase from *Escherichia coli* K-235. J. Biol. Chem. 246, 2543-2550.

Kyte, J. and Doolittle, R. F. (1982) A simple method for displaying the hydropathic character of a protein. J. Mol. Biol. 157, 105-132.

Lipman, D. J. and Pearson, W. R. (1985) Rapid and sensitive protein similarity searches. Science 227, 1435-1441.

Masson, L. and Holbein, B. E. (1985) Role of lipid intermediate in the synthesis of serogroup B *Neisseria meningitidis* capsular polysaccharide. J. Bacteriol.. 161, 861-867.

Merker, R. I. and Troy, F. A. (1990) Biosynthesis of the polysialic acid capsule in *Escherichia coli* K1. Cold inactivation of sialic acid synthase regulates capsule expression below 20°C. Glycobiology 1, 93-100.

Owen, P. and Kaback, H. R. (1978) Molecular structure of membrane vesicles from *Escherichia coli.* Proc. Natl. Acad. Sci. USA 75, 3148-52.

Orskov, I., Sharma, V. and Orskov, F. (1976) Genetic mapping of the K1 and K4 antigens of K(L) antigens and K antigens of 08:K27(A), 08:K8(L) and 09:K57(B). Acta Pathol. Microbiol. Scand. Sect. B 84, 125-131.

Pavelka, M. J., Wright, L. F. and Silver, R. P. (1991) Identification of two genes, kpsM and kpsT, in region 3 of the polysialic acid gene cluster of *Escherichia coli* K1. J. Bacteriol. 173, 4603-4610.

Roberts, I. S., Mountford, R., Hodge, R., Jann, K. B. and Boulnois, G. J. (1988) Common organization of gene clusters for production of different capsular polysaccharides (K antigens) in *Escherichia coli.* J. Bacteriol. 170, 1305-1310.

Rohr, T. E. and Troy, F. A. (1980) S.tructure and biosynthesis of surface polymers containing polysialic acid in *Escherichia coli.* J. Biol. Chem. 255, 2332-2342.

Silver, R. P., Finn, C. W., Vann, W. F., Aaronson, W., Schneerson, R., Kretchmer, P. J. and Garon, C. F. (1981) Molecular cloning of the K1 capsular polysaccharide genes of *E. coli.* Nature 289, 696-698.

Silver, R. P. and Vimr, E. R. (1990) Polysialic acid capsule of *Escherichia coli* K1, In Iglewski, B. H. and Clark, V. L. (eds.) The bacteria: a treatise on structure and function, XI, Molecular basis of bacterial pathogenesis. Academic Press, San Diego, pp. 39-60.

Steenbergen, S. M. and Vimr, E. R. (1990) Mechanism of polysialic acid chain elongation in *Escherichia coli* K1. Mol. Microbiol. 4, 603-611.

Troy, F. A. (1992) Polysialylation: from bacteria to brains. Glycobiology 2, 5-23.

Troy, F. A., Janas, T., Janas, T. and Merker, R. I. (1991) Vectorial translocation of polysialic acid chains across the inner membrane of neuroinvasive *E. coli* K1. FASEB J. 5, A1548.

Troy, F. A., Janas, T. and Merker, R. I. (1990a) Topology of the poly-a-2,8-sialyltransferase in *E. coli* K1 and energetics of polysialic acid chain translocation across the inner membrane. Glycoconj J. 7, 383.

Troy, F. A., Janas, T. and Merker, R. I. (1990b) Topology of the polysialyltransferase complex in the inner membrane of *E. coli* K1. FASEB J. 4, 3189.

Troy, F. A., Janas, T., Janas, T. and Merker, R. I. (1990c) Transmembrane translocation of polysialic acid chains across the inner membrane of neuroinvasive *E. coli* K1. Glycoconj J. 8, 152.

Troy, F. A. and McCloskey, M. A. (1979) Role of a membranous sialyltransferase complex in the synthesis of surface polymers containing sialic acid in *Escherichia coli*: temperature-induced alteration in the assembly process. J. Biol. Chem. 254, 7377-7387.

Troy, F. A., Vijay, I. K. and Tesche, N. (1975) Role of undecaprenyl phosphate in synthesis of polymers containing sialic acid *in Escherichia coli.* J. Biol. Chem. 250, 156-163.

Vann, W. F., Silver, R. P., Abeijon, C., Chang, K., Aaronson, W., Sutton, A., Finn, C. W., Lindner, W. and Kotsatos, M. (1987) Purification, properties, and genetic location of *Escherichia coli* cytidine 5'-monophosphate N-acetylneuraminic acid synthetase. J. Biol. Chem. 262, 17556-17562.

Vimr, E. R., Merker, R. I. and Troy, F. A. (1983) Genetic mechanisms regulating synthesis of capsular polysialic acid in *Escherichia coli* K1 strains. Fed. Proc. 42, 2164.

Vimr, E. R. (1991) Map position and genomic organization of the kps cluster for polysialic acid synthesis in *Escherichia coli* K1. J. Bacteriol. 173, 1335-1338.

Vimr, E. R., McCoy, R. D., Vollger, H. F., Wilkison, N. C. and Troy, F. A. (1984) Use of prokaryotic-derived probes to identify poly(sialic acid) in neonatal neuronal membranes. Proc. Natl. Acad. Sci. USA 1971-1975.

Vimr, E. R. and Troy, F. A. (1985) Identification of an inducible catabolic system for sialic acids (nan) in *Escherichia coli.* J. Bacteriol. 164, 845-853.

Weisgerber, C., Hansen, A. and Frosch, M. (1991) Complete nucleotide and deduced protein sequence of CMP-NeuAc: poly-a-2,8 sialosyl sialyltransferase of *Escherichia coli* K1. Glycobiology 1, 357-365.

Weisgerber, C. and Troy, F. A. (1990) Biosynthesis of the polysialic acid capsule in *Escherichia coli* K1. The endogenous acceptor of polysialic acid is a membrane protein of 20 kDa. J. Biol. Chem. 265, 1578-87.

Whitfield, C. and Troy, F. A. (1984) Biosynthesis and Assembly of the polysialic acid capsule in *Escherichia coli* K1: Activation of sialyl polymer synthesis in inactive sialyltransferase complexes requires protein synthesis. J. Biol. Chem. 259, 12776-12780.

Wyle, F. A., Artenstein, M. S., Brandt, B. L., Tramont, E. C., Kasper, D. L., Altieri, P. L., Berman, S. L. and Lowenthal, J. P. (1972) Immunologic responses of man to group B meningococcal polysaccharide vaccines. J. Infect. Dis. 126, 514-522.

THE EXPORT OF CAPSULAR POLYSACCHARIDES BY ESCHERICHIA COLI.

C. Pazzani[1], A. Smith[1], D. Bronner[2], K. Jann[2], G. J. Boulnois[1] and I. S. Roberts[1].

[1] Department of Microbiology, University of Leicester, P. O. Box 138, Medical Science Building, Leicester, LE1 1 9HN. [2] Max Planck Institut fur Immunobiologie, Freiburg i. Br., Germany.

SUMMARY: Escherichia coli can express a wide range of chemically diverse capsular polysaccharides or K antigens on its cell surface. On the basis of a number of biochemical and genetic criteria it is possible to divide these capsular polysaccharides into two groups, I and II. The gene clusters encoding for the biosynthesis of a number of chemically diverse group II polysaccharides have been cloned and subjected to a detailed molecular analysis. These studies have demonstrated that there is a conserved genetic organisation between different group II capsule gene clusters and that the functions involved in the export of these polysaccharides appear to be conserved. In this paper we review the recent results of our studies on the export of group II polysaccharides by E. coli.

INTRODUCTION

Many Escherichia coli isolates are surrounded by a capsular polysaccharide (K antigen) of which over 70 chemically different structures have been observed (Orskov et al., 1977). The genetic basis for this diversity is not clear. On the basis of chemical structure, size, mode of expression and chromosomal location of genetic determinants the capsular polysaccharides can be divided into two broad groups, I and II (Jann and Jann, 1987). A given isolate produces a single K antigen and co-expression or switching of capsule types either within or between groups has not been observed (Jann, 1985).

Group I K antigens occur as thick copious capsules and are restricted to 08 and 09 and occasionally 020 and 0101 serotypes (Jann and Jann, 1990). They have a low charge density and are expressed at all growth temperatures. Some group I polysaccharides are anchored in the outer membrane by linkage to core-lipid A and can be thought of as extended lipopolysaccharides (LPS) (Jann and Jann, 1987 and 1990). Genetic determinants for group I capsular polysaccharide production map near the his and rfb (O antigen synthesis) loci (Ørskov et al., 1977) but have not been analysed in detail. Studies using the K27 antigen indicate that a second *trp*-linked locus is necessary for surface expression of a complete group I capsule, though this is not the case for the K30 antigen (Laasko et al., 1988; Schmidt et al., 1977).

The group II capsules have a higher charge density than group I polysaccharides and are not expressed at low growth temperatures (17-20°C). Group II capsules are co-expressed with a variety of O antigens (excluding the four above) (Jann and Jann, 1987). In all group II K antigens so far analysed the reducing end of the polysaccharide is substituted with phosphatidic acid and this substitution may act as the membrane anchor for group II capsules (Jann and Jann, 1990; Schmidt et al., 1982). Unlike group I, group II capsule producing strains have a high CMP-KDO (2-keto-3-deoxy mannooctulonic acid) synthetase activity at the capsule permissive temperature (Finke et al., 1989 and 1990). Genetic determinants for the expression of some group II capsules have been mapped at 64 minutes (Vimr, 1991) near serA on the E. coli chromosome (Ørskov et al., 1976) and have been termed kps (Silver et al., 1984; Vimr et al., 1989), formerly kpsA (Ørskov and Nyman, 1974). Gene clusters encoding several chemically different group II K antigens have been cloned and shown to have a common organisation (Roberts et al., 1986, 1988). The group II capsule gene clusters analysed to date consist of three functional regions. Regions 1 and 3 (kps) are homologous between different group II capsule gene clusters and are thought to encode products which perform identical functions in the expression of chemically different group II K antigens (Boulnois et al., 1987; Boulnois and Roberts, 1990; Roberts et al., 1988). Region 1 is about 7kb and, in part, encodes proteins for the transport of mature, phosphatidic acid-linked polysaccharide from the periplasm to the cell surface. In addition, the structural gene encoding CMP-KDO synthetase is located within region 1 (Pazzani, C., manuscript in preparation). The role of this enzyme is unclear but it has been postulated to play a part in the initiation of polymer biosynthesis (Finke et al., 1989) because KDO has been found at the terminus of some group II capsules which do not have KDO in the repeat unit (Jann and Jann, 1990). Region 3 is 1.6kb and encodes two proteins believed to be involved in the translocation of group II polysaccharide across the inner membrane (Kroncke et al., 1990; Pavelka et al., 1991; Smith *et al.*, 1990). Region 2 is unique for each group II polysaccharide and is positioned between the common regions 1 and 3 (Boulnois and Jann, 1990). The available evidence suggests that region 2 encodes enzymes for the synthesis of the group II polysaccharide in question (Boulnois et al., 1987, Roberts et al., 1988). In the case of the K1 antigen gene cluster, region 2 encodes for sialic acid synthesis, activation and polymerisation (Vimr et al., 1989). The size of region 2 broadly correlates with the chemical complexity of the polysaccharide (Drake et

al., 1990). An isolate expressing a given group II capsule carries region 2 determinants for the biosynthesis of that polysaccharide only and silent region 2 determinants are not present on the E. coli chromosome (Roberts, M. et al., 1988).
We have recently completed the nucleotide sequence of the entire K5 capsule gene cluster. Here we review what this has revealed about possible processes involved in the export of group II capsular polysaccharides by E. coli.

ANALYSIS OF REGION 3.

Mutations in region 3 result in cytoplasmic polysaccharide of lower molecular weight than the cell surface material. This cytoplasmic polysaccharide is unlinked to the lipid (Kroncke, et al., 1990), a phenotype consistent with the synthesis of polysaccharide occurring on the inner face of the cytoplasmic membrane (Boulnois and Jann, 1990). The determination of the complete nucleotide sequence of region 3 from the cloned K5 and K1 antigen gene clusters has revealed a single transcriptional unit comprised of two genes termed kpsM and kpsT (Pavelka, et al., 1991; Smith et al., 1990).

1) The KpsM protein.

The hydropathy profile of the predicted amino acid sequence of the KpsM protein is characteristic of an integral membrane protein with six potential membrane-spanning domains. The integral membrane components of the periplasmic binding protein-dependent transport systems generally consist of five or six potential membrane-spanning alpha-helices separated by short stretches of hydrophilic sequence (Hiles et al., 1987). Thus the kpsM gene may encode the integral membrane component of an analogous putative polysaccharide transport system. The 30.2 KDa BexB protein encoded by the Haemophilus influenzae capsule gene cluster, has also been identified as a candidate integral membrane protein involved in capsule expression (Kroll et al., 1990). Comparison of the putative amino acid sequence of the KpsM and BexB proteins using the Lipman and Pearson algorithm (1985) revealed 26.0% identity. Allowing for conservation amino acid substitutions by the criterion of the algorithm, homology is extended to 69.0%. Although the sequence homology between the KpsM and BexB proteins is relatively modest, their hydropathy profiles are almost identical (Smith et al., 1990). This may indicate that the KpsM and BexB proteins have a similar function in capsule biogenesis, possibly mediating the transport of capsular polysaccharide across the inner membrane in a carrier or pore-mediated mechanism.

2) The KpsT protein.

Analysis of the putative amino acid sequence of the KpsT protein revealed potential adenine nucleotide binding fold sequences (Walker et al., 1982). Such sequences have been identified in the inner membrane ATP-binding components of enterobacterial periplasmic binding protein-dependent transport systems (Higgins et al., 1990). These transport systems are multicomponent, consisting of a periplasmic binding protein and a complex of several inner-membrane-associated proteins (Higgins et al., 1990). The latter generally consists of two highly hydrophobic integral membrane proteins, which are proposed to mediate the translocation of the substrate across the membrane, and two relatively hydrophilic membrane proteins which bind ATP and are believed to couple ATP hydrolysis to the transport process (Higgins et al., 1990). Periplasmic binding protein-dependent transport systems are generally involved in the uptake of nutrients into the bacterial cell. However, a number of proteins have been identified which have homology with this family of inner membrane transporters, but which are involved in the export of material from prokaryotic and eukaryotic cells. This group of export proteins appears to have a diverse range of substrates and includes the eukaryotic Mdr protein responsible for the export of cytotoxic drugs from tumour cells (Gros et al., 1986), the E. coli HlyB protein involved in the secretion of haemolysin (Holland et al., 1990) and NdvA protein of Rhizobium meliloti (Stanfield et al., 1988) and ChvA protein of Agrobacterium tumefaciens (Cangelosi et al., 1989) which are implicated in the export of cyclic ß-(1-2)glucan. It is interesting to note that mutations in the chvA gene result in cytoplasmic ß-(1-2)glucan, which has a lower molecular weight than the wild-type glucan (Cangelosi et al., 1989), a phenotype analogous to E. coli K5 Region 3 mutants (Kroncke et al., 1990). Based on the observation that the KpsT protein has a putative ATP-binding site, we proposed that it may be a subunit of a similar export system, coupling ATP hydrolysis to the transport of capsular polysaccharide across the inner membrane.

Region 1 of the capsulation (cap) locus of H. influenzae is proposed to be essential for the energy-dependent export of capsular polysaccharide (Kroll et al., 1988; 1989) and thus it may have a similar role in capsule biogenesis to Region 3 of the E. coli kps locus. Four genes have been identified in Region 1 of the cap locus, bexA, bexB, bexC and bexD (Kroll et al., 1988; 1990) and analysis of the encoded proteins has revealed the presence of potential adenine nucleotide binding fold sequences in the 24.7 KDa BexA protein (Kroll *et* al., 1988). Comparison of the putative amino acid sequence of the KpsT and BexA proteins, using the sequence alignment program devised by Lipman and Pearson (1985), revealed 45.2% homology. Taking into account conservative amino acid substitutions, homology is extended to 89.4%. The two cysteine residues of the KpsT protein align exactly with two of the five cysteine residues of the BexA protein and the hydropathy profiles of the two proteins, calculated by the method of Kyte and Doolittle (1982) using an 11-residue window, are very similar. Thus the KpsT and BexA proteins not only have extensive primary sequence homology, but may also have a similar secondary structure. This indicates that the two proteins may have a common role in the export of capsular polysaccharide across the inner membrane, and the present of an ATP-binding site would suggest that they both act as energy-coupling subunits for this process.

Therefore Region 3 of the E. coli K5 kps locus appears to consist of an operon of two genes: kpsM and kpsT. We proposed that the two putative encoded proteins may act together in a capsular polysaccharide export complex analogous to the inner-membrane complex of the periplasmic binding protein-dependent transport systems. The KpsT protein may couple ATP hydrolysis to the KpsM protein-mediated transport of the (growing) polysaccharide across the inner membrane. The archetypal transport system appears to require two hydrophobic domains encoded as two single polypeptides or as a larger two-domain protein. Similarly, two ATP-binding domains are generally present (Higgins *et al.*, 1990). To fulfil these criteria, each of the KpsT and KpsM proteins may be present at homodimers in the putative export complex.

ANALYSIS OF REGION 1.

Previous studies on the role of region 1-encoded functions exploited deletion derivatives of the cloned K1 antigen gene cluster in which almost the entire region 1 was deleted (Boulnois et al., 1987). Such mutations result in phosphatidic acid-linked polysaccharide in the periplasmic space (Boulnois et al., 1987), indicating, at least in part, a role for the region 1 gene products in the translocation of polysaccharide onto the cell surface. Studies on the cloned K1 and K7 antigen gene clusters identified at least five proteins encoded by this region (Boulnois and Jann, 1990, Roberts et al., 1988), and one, a protein of 60 KDa, is located within the periplasmic space (Silver et al., 1987).

Strains expressing group II K antigens have increased levels of CMP-KDO synthetase activity at the capsule permissive temperature (37°C) compared to either unencapsulated E. coli strains or those expressing group I K antigens (Finke et al., 1990). In contrast, at non-permissive temperature (18°C), the level of CMP-KDO synthetase activity is low and essentially constant irrespective of encapsulation and reflects the presence of a CMP-KDO synthetase encoded by kdsB which is involved in LPS biogenesis (Finke et al., 1990; Goldman and Kohlbrenner, 1985). These observations together with the finding of KDO at the reducing termini of a number of group II polysaccharides which lack this sugar in the repeat unit, has led to the hypothesis that CMP-KDO synthetase plays an important role in the expression of group II polysaccharides (Finke et al., 1989). Studies using subcloned fragments of the cloned K5 antigen gene cluster indicated that the elevated CMP-KDO synthetase levels correlates with strains harbouring region 1 (Finke et al., 1989). Whether the elevated levels of CMP-KDO synthetase are due to a structural gene within region 1 encoding for a capsule specific CMP-KDO synthetase or due to increased expression of the kdsB gene mediated by a regulator encoded within region 1 is unknown.

Recently the complete nucleotide sequence of region 1 from the K5 antigen gene cluster has been determined (Pazzani, Jann, Boulnois and Roberts, unpublished results). Analysis of this nucleotide sequence indicated that region 1 consisted of five open reading frames (ORF) organised in a way consistent with a single transcriptional unit with the genes being organised 5' kpsE-KpsD-KpsU-KpsC-KpsS 3'

1) The KpsE protein.

The hydropathy plot (Kyte and Doolittle, 1982) of the predicted amino acid sequence of the 43 KDa KpsE protein revealed a relatively hydrophilic protein with, in addition to the signal sequence, N-terminal and C-terminal hydrophobic domains. Since both of these domains are approximately 20 amino acids they are of sufficient size to span the membrane and potentially act as a membrane anchors for this protein. Comparisons using the sequence alignment program of Lipman and Pearson (1985) between the predicted amino acid sequences of the KpsE protein and the BexC protein of the H. influenzae capsule gene cluster revealed 28.1% identity. Taking conservative amino acid substitutions into account the homology is 73.2%. A similar analysis between the KpsE protein and the CtrB protein of the Niesseria meningitidis capsule gene cluster revealed 26.8% identity and 73.2% homology.

The hydropathy plots of the CtrB, BexC and KpsE proteins are similar having hydrophobic N and C-termini (Frosch et al., 1991; Kroll et al., 1990) which could be a membrane spanning domains anchoring these protein at a number of sites. The Ctrb protein has been postulated to be an inner membrane protein with a periplasmic domain, since Ctrb-PhoA$^+$ fusion proteins are associated with the inner membrane fraction (Frosch et al., 1991). Likewise, Pho$^+$ fusions to the BexC protein suggests that this protein has periplasmic domain although in this case the site of the fusion protein was not localised by cell fractionation (Kroll et al., 1990). Alternatively, the observation that both BexC and KpsE have putative N-terminal signal sequences might suggest that these proteins are exported into the periplasmic space and are associated with the outer rather than inner membrane. If this is the case then this might indicate a role for these proteins in the later stages of polysaccharide export.

2) The KpsD protein.

The second ORF is 1974 bp long and encodes a protein of 558 amino acids with a predicted molecular weight of 60514 Da. This protein corresponds to the 60 KDa periplasmic protein termed KpsD identifed previously as a region 1 gene product important in the export of group II polysaccharides (Silver et al., 1987).

Analysis of the predicted amino acid sequence of the KpsD protein revealed a typical signal sequence in which the terminal methionine residue is followed by two positively charged lysine residues and a long stretch of uncharged amino acids terminating in an aspartic acid residue. No significant homology could be detected between the predicted amino acid sequence of this protein and the BexA, B, C and D proteins of H. influenzae or the CtrA, B, C or D proteins of N. meningitidis. However, protein data base searches revealed that this protein and the PgpB protein of E. coli are 25% identical over 68 amino acids, and taking conservative amino acid changes into account are 67% homologous. The PgpB protein of E. coli is phosphatidyl glycerophosphate (PGP) B phosphatase which is capable of hydrolysing PGP, phosphatidic acid (PA) and lysophosphatidic acid (LPA) (Icho, 1988). The homology between the KpsD protein and the PgpB protein of E. coli although relatively low (25% identity over 68 amino acids) is perhaps significant since the homology is localised to the hydrophilic domain of the PgpB protein which is thought to interact with phosphatidic acid (Icho, 1988). It is known that group II K antigens are linked to phosphatidic acid at the reducing terminus (Schmidt et al., 1982). Therefore, the KpsD protein may interact in the periplasm with the polysaccharide via phosphatidic acid.

3) The KpsU protein.

The third ORF is 738 bp and encodes for a protein termed KpsU of 246 amino acids with a predicted molecular weight of 27128 Da. Protein sequence database searches using the predicted amino acid sequence of the KpsU protein revealed 44.3% identity with the CMP-KDO synthetase protein encoded by the E. coli kdsB gene and taking conservative amino acid changes into account the two proteins are 70.7% homologous. Comparison of the respective nucleotide sequences demonstrated 63.5% identity over 540 bp. The predicted sequence of the KpsU protein was also 21.1% identical and 66.1% homologous to CMP-NANA synthetase encoded by the K1 antigen region 2 (Zapata et al., 1989). Both enzymes catalyse the generation of cytidine monophosphate sugar nucleotides. The finding that KpsU protein is highly homologous to the CMP-KDO synthetase enzyme encoded by the kdsB gene suggests that the KpsU is a functional CMP-KDO synthetase enzyme. The KdsB enzyme is thought to catalyse the rate-limiting step in KDO incorporation into lipopolysaccharide (Goldman and Kohlbrenner, 1985) and is present at low levels in all E. coli strains independent of encapsulation (Finke et al., 1990). The relatively high DNA homology of 63.5% between the kpsU and kdsB genes suggests that the two genes may be the product of a gene duplication. The identification of a structural gene within region 1 encoding for a second CMP-KDO synthetase enzyme explains the increased levels of this enzyme at capsule permissive temperatures in strains expressing group II K antigens and is in keeping with the elevated levels of CMP-KDO synthetase activity in strains carrying the subcloned region 1 of the K5 antigen gene cluster (Finke et al., 1989). The observation that KDO is found at the reducing termini of a number of group II polysaccharides may indicate an important role for this enzyme in capsule biogenesis. Why a second CMP-KDO synthetase enzyme is encoded within region 1, in addition to the CMP-KDO synthetase encoded by the kdsB gene, which functions in LPS biogenesis is not clear. One possibility is that it allows the independent regulation of the expression of group II capsule K antigens and LPS. For instance, group II capsule expression is regulated by environmental stimuli such as temperature and CMP-KDO synthetase (KpsU) activity is similarly regulated (Finke et al., 1990). Alternatively the presence of two enzymes may reflect biochemical differences between them which are specific to their roles in LPS and group II capsule expression respectively. The recent purification of the KpsU protein (C. Rosenou and K Jann, unpublished results) should help resolve this question.

4) The KpsC protein.

There are a number of potential ATG initiation codons for the fourth ORF which would generate putative proteins ranging in molecular weight from 75684 to 55000 Da. The failure to visualise a candidate polypeptide in the minicell experiments (Pazzani, Boulnois and Roberts unpublished results) for this ORF make assignment of the likely start point for translation difficult. However, a protein of 74-77 KDa termed KpsC has previously been identified as being encoded by region 1 of the K1, K5 and K7 antigen gene clusters (Boulnois and Jann, 1990; Roberts et al., 1986; Silver et al., 1983; Silver et al., 1987). Therefore it is likely that the ORF encodes for a protein of predicted molecular weight of 75684 Da. The predicted sequence of the KpsC protein together with hydropathy plots and Gravy score analysis are consistent with a soluble, cytosolic protein. Protein database searches failed to identify significant homologies apart from low level homology to a number of DNA binding proteins, this is likely to be a reflection of the relatively basic nature of the protein. The role of this protein in polysaccharide capsule production is as yet unclear, but the basic nature of the protein may allow binding to the negatively charged polysaccharide molecule via ionic interactions.

5) The KpsS protein.

The final ORF also has a number of potential ATG initiation codons generating putative proteins ranging in predicted molecular weight from 46328 to 30000 Da. No proteins with molecular weights corresponding to any of those predicted from the sequence analysis were identified in minicell analysis of the relevant clone (Pazzani, Boulnois and Roberts, unpublished results). However only the ATG initiation codon generating a protein of molecular weight of 43987 Da has an appropriate SD sequence 5' to the ATG. The size of this protein corresponds well with the a 42 KDa protein identified previously by minicell analysis to be encoded by the region 1 of the K5 and K7 antigen gene clusters (Boulnois and Jann, 1990: Roberts et al., 1986) and with a similar molecular weight protein termed KpsS from region 1 of the K1 antigen gene cluster (Silver et al., 1983; 1987).

Analysis of the predicted amino acid sequence of KpsS protein together with data base searches failed to shed light on the possible role of this protein in the expression of the K5 antigen. Mutations in the kpsS gene of the K1 antigen gene cluster result in a decrease in endogenous sialyltransferase activity (Vimr *et al.*, 1989) and it has been postulated that the KpsS protein interacts with both the sialyltransferase enzyme and other K1 proteins to form a multicomponent enzyme complex associated with the inner membrane (Vimr *et al.*, 1989). A Tn1000 transposon insertion in the kpsS gene on plasmid pGB118 results in K5 polysaccharide within the cytoplasm but not specifically associated with the inner membrane (D. Bronner and K Jann unpublished results). This observation might support a role for the KpsS protein in interacting with the polysaccharide biosynthetic apparatus and the inner membrane. This is only one interpretation, but what is clear is that the biosynthesis of a bacterial capsule onto the cell surface is a complex, dynamic process involving the interaction of a large number of proteins encoded by the capsule gene cluster.

REGULATION OF REGION 1 EXPRESSION.

Expression of group II capsules in E. coli is regulated by temperature, with no capsule expression below 20°C (Kroncke et al., 1990). In the case of K1 it has been proposed that temperature dependency is mediated by reversible cold-inactivation of the enzymes involved in the synthesis and polymerisation of sialic acid (Merker and Troy, 1990).

However, this cannot be the case for the K5 region 1 functions since, extracts from cells grown at non permissive temperatures (18°C) for capsule expression have low levels of KpsU (CMP-KDO synthetase) activity when assayed at 37°c compared to extracts from cells grown at 37°C. Therefore changes in the levels of KpsU activity most likely reflect increased synthesis of KpsU at 37°C. Whether this regulation is at the level of transcription or translation is as yet unclear. Sequences 5' to kpsE and around the putative promoter of the region 1 operon are important in temperature dependent regulation of region 1 functions, since, plasmid constructs in which these sequences have been deleted no longer show temperature dependent expression of KpsU activity. Experiments are now in progress in our laboratories to address the question of the environmental regulation of E. coli group II K antigen expression.

ACKNOWLEDGEMENTS

Work in the authors (GJB, ISR) laboratory was supported by grants from both the Medical Research Council and the Science and Engineering Research Council of the UK. During this study GJB was a Lister Institute-Jenner Research Fellow. Work in KJ's laboratory was supported by grant Ja 115/28 from the Deutsche Forschungsgemeinschaft.

REFERENCES

Boulnois, G.J., Roberts, I.S., Hodge, R., Hardy, K.R.,Jann, K.B. and Timmis, K.N. (1987) Mol. Gen. Genet. 208:242-246.
Boulnois, G.J. and Jann, K. (1990) Mol. Microbiol. 3:1819-1823.
Boulnois, G. J. and Roberts, I.S. (1990) Curr. Top. Microbiol. Immunol. 150:1-18.
Cangelosi, G.A., Martinetti, G., Leigh, J.A., Lee, C.C., Theines, C. and Nester, E. W. (1989) J. Bacteriol. 171:1609-1615.
Drake, C.R., Roberts, I.S., Jann, B., Jann, K. and Boulnois, G.J. (1990) FEMS Microbiol. Lett. 66:227-230.
Finke, A., Roberts, I.S., Boulnois, G.J., Pazzani, C. and Jann, K. (1989) J. Bacteriol. 171:3074-3075.
Finke, A., Jann, B. and Jann, K. (1990) FEMS Microbiol. Lett. 69:129-134.
Frosch, M., Edwards, U., Bousset, K., Krausse, B. and Weisgerber, C. (1991) Mol. Microbiol. 5:1251-1263.
Goldman, R.C. and Kohlbrenner, W.E. (1985) J. Bacteriol. 163:256-261.
Gros, P., Croop, J. and Housman, D. (1986) Cell 27:371-380.
Higgins, C.F., Hyde, S.C., Mimmack, M.M., Gileadi, U., Gil, D.R. and Gallagher, M.P. (1990) J. Bioenergetics and Biomembranes. 22:571-592.
Hiles, I.D., Gallagher, M.P., Jamieson, D.J. and Higgins, C.F. (1987)
J. Mol. Biol. 195:125-142.
Holland, I.B., Kenny, B. and Blight, M. (1990) Biochemie 72:131-141.
Icho, T. (1988) J. Bacteriol. 170:5117-5124.
Jann, K. (1985) In: The Virulence Of *Escherichia coli* (M. Sussman Ed), Academic Press Inc., New York, pp. 156-176.
Jann, K. and Jann, B. (1987) Rev. Infect. Dis. 9:5517-5526.
Jann, B. and Jann, K. (1990) Curr. Top. Microbiol. Immunol. 150:19-42.

Kroll, J.S., Hopkins, I. and Moxon, R. (1988) Cell 53:347-356.
Kroll, J.S., Zamze, S., Loynds, B. and Moxon, E.R. (1989) J. Bacteriol. 171:3343-3347.
Kroll, J.S., Loynds, B., Brophy, L.N. and Moxon, E.R. (1990) Mol. Microbiol. 4:1853-1862.
Kroncke, K.D., Boulnois, G.J., Roberts, I.S., Bitter-Suermann, D., Golecki, J.R., Jann, B. and Jann, K. (1990) J. Bacteriol. 172:1085-1091.
Kyte, J. and Doolittle, R.F. (1982) J. Mol. Biol. 157:105-132.
Laasko, D.M., Homonylo, M.K., Wilmot, S.J. and Whitfield, C. (1988) Can. J. Microbiol. 34:987-992.
Lipman, D.J. and Pearson, W.R. (1985) Science 227: 1435-1441.
Merker, R.I. and Troy, F.A. (1990) Glycobiology 1:93-100.
Orskov, I. and Nyman, K. (1974) J. Bacteriol. 120:43-51.
Orskov, I., Sharma, V. and Orskov, F. (1976) Acta. Pathol. Microbiol. Scand. 84:125-131.
Orskov, I., Orskov, F., Jann, B. and Jann, K. (1977) Bacteriol. Rev. 41:667-710.
Pavelka, M.S., Wright, L.F. and Silver, R.P. (1991) J. Bacteriol. 173:4603-4610.
Roberts, I.S., Mountford, R., High, N., Jann, K., Timmis, K.N. and Boulnois, G.J. (1986) J. Bacteriol. 168:1228-1233.
Roberts, I.S., Mountford, R., Hodge, R., Jann, K. and Boulnois, G.J. (1988) J. Bacteriol. 170:1305-1310.
Roberts, M., Roberts, I.S., Korhonen, T.K., Jann, K., Bitter-Suermann, D., Boulnois, G.J. and Williams, P.H. (1988) J. Clin. Microbiol. 26:385-387.
Schmidt, G., Jann, B. and Jann, K. (1977) J. Gen. Microbiol. 100:355-361.
Schmidt, G. and Jann, K. (1982) FEMS Microbiol. Lett. 14:69-74.
Silver, R.P., Vann, W.F. and Aaronson, W. (1983) J. Bacteriol. 157:568-575.
Silver, R.P., Aaronson, W. and Vann, W.F. (1987) J.Bacteriol. 169:5489-5495.
Silver, R.P., Aaronson, W. and Vann, W.F. (1988) Rev. Infect. Dis. 10:282-286.
Smith, A.N., Boulnois, G.J. and Roberts, I.S. (1990) Mol. Microbiol. 4:1863-1869.
Stanfield, S.W., Ielpi, L., O'Brochta, D., Helinski, D.R. and Ditta, G.S. (1988) Proc. Natl. Acad. Sci. USA 74:5463-5467.
Vimr, E.R., Aaronson, W. and Silver, R.P. (1989) J. Bacteriol. 171:1106-1117.
Vimr, E.R. (1991). J.Bacteriol. 173:1335-1338.
Walker, J.E., Saraste, M., Runswick, M.J. and Gay, N.J. (1982) EMBO J. 1:945-951.
Zapata, G., Vann, W.F., Aaronson, W., Lewis, M.S. and Moos, M. (1989) J. Biol. Chem. 264:12769-14774.

Polysialic Acid
J. Roth, U. Rutishauser and F. A. Troy II (eds.)

STRUCTURE AND FUNCTION OF ENZYMES IN SIALIC METABOLISM IN POLYSIALIC PRODUCING BACTERIA

Willie F. Vann[1], Gerardo Zapata[1], Ian Roberts[2], Graham Boulnois[2], and Richard P. Silver[3]

[1]Laboratory of Bacterial Polysaccharides, Center for Biologics Evaluation and Research, Bethesda, Maryland, USA
[2]Department of Microbiology, University of Leicester, Leicester, ENGLAND
[3]Department of Microbiology, University of Rochester, Rochester, N.Y., USA

SUMMARY: The enzymes necessary for the synthesis and activation of N-acetylneuraminic acid in E. coli K1 are encoded by 3 genes, neuA, neuB, and neuC. All of the genes have been sequenced. Only the neuA gene has been assigned an enzymatic function, namely CMP-neuAc synthetase. A neuAc synthesis reaction has been reported in both E. coli and Neisseria meningitidis, but has not been correlated to a specific gene. The amino acid sequence of CMP-neuAc synthetase from E. coli and Neisseria meningitidis share significant regions of homology but differ in molecular weight, pI, and specificity activity. Sequence homology suggest conserved basic residues. E. coli CMP-neuAc synthetase catalyzes the transfer of the β-neuAc anomeric oxygen to the α-phosphorus of CTP forming CMP-β-neuAc. The enzyme contains 2 catalytically non-essential sulfhydryls. Stability studies suggest that inactivation of the enzyme by denaturation results from subtle changes in protein structure. The neuA (CMP-neuAc synthetase) and neuC genes are part of the same operon. Both E. coli and Neisseria meningitidis have a neuC like gene based on sequence homology. The E. coli enzyme has been partially purified.

The capsular polysaccharides of several pathogenic bacteria are homopolymers of N-acetylneuraminic acid. α(2-8)polysialic acid is produced by Escherichia coli K1, (Dewitt, 1961; McGuire, 1964) Neisseria meningitidis Group B (Bhattacharjee, 1975; Orskov, 1979), Morexella nonliquifacians (Bovre, 1983; Devi, 1991), and Pasturella haemolytica A2 (Adlam, 1987). α(2-9) polysialic acid is produced by N. meningitidis Group C (Bhattacharjee, 1975) and α(2-8) α(2-9)polysialic acid by E. coli K92 (Egan, 1977). These bacteria cause such invasive diseases as meningitis, septicemia, and urinary tract infections in humans. The genes for the biosynthesis of these polymers in E. coli and N. meningitidis have been cloned and expressed (Silver, 1981; Echarti, 1983; Frosch, 1989). A map of the E. coli K1 gene cluster is given in Figure 1. The genes in region 2 are specific to the synthesis of polysialic acid and are necessary for neuAc synthesis, activation, and polymerization (Roberts, 1988;Vimr, 1989). Vimr et. al. (1989) have identified 3 genes in region 2 which may be involved in sialic acid synthesis or activation, neu A, neu B, and neu C. The neu D gene also in this region has been shown by Silver et. al. (personal communication) not to be required for neuAc synthesis.

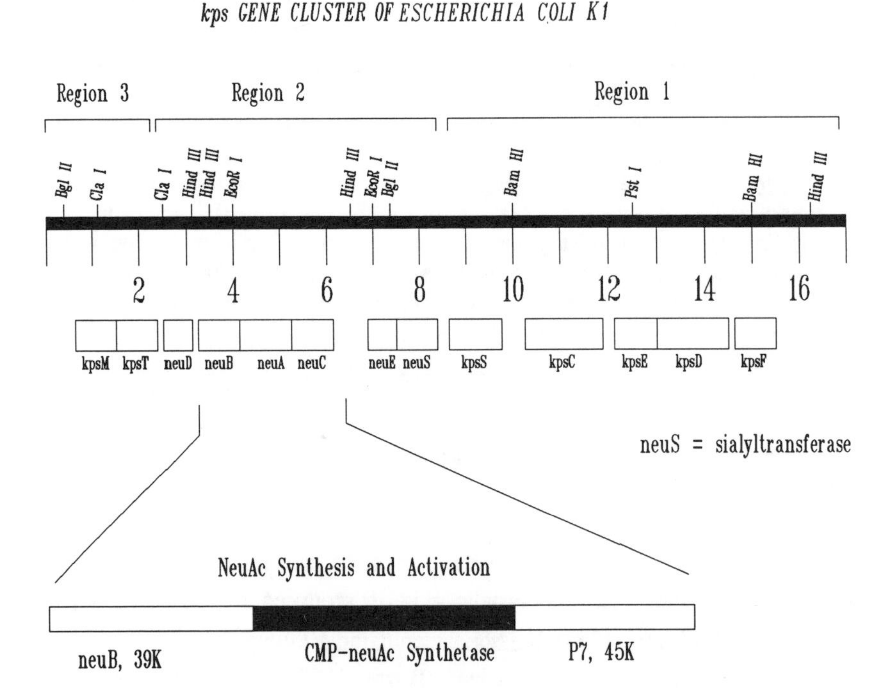

Figure 1. Location of neuAc synthesis and activation genes in E. coli K1 gene cluster.

N-ACETYLNEURAMINIC ACID SYNTHESIS

NeuAc is synthesized in mammalian tissues by condensing N-acetylmannosamine-6-phosphate with phosphoenolpyruvate to form N-acetylneuraminic acid-9-phosphate (Watson, 1966). N-acetylneuraminic acid-9-phosphate is converted to neuAc by the action of a phosphatase. This differs from the reactions demonstrated in the polysialic acid producing bacteria N. meningitidis group B. Blacklow and Warren et. al (1962) demonstrated that manAc and PEP are condensed directly to form neuAc in these bacteria. The N. meningitidis neuAc synthase has not been purified to homogeneity although partial purification has been reported (Blacklow, 1962; Brossmer, 1980). A chromatogram of an ammonium sulfate fraction of N. meningitidis B on ion exchange is illustrated in Figure 2.

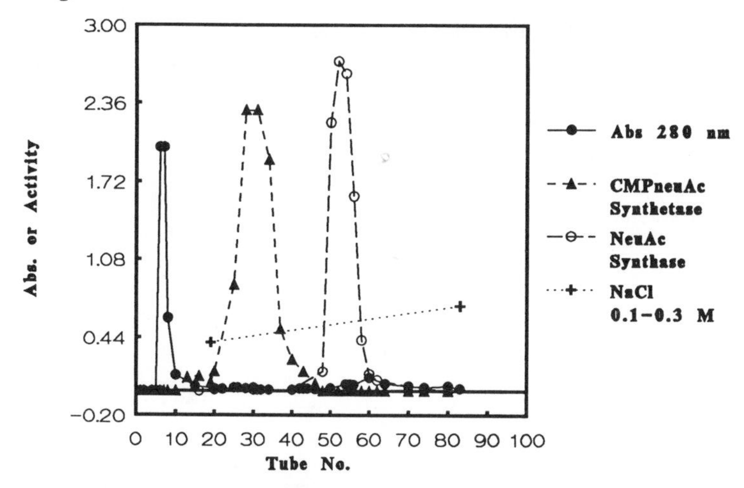

Figure 2. Fractionation of N. meningitidis CMP-neuAc synthetase and neuAc synthase on Fast Flow Q-Sepharose. A 50-80% ammonium sulfate fraction was loaded onto Fast Flow Q-Sepharose in 0.05 M bicine, 1 mM DTT, 0.1 M NaCl, pH 7.6 and eluted with a gradient of 0.1 to 0.3 M NaCl.

The N. meningitidis neuAc synthase is an acidic enzyme as judged by its behavior on Q-Sepharose. Although, the active fraction from the Q-Sepharose column can be further purified on Phenyl Sepharose, homogeneous protein was not obtained. Purification of this enzyme will await cloning of the neuAc synthase gene into an expression vector.

The gene for the meningococcal neuAc synthase was located to a region on the capsule synthesis gene cluster designated the A-region. This region described by M. Frosch et. al. (1991) is common to both polysialic producing strains. Further characterization of the gene encoding neuAc synthase in Neisseria has not been reported.

A similar synthase activity has been reported by Merker and Troy (1991) in E. coli strains producing the K1 polysialic capsule. This activity catalyzes the Ca^{++} dependent synthesis of neuAc from manAc and PEP. Vimr et. al. (1989) reported complementation studies which suggest that 2 genes may be necessary for sialic synthesis, neu B and neu C. This synthase activity has not been conclusively associated with strains harboring plasmids containing either of these genes. The plasmid pSR32 (Silver, 1984; personal communication) contains kps M and T and the neu A, B, C and D genes (Figure 1.) . Cells harboring this plasmid accumulate CMP-neuAç as measured by, TLC, Dionex HPAE chromatography or thiobarbituric acid assay of ethanol extracts. Recently, Silver and Wright (unpublished results) constructed plasmids containing only functional A, B and C genes.

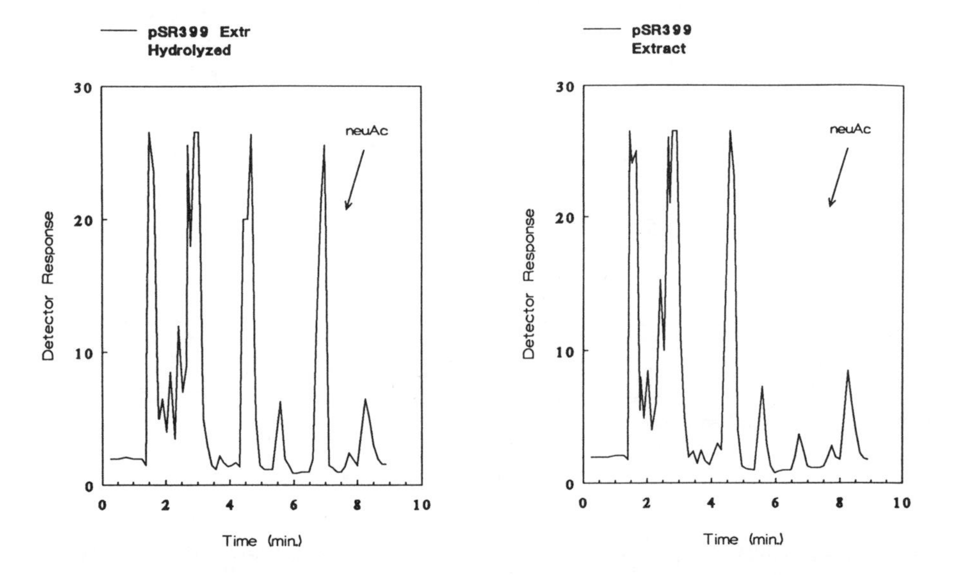

Figure 3. Detection of neuAc by high pH ion exchange chromatography of ethanol extracts from EV80:pSR399. Lysates of overnight cell cultures were extracted with 50% ethanol as described (Vimr and Troy, 1985) and injected onto a Dionex Carbopac-PA1 before and after hydrolysis for 30 min. with 0.2 M trifluoroacetic acid at 37 ^{o}C.

Analysis of extracts of cells harboring one of these plasmids, pSR399 is illustrated in Figure 3. These cells produce neuAc in a free and acid labile form (i.e. CMP-neuAc). This confirms the complementation experiments which indicated that only neu B and neu C are necessary for neuAc synthesis.

N-ACETYLNEURAMINIC ACID ACTIVATION

As in mammalian tissues, neuAc is activated to CMP-neuAc in polysialic producing bacteria by a CMP-neuAc synthetase. CTP and neuAc are condensed with the formation of pyrophosphate (Roseman, 1962;Warren, 1962). It has been reported that the production of E. coli CMP-neuAc synthetase is cold sensitive and suggested that CMP-neuAc synthetase may be an important enzyme in the regulation of polysialic synthesis in E. coli cells grown at 15-20 ^{o}C (Gonzalez-Clemente, 1989). The E. coli K1 enzyme is the product of the neu A gene and has been purified to homogeneity. The amino acid sequence was determine from the neuA nucleotide sequence. Recently, an improved method for the purification of recombinant E. coli K1 CMP-neuAc synthetase has been described based on dye binding chromatography (Shames, 1991; Ichikawa, 1991). An interesting mutant enzyme has been constructed in the laboratory of Dr. Chi H. Wong (Liu, 1992). The enzyme contains a peptide tag to facilitate screening and is fully active. The native enzyme is a 48,621 molecular weight protein with tendency to form dimers. It contains 418 amino acids (Zapata, 1989).

The gene for the CMP-neuAc synthetase of N. meningitidis is located in the A region of the Group B gene cluster (Frosch, 1989;Boulnois, 1990). The gene has been cloned into the expression vector pT7-6 and purified to homogeneity by hydrophobic interaction, ion exchange, and dye binding chromatography. The estimated molecular weight by FPLC is 40,000 and 28,000 by PAGE. The amino terminal sequence of the purified protein agrees with the derived sequence of an open reading frame in the A region encoding a 24,000 MW protein. Brossmer et. al. (1980) found that partially purified meningococcal CMP-neuAc synthetase eluted on Sephadex G-150 with a calculated molecular weight of 37,000. Like the E. coli enzyme the Neisseria CMP-neuAc synthetase may exist in the native state as a dimer. The amino terminus is homologous to that of the E. coli CMP-neuAc synthetase (Zapata, 1989) and CMP-KDO synthetase (Goldman, 1985; Roberts, unpublished results) suggesting a conserved sequence in the amino terminus of CMP-neuAc synthetases.

The catalytic mechanism of the E. coli enzyme was investigated by steady state kinetics and ^{13}C labeled substrates (Ambrose, 1992). Lineweaver-Burke plots of initial velocity reactions suggest an ordered addition mechanism (Figure 4).

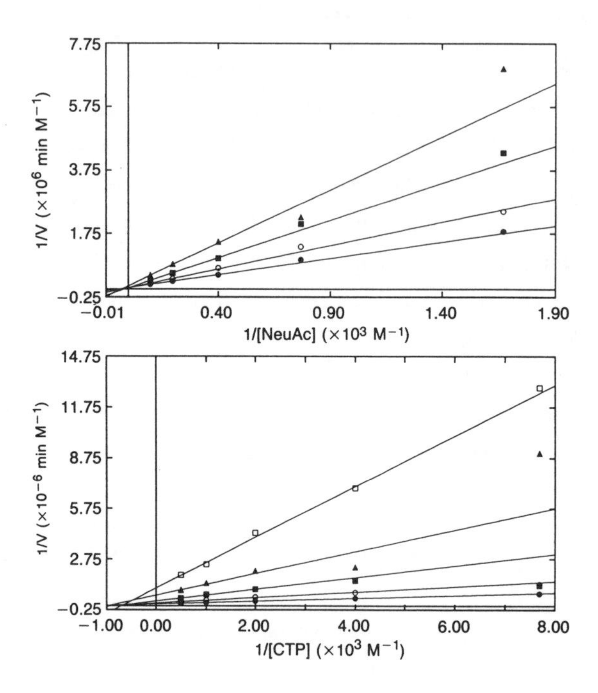

Figure 4. Inverse plots of initial velocity versus substrate for E. coli CMP-neuAc synthetase. Enzyme was assayed in a 5x5 matrix of substrate concentrations varying CTP from 0.25 mM to 2 mM and neuAc from 0.6 mM to 10 mM as decribed (Vann, 1987; Ambrose, 1992).

The absence of characteristic Ping-Pong kinetics also suggest that the catalytic reaction does not involve a covalent intermediate. Substrate inhibition has been observed at high substrate concentrations. The anomeric specificity of the enzyme was determined using [^{13}C]-2-neuAc (Ambrose, 1992; Kohlbrenner, 1985) to be toward the β-anomer. In a related experiment, the change in the chemical shift of the anomeric carbon induced by ^{18}O exchange was used to determine the fate of the anomeric oxygen. The anomeric oxygen was not exchange with solvent $H_2{}^{18}O$ during the catalytic reaction. The conclusions from these experiments are summarized in Figure 5.

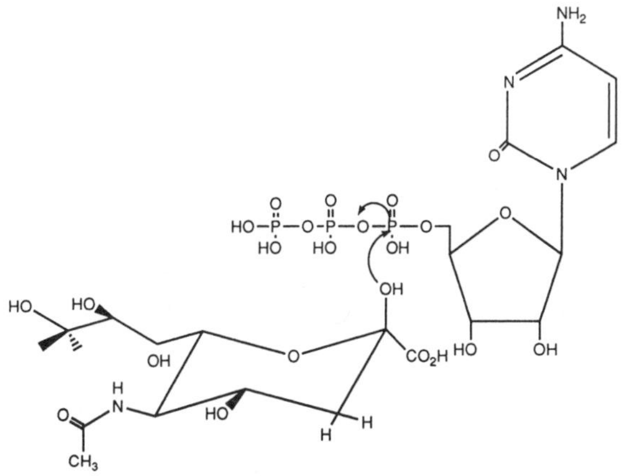

Figure 5. Proposed mechanism of E. coli CMP-neuAc synthetase.

E. coli CMP-neuAc synthetase catalyzes the transfer of the anomeric oxygen of β-neuAc to the α-phosphorus of CTP via an Sn2 type reaction. Both the substrate and the product are in the β-configuration. This implies that α-linked polysialic acid must be polymerized with a net inversion of configuration.

Chemical modification of E. coli CMP-neuAc Synthetase: Chemical modification of the E. coli CMP-neuAc synthetase suggest the presence of arginine, cysteine, and lysine residues which affect enzyme activity after covalent modification. Analysis of the sequence homology of CMP-neuAc synthetases of E. coli and N. meningitidis, and the CMP-KDO synthetase of E. coli suggest the presence of conserved basic residues in the amino terminus of the enzyme. The enzyme is inactivated by the aldehyde reagents CTP dialdehyde and pyridoxal-phosphate. The substrate CTP exhibits a protective effect in the presence of both of these reagents (Vann, unpublished results).

Although, the enzyme does not require the presence of thiols for activity it is inactivated by sulfhydryl specific reagents. The enzyme only contains 2 cysteine residues which allowed a more detailed study of the role of sulfhydryls in catalytic function using a combination of chemical modification and site directed mutagenesis.

These residues have differing accessibility to solvent and have diverse reactivity toward sulfhydryl reagents. The reagent dithioldipyridine reacts with 1 residue in the native protein. The rate of reaction of this residue is increased by either SDS or heat (Zapata, 1992a).

Site Directed Mutagenesis of E. coli CMP-neuAc Synthetase: Site directed mutagenesis of the cysteine residues cys-129 to serine or cys-329 to threonine demonstrate that these amino acid side chains are not essential for catalytic activity. Both of these mutant enzymes are active. This also means that a dissulfide bond is not essential for the formation of a catalytically active enzyme. These sulfhydryls are important for the stability of the protein as evident from heat denaturation experiments (Zapata, 1992a).

The degree of exposure of cysteine residues was investigated using mutants. The reagent N-ethylmaleimide reacts readily with native and mutant enzymes only above 40-42 ^{o}C. An investigation of the reactivity of [^{14}C]-NEM with wild type, mutant cys129ser, and cys329thr demonstrates that residue 329 reacts more readily. Residue 129 appears to be buried in the protein structure (Zapata, unpublished results).

Denaturation of CMP-neuAc synthetase: Our studies on the denaturation pathway of this enzyme are a direct result of determining the degree of exposure of cysteine side chains and the effect of site directed mutagenesis on the structure of the protein. Circular dichroism of the native E. coli enzyme suggest a structure with 37% α-helix which is in general agreement with predictions from the nucleotide derived amino acid sequence. Some denaturing conditions which lead to loss of enzymatic activity result in aggregation without major changes in the circular dichroism or fluorescence spectra. Some of the conditions which can lead to aggregation are listed in Table I.

Table I. Conditions which Promote Aggregation of E. coli CMP-neuAc Synthetase

Temperature	46 ^{o}C
Proteolysis	Clostripain
Urea	3-4M
Guanidium hydrochloride	2M

Complete denaturation as judged by circular dichroism and fluorescence spectra requires 5 M guanidinium hydrochloride. Our interpretation of these findings is illustrated in Figure 6.

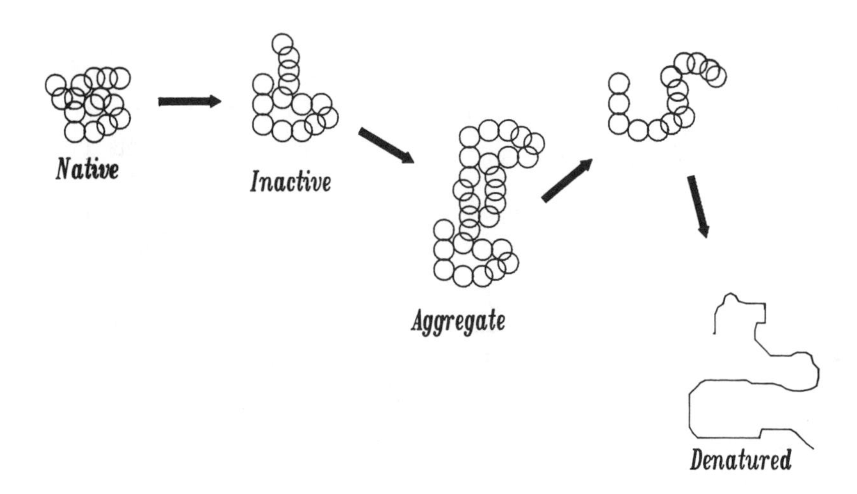

Figure 6. Proposed Denaturation Pathway of E. coli CMP-neuAc Synthetase (Vann, unpublished results).

Whether these properties are general to bacterial CMP-neuAc synthetases awaits their purification and characterization. Antibodies have been prepared to the E. coli enzyme which may be helpful in distinguishing structural similarities.

Neu C GENE PRODUCT, P7

Silver et. al. (1984) demonstrated that neuC is required for the synthesis of extracellular polysialic acid. E. coli with an insertion mutation in the neuC gene require the addition of neuAc to the growth media for formation of polysialic capsule. These strains are not sensitive to K1 specific phage (Silver, 1984) sensitivity nor do they produce characteristic halos on capsule specific antiserum agar plates (Zapata, 1992). NeuAc has not been detected in strains harboring the plasmid pSR50 which contains an insertion mutation in the neuC gene. A definition of the exact function of this protein will be facilitated by purification of P7 to homogeneity.

The nucleotide sequence of neuC has been determined (Zapata, 1992b). It is apparently part of an operon. The first codon of neuC overlaps with the last codon of the neuA gene. A putative transcription terminator sequence follows the neuC gene. A chimeric gene was constructed with the amino terminal coding region of neuC and the complete β-galactosidase gene. In such a construct β-galactosidase activity was induced in concert with CMP-neuAc synthetase activity when the IPTG controlled tac promoter was placed upstream of the neuA gene. Two possible start codons were detected in an open reading frame for a 44 k protein. A P7-β-galactosidase chimeric protein was constructed and purified. The amino terminal sequence of this protein confirm the nucleotide sequence and suggest that the first start codon is translated (Zapata, 1992).

Antibodies were prepared to a nucleotide derived amino terminal sequence of P7. The antibodies were affinity purified using the synthetic peptide and subsequently used to detect the native protein in chromatography fractions. The native protein has also been partially purified by dye binding chromatography of ammonium sulfate fractions of E. coli JM109:pWA1 lysates. The plasmid pWA1 contains the neuA and neuC genes under the control of the tac promoter. The protein migrates on SDS polyacrylamide gels with a molecular weight of 44,000 and on isoelectric focusing gels with a pI = 7.5 (theoretical = 7.2). The amino terminal sequence of this protein agrees with that predicted by the nucleotide sequence (Vann, unpublished results).

Recently, Boulnois et. al (unpublished results) have identified a neuC like gene in the meningococcal Group B cluster. This protein is adjacent to the gene encoding CMP-neuAc synthetase. Perhaps the arrangement of neuAc metabolic genes in polysialic acid producing bacteria is similar.

The synthesis and activation of neuAc is key to the biosynthesis of polysialic acid. Purification and characterization of the enzymes involved are a first step in understanding their role in the control of polysialic acid biosynthesis. CMP-neuAc is a feedback inhibitor of UDP-glcNAc epimerase in mammals (Kikuchi, 1973). Such feedback control has not been demonstrated in bacteria. Two laboratories have suggested that temperature may play a role in regulating neuAc synthesis (Merker and Troy, 1991) and activation (Gonzalez-Clemente, 1989). Polysialic acid synthesis may be down regulated at 15-20 ^{o}C as a consequence of the sensitivity of these enzyme activities to growth of bacteria at low temperature.

ACKNOWLEDGEMENTS

The authors wish to thank Dr. Eric Vimr, Depart. of Veterinary Pathobiology, University of Illinois, Urbana, IL for kindly supplying the bacteria strains EV5 and EV80.

REFERENCES

Adlam, C, Knight, J.M., Mugridge, A., Williams, J. M., and Lindon, J. C. (1987) FEMS Microbiol. Lett. 42: 23-25.

Ambrose, M. G., Freese, S. J., Reinhold, M. S., Warner, T. G. and Vann, W. F. (1992) Biochemistry 31: 775-780.

Bhattacharjee, A. K., Jennings, H. J., Kenny C. P., Martin, A., and Smith, I. C. P. (1975) J. Biol. Chem. 250: 1926-1932.

Blacklow, R. S., and Warren, L. (1962) J. Biol. Chem. 237: 3520-3526.

Boulnois, G. J., and Roberts, I. S. (1990) Current Topics in Microbiology and Immunology 150: 1-18.

Bovre, K., Byrn, K., Closs, O., Hagan, N., and Froholm, L. O. (1983) NIPH Ann. 6: 65-73.

Brossmer, R., Rose, U., Kasper, D., Smith, T. L., Grasmuk, H., and Unger, F. M. (1980) Biochem. Biophys. Res. Commun. 96: 1282-1289.

Devi, S., Schneerson, R., Egan, W., Vann, W. F., Robbins, J. B., and Shiloach, J. (1991) Infect. Immun. 59: 732-736.

Dewitt, C. W. and Rowe, J. A. (1961) J. Bacteriol. 82: 838-848.

Echarti, C., Hirschel, B., Boulnois, G. J., Varley, J. M., Waldvogel, F., and Timmis, K. N. (1983) Infect. Immun. 41: 56-60.

Egan, W., Liu, T.-Y., Dorow, D., Cohen, J. S., Robbins, J. D., Gotschlich, and Robbins, J. B. (1977) Biochemistry 16: 3687-3692.

Frosch, M. Edwards, U., Bousset, K., Krausse, B. and Weisgerber, C. (1991) Mol. Microbiol. 5: 1251-1263.

Frosch, M., Weisgerber, C. and Meyer, T. F. (1989) Proc. Natl. Acad. Sci USA 86: 1669-1673.

Goldman, R. C., Bolling, T. J., Kohlbrenner, W. E., Kim, Y., and Fox, J. L. (1986) J. Biol. Chem. 261: 15831-15835.

Gonzalez-Clemente, C., Luengo, J. M., Rodriguez-Aparicio, L. B., and Reglero, A. (1989) FEBS Letters 250: 429-432.

Ichikawa, Y., Shen, G.-J., and Wong, C.-H. (1991) J. Am. Chem. Soc. 113: 4698-4700.

Kohlbrenner, , W. E., and Fesik, S. W. (1985) J. Biol. Chem. 260: 14695-14700.

Kikuchi, K. and Tsuiki, S. (1973) Biochim. Biophys. Acta 327: 197-206.

Liu, J. L.-C., Shen, G.-J., Ichikawa, Y., Rutan, J. F., Zapata, G., Vann, W. F., and Wong, C.-H. (1992) J. Am. Chem. Soc. 114: 3901-3910.

McGuire, E. J. and Binkley, R. (1964) Biochemistry 3: 247-251.

Merker, R. I. and Troy, F. A. (1991) Glycobiol. 1: 93-100.

Orskov, F., Orskov, I., Sutton, A., Schneerson, R., Lin, W.-L., Egan, W., Hoff, G. E., and Robbins, J. B. (1979) J. Exp. Med. 149: 669-685.

Roberts, I. Mountford, R., Hodge, R., Jann, K., and Boulnois, G. J. (1988) J. Bacteriol. 170: 1305-1310.

Roseman, S. (1962) Proc. Natl. Acad. Sci. 48: 437-441.
Silver, R. P., Finn, C. W., Vann, W. F., Aaronson, W., Schneerson, R., Kretchmer, P. J., and Garon, C. F. (1981) Nature 289: 167-172.
Silver, R. P., Vann, W. F., and Aaronson, W.(1984) J. Bacteriol. 157: 568-575.
Troy F. A. (1992) Glycobiol. 2: 5-23.
Vann, W. F., Silver, R. P., Abeijon, C., Chang, K., Aaronson, W., Sutton, A., Finn, C. W., Lindner, W., and Kotsatos, M. (1987) J. Biol. Chem. 262: 17556-17562.
Vimr, E. R., Aaronson, W., and Silver, R. P. (1989) J. Bacteriol. 171: 1106-1117.
Vimr, E. R. and Troy, F. A. (1985) J. Bacteriol. 164: 845-853.
Warren, L. and Blacklow, R. S. (1962) J. Biol. Chem. 237: 3527-3534.
Zapata, G., Vann, W. F., Aaronson, W., Lewis, M. S., and Moos, M. (1989) J. Biol. Chem. 264: 14769-14774.
Zapata, G., Crowley, J., and Vann, W. F. (1992a) FASEB J. 6: 1221.
Zapata, G., Crowley, J., and Vann, W. F. (1992b) J. Bacteriol. 174: 315-319.

Polysialic Acid
J. Roth, U. Rutishauser and F. A. Troy II (eds.)

CHARACTERIZATION OF BACTERIOPHAGE E ENDO-SIALIDASE SPECIFIC FOR ALPHA-2,8-LINKED POLYSIALIC ACID

Graham S. Long, Peter W. Taylor* and J. Paul Luzio

University of Cambridge, Department of Clinical Biochemistry, Addenbrooke's Hospital, Hill's Road, Cambridge CB2 2QR, U.K. and *Research Centre, Ciba-Geigy Pharmaceuticals, Wimblehurst Road, Horsham, West Sussex, RH12 4AB, U.K.

SUMMARY

Bacteriophage E specifically recognizes and infects strains of Escherichia coli which display the α-2,8-linked polysialic acid capsular K1 marker (Gross et al., 1977). Initial absorption of the phage to the host bacterium is thought to be achieved by virtue of the endo-sialidase activity associated with the tail spikes of the phage particle.

In order to obtain information about the structure and primary sequence of the enzyme, milligram quantities of pure endo-sialidase were prepared using a modification of the protocol described by Tomlinson and Taylor (1985). The intact native enzyme has a molecular weight in excess of 200kDa, as judged by SDS-polyacrylamide gel electrophoresis and molecular sieve chromatography. The holoenzyme can be reduced to a single subunit of approx. 74kDa.

Purified 74kDa subunit has been directly analyzed for N-terminal sequence and subjected to cyanogen bromide treatment to yield smaller protein fragments. A partial N-terminal sequence for the 74kDa subunit has been obtained, although substantial N-terminal blockage of this protein was observed. The partial amino acid sequences of five separate cyanogen bromide digest fragments have been determined. All six protein sequences thus far obtained do not show a significant identity with any other protein or DNA sequences listed in the EMBL database (release 29, December 1991). The sequences have been used to construct synthetic oligonucleotide probes to allow molecular cloning of the enzyme. The radiolabelled probes have been shown to preferentially hybridize to purified bacteriophage E DNA by Southern blotting. Restriction fragments of bacteriophage E DNA have been identified which potentially encode the endo-sialidase gene.

The presence of elevated levels of cell-surface polysialic acid expression has been shown in tumour cell lines and human tumour tissue (Livingston et al., 1988; Roth et al., 1988a). Bacteriophage E endo-sialidase has the potential to be used as a probe for neoplastic tissues displaying this marker by virtue of its apparent absolute specificity for α-2,8-linked polysialic acid (Tomlinson and Taylor, 1985).

INTRODUCTION

Polysialic acid (PSA) or α-2,8-linked poly-N-acetylneuraminic acid is an unusual carbohydrate expressed on the cell surface of certain tumour cells, cell lines and some pathogenic bacteria (Roth et al., 1988a;b; Figarella-Branger et al., 1990; Moolenaar et al., 1990; Heitz et al., 1990; Livingston et al., 1988). An endo-sialidase hydrolyzing α-2,8-polysialosyl linkages with a high degree of specificity has been identified in bacteriophage E. This bacteriophage is a member of a family of similar phages originally isolated from sewage to help in the clinical identification of K1 serotype strains of E.coli (Gross et al., 1977) which were recognized as being responsible for a high rate of mortality in cases of neonatal meningitis (Robbins et al., 1974; Bhattachorjee et al., 1975; Orskov et al., 1977). In addition to bacteriophage E endo-sialidase there have been several other reports of similar α-2,8-specific endo-sialidases associated with K1 coliphages (Kwiatkowski et al., 1982; Finne and Makela, 1985; Hallenbeck et al., 1987; Pelkonen et al., 1989). It was realized that there may be potential in using the enzyme for the diagnosis and therapy of K1 meningitis, septicaemia or bactericaemia by virtue of the enzyme's specificity for α-2,8-sialosyl linkages (Taylor, 1987). One report demonstrated a dramatically effective treatment of E.coli K1 injected intramuscularly or intracerebrally in mice by administration of a low titer of anti-K1 phage intramuscularly (Williams-Smith and Huggins, 1982). A single dose of phage was significantly more effective than multiple doses of different antibiotics. Murine models have also been successfully treated against lethal E.coli K1 injection by administration of purified bacteriophage E endo-sialidase (Taylor, 1987).

Purified endo-sialidase has been successfully used as a probe for cell surface PSA expression, and it has become apparent that PSA is found in mammals almost exclusively in association with neuronal cell adhesion molecule (N-CAM). The level of PSA expression on N-CAM appears to be developmentally regulated with a high sialic acid content embryonic form and a low sialic acid content adult form (Hoffman et al., 1982). A regression to the highly polysialylated weakly adhesive form is seen in certain malignancies. Thus, PSA appears to be a tumour associated cell surface marker in adult tissue. Indeed, PSA has been postulated to be an oncodevelopmental marker in human kidney (Roth et al., 1988a;b). Thus, endo-sialidase has the potential to be used as a probe for neoplastic tissues displaying the putative polysialic acid oncodevelopmental marker.

There is recent evidence to suggest that PSA may also be a marker for metastatic potential. In Wilm's kidney tumour, PSA is associated with cells on the periphery of the tumour mass which show an invasive nature (Roth et al., 1988b). In one patient with a Wilm's tumour, cells expressing PSA were detected in the blood stream and at the site of a secondary tumour growth in the lung (Roth et al., 1988b). Roth and Zuber (1990) have proposed that PSA may modulate the behaviour and invasive potential of Wilm's tumour cells. Following on from these studies endo-sialidase may be used to further elucidate the relevance of PSA in tumour metastasis and possibly to investigate the mechanisms of metastasis for some phenotypes of tumour cells.

The present study was aimed at the further characterization and molecular cloning of bacteriophage E endo-sialidase.

ENZYME PURIFICATION

Purified bacteriophage E endo-sialidase was prepared in milligram quantities using a modification of the protocol described by Tomlinson and Taylor (1985). The α-2,8-polysialic acid positive host bacterium used was Escherichia coli K1 LP1674/ATCC No. 53351 (serotype O7:K10), a bacterial strain originally isolated from a patient with a urinary tract infection (Taylor, 1976). Lysates of mid-log phase bacterial cultures infected with bacteriophage E were subjected to fractional ammonium sulphate precipitation, and the fraction corresponding to 30-60% saturation of ammonium sulphate was collected. Intact phage particles were separated from freely soluble endo-sialidase by caesium chloride density gradient ultracentrifugation, and the soluble enzyme was then purified to homogeneity by gel filtration and anion exchange chromatography (Table I).

Table I. Purification of endo-sialidase

STEP	VOL (ml)	TOTAL PROTEIN (mg)	TOTAL ENZYME ACTIVITY (mU)	SPECIFIC ACTIVITY (mU/mg)	PURITY (fold)
Bacterial lysate	2000	4820	10500	2	1
Ammonium sulphate	95	295	1190	4	2
Cs Cl gradient	46	240	1410	6	3
S300 gel	338	22.3	3170	142	71
Q Sepharose anion exchange	117	12.2	3560	292	146
DEAE Sephadex A-25 anion exchange	30	2.45	3910	1590	797

Total enzyme activity was found to increase with each purification step following ammonium sulphate precipitation of protein. Salts have been found to interfere with the thiobarbituric acid assay of free sialic acids used to monitor the purification, (especially ammonium sulphate and caesium chloride). It is also possible that molecules which inhibit the activity of the endo-sialidase enzyme are removed with each subsequent purification step thus allowing a disproportionate increase in activity.

CHARACTERIZATION OF ENDO-SIALIDASE

Structure of bacteriophage E endo-sialidase :

Using SDS-PAGE the intact native enzyme was found to have a molecular weight equivalent to a protein in excess of 200kDa, and using gel filtration a molecular weight in excess of 300kDa. Heat treatment at 100°C for 5 minutes in the presence of SDS gave rise to a single protein subunit of 74kDa. When reducing agents were added, the same result was obtained if the enzyme was heat treated at 100°C, but the integrity of the intact enzyme was maintained if the sample was kept at room temperature. It would appear that the subunit complex is a heat labile structure rather than a disulphide-bonded moiety. The molecular weight determinations suggest that the enzyme is either a trimer or a tetramer of the 74kDa subunit.

Enzyme kinetics :

Determination of the K_m of bacteriophage E endo-sialidase for two different preparations of E.coli K1 PSA substrate gave identical estimates of 10.6μM. These values were calculated on the assumption that the mean molecular weight of each polymer preparation was equivalent to an oligosaccharide of 175 units of sialic acid ($M_r \sim 54000$).

Inhibitor studies :

It has been reported that DNA and other polyanions inhibit the activity of endo-sialidase K1F (Hallenbeck et al., 1987). In the present study weak inhibition ($\leq 20\%$) of bacteriophage E endo-sialidase was also observed with pGEX plasmid DNA (0.1mg/ml) and heparin (250 units/ml).

The neuraminic acid derivative 2-desoxy-2,3-dehydro neuraminic acid (DD-NANA) has been shown to be a potent inhibitor of many exo-sialidases (Van Der Horst et al., 1990). 1mM DD-NANA did not significantly inhibit endo-sialidase activity whereas the same concentration of DD-NANA caused 97% inhibition of Vibrio cholera type II sialidase.

PEPTIDE SEQUENCING OF BACTERIOPHAGE E ENDO-SIALIDASE

The purest enzyme preparations were consistently shown to produce a single 74kDa reduced protein band using SDS-PAGE analysis. Therefore, this subunit was purified by electroelution from SDS-PAGE gels, and partial peptide sequences were obtained by Edman degradation analysis of the intact 74kDa subunit and for five separate cyanogen bromide fragments (Table II).

The strength of the sequence signal obtained for the 74kDa subunit indicated an amino acid yield of less than 5% of that analyzed. This suggested that the protein was N-terminally blocked. Thus, the amino acid sequence obtained was either authentically that of the intact 74kDa subunit, or represented a minor protein contaminant.

All six protein sequences failed to show a significant identity with any other protein or DNA sequences listed in the EMBL (release 29, December 1991) and other major databases, i.e. SwissProt (release 21, March 1992), GenBank (release 67.0, 15 March 1991), NBRF (National Biomedical Research Foundation; release 35.0, 5 May 1989), NewEMBL (Daresbury laboratory collection of daily updates) and VecBase (release 3, 5 August 1987). Also, the consensus sequence XXSXDXGXTWXX (Roggentin, P. *et al.*, 1989) common to bacterial sialidases was not apparent in any of the bacteriophage E endo-sialidase partial peptide sequences determined so far.

Table II. Sequenced peptide fragments of bacteriophage E endo-sialidase

ENDO-SIALIDASE FRAGMENT	Mr (kDa)	No. OF AMINO ACIDS SEQUENCED	AMINO ACID YIELD (%)
Intact reduced subunit	74.0	25	< 5
CNBr digest fragment 1	15.5	20	> 80
CNBr digest fragment 2	9.4	20	> 80
CNBr digest fragment 3	8.5	9	< 50
CNBr digest fragment 4	29.0	14	> 80
CNBr digest fragment 5	23.0	15	> 80

MOLECULAR CLONING OF BACTERIOPHAGE E ENDO-SIALIDASE

The cloning strategy adopted was to obtain sufficient pure enzyme to allow N-terminal peptide sequencing and then to design synthetic oligonucleotide probes to identify DNA fragments and clones which encode bacteriophage E endo-sialidase. Five synthetic oligonucleotide probes have been constructed from the partial peptide sequences (Table III) with reference to E.coli codon usage tables. These have been end-labelled with [γ-^{32}P]-ATP and used to probe Southern blots of purified bacteriophage E genomic DNA restriction digests. Restriction fragments have been identified which potentially encode sequence for the 74kDa subunit of bacteriophage E endo-sialidase. An 8.5kb HindIII fragment, a 2.2kb XbaI x PstI fragment and a 1.9kb HindIII x PstI fragment are currently being cloned into Bluescript SK+ phagemid. Of these fragments only the 8.5kb HindIII fragment reacted with more than one of the oligonucleotide probes. Surplus sequence within each fragment not encoding endo-sialidase can be excised from the insert of positive recombinants by the technique of nested deletion (Sambrook et al., 1989).

Table III. Synthetic oligonucleotide probes

PROBE	ENZYME FRAGMENT	NO. OF NUCLEOTIDES	DEGENERACY
1	74kDa subunit	18	24
2	15.5kDa CNBr	29	192
3	9.4kDa CNBr	29	256
4	29.0kDa CNBr	23	64
5	23.0kDa CNBr	17	16

FUTURE WORK

At present the primary objective is to complete the cloning and sequencing of the DNA fragments which selectively hybridize to the oligonucleotide probes. Once the full primary nucleotide sequence of bacteriophage E endo-sialidase has been determined, it will be expressed in suitable bacterial expression systems.

It is anticipated that through a greater understanding of the structure-function relationship of endo-sialidase, it will be feasible to develop rational techniques for therapy and monitoring of disease states mediated by α-2,8-polysialic acid expression.

ACKNOWLEGEMENTS

This work was funded by Ciba-Geigy Pharmaceuticals. The DD-NANA sialidase inhibitor was kindly provided by Professor R. Brossmer of Heidelberg University.

REFERENCES

Bhattacharjee, A. K., Jennings, H. J., Kenny, C. P., Martin, A. and Smith, I. C. P. (1975) Structural determination of the sialic acid polysaccharide antigens of Neisseria meningitidis serogroups B and C with carbon 13 nuclear magnetic resonance. J. Biol. Chem. 250, 1926-1932

Figarella-Branger, D. F., Durbec, P. L. and Rougan, G. N. (1990) Differential spectrum of expression of neural cell adhesion molecule isoforms and L1 adhesion molecules on human neuroectodermal tumours. Cancer Res. 50, 6364-6370

Finne, J. and Makela, P. H. (1985) Cleavage of the polysialosyl units of brain glycolipids by a bacteriophage endosialidase : Involvement of a long polysaccharide segment in molecular interactions of polysialic acid. J. Biol. Chem. 260, 1265-1270

Gross, R. J., Cheasty, T. and Rowe, B. (1977) Isolation of bacteriophages specific for the K1 polysaccharide antigen of Escherichia coli. J. Clin. Microbiol. 6, 548-550

Hallenbeck, P. C., Vimr, E. R., Yu, F., Bussler, B. and Troy, F. A. (1987) Purification and properties of a bacteriophage-induced endo-N-acetylneuraminidase specific for poly-α-2,8-sialosyl carbohydrate units. J. Biol. Chem. 262, 3553-3561

Heitz, P. U., Komminoth, P., Lackie, P. M. Zuber, C. and Roth, J. (1990) Demonstration of polysialic acid and N-CAM in neuroendocrine tumours. Verh. Dtsch. Ges. Pathol. 74, 376-377

Hoffman, S., Sorkin, B. C., White, P. C., Brackenbury, R., Mailhammer, R., Rutihauser, U., Cunningham, B. A. and Edelman, G. M. (1982) Chemical characterization of a neural cell adhesion molecule purified from embryonic brain membranes. J. Biol. Chem., 257, 7720-7729

Kwiatkowski, B., Boschek, B., Thiele, H. and Stirm, S. (1982) Endo-N-acetylneuraminidase associated with bacteriophage particles. J. Virol. 43, 697-704

Livingston, B. D., Jacobs, J. L., Glick, M. C. and Troy, F. A. (1988) Extended polysialic acid chains (n>55) in glycoproteins from human neuroblastoma cells. J. Biol. Chem. 263, 9443-9448

Moolenaar, C. E., Muller, E. J., Schal, D. J., Figdor, C. G. Bock, E., Bitter-Suermann, D. and Michalides, R. J. (1990) Expression of neural cell adhesion molecule-related sialoglycoprotein in small cell lung cancer and neuroblastoma cell lines H69 and CHP-212. Cancer Res. 50, 1102-1106

Orskov, I., Orskov, F., Jann, B and Jann, K. (1977) Serology, chemistry and genetics of O and K antigens of Escherishia coli. Bacteriol. Rev. 41, 667-710

Pelkonen, S., Pelkonen, J. and Finne, J. (1989) Common cleavage patterns of polysialic acid by bacteriophage endosialidases of different properties and origins. J. Virol. 63, 4409-4416

Robbins, J. B., McCracken, G. H., Gotschlich, E. C., Orskov, F., Orskov, I. and Hanson, L. A. (1974) Escherichia coli K1 capsular polysaccharide associated with neonatal meningitis. N. Engl. J. Med. 290, 1216-1220

Roggentin, P., Rothe, B., Kaper, J. B., Galen, J., Laurisuk, L., Vimr, E. R. and Schauer, R. (1989) Conserved sequences in bacterial and viral sialidases. Glycoconjugate J. 6, 349-353

Roth, J., Zuber, C., Wagner, P., Taatjes, D. J., Weisgerber, C., Heitz, P. U., Goridis, C. and Bitter-Suermann, D. (1988a) Reexpression of poly(sialic acid) units of the neural cell adhesion molecule in Wilm's tumour. Proc. Natl. Acad. USA 85, 2999-3003

Roth, J., Zuber, C., Wagner, P., Blaha, I., Bitter-Suermann, D. and Heitz, P. U. (1988b) Presence of the long chain form of polysialic acid of the neural cell adhesion molecule in Wilm's tumour. Am. J. Pathol. 133, 227-240

Roth, J. and Zuber, C. (1990) Immunoelectron microscopic investigation of surface coat of Wilm's tumour cells. Dense lamina is composed of highly sialylated neural cell adhesion molecule. Lab. Invest. 62, 55-60

Sambrook, J., Fritsch, E. F. and Maniatis, T. (1989) Molecular Cloning, A Laboratory Manual. Second Edition, (Irwin, N., Ford, N., Nolan, C., Ferguson, M. and Ockler, M., eds.), vol. 2, pp. 13.34-13.41, Cold Spring Harbor Laboratory Press

Taylor, P. W. (1976) Immunochemical investigations on lipopolysaccharides and acidic polysaccharides from serum-sensitive and serum-resistant strains of Escherichia coli isolated from urinary tract infections. J. Med. Microbiol. 9, 405-421

Taylor, P. W. (1987) Enzymatic detection of bacterial capsular polysaccharide antigens. United States Patent No. 4695541

Tomlinson, S. and Taylor, P. W. (1985) Neuraminidase associated with coliphage E that specifically depolymerizes the Escherichia coli K1 capsular polysaccharide. J. Virol. 55, 374-378

Van Der Horst, G. T., Mancini, G. M., Brossmer, R., Rose, U. and Verheijen, F. W. (1990) Photoaffinity labelling of a bacterial sialidase with an aryl azide derivative of sialic acid. J. Biol Chem. 265, 10801-10804

Williams Smith, H. and Huggins, M. B. (1982) Successful treatment of experimental Escherichia coli infections in mice using phage : its general superiority over antibiotics. J. Gen. Microbiol. 128, 307-318

Polysialic Acid
J. Roth, U. Rutishauser and F. A. Troy II (eds.)

AFFINITY LABELING REAGENTS FOR THE CHARACTERIZATION AND ANALYSIS OF SIALIDASES AND SIALIC ACID METABOLIC PROCESSES

Thomas G. Warner

Genentech, Inc. Department of Medicinal and Analytical Chemistry, S. San Francisco, CA 94080

SUMMARY: Derivatives of sialic acid containing photoreactive functional groups have been prepared and tested as photoaffinity labeling reagents. Studies were carried out to determine the feasibility of using these molecules to identify and characterize sialidases and other proteins involved in sialic acid metabolism. Labeling of the human sialidase in a crude preparation from placenta with a radioactive (^{3}H) probe was consistent with the protein having a minimum molecular weight of about 60 kDa. Additional probes were synthesized to incorporate ^{125}I in order to give greater sensitivity of detection than that obtained with the tritiated compounds. The iodinated labeling reagents were evaluated by testing their limits of detection of the C. perfringens sialidase in a crude cell homogenate. It was estimated that as little as 0.6-6 pmol of sialidase could be detected when it was about 0.02-0.2 % of the total protein present in this complex protein mixture. The photoprobes were also utilized to characterize the substrate interactive domain of the highly purified Salmonella typhimurium LT 2 sialidase. A 9.6 kDa peptide near the carboxy terminus was specifically labeled. This peptide had a predicted structure which consisted of alternating segments of β-sheets connected with loops. The predicted secondary structure was apparently similar to that observed for the active sites of viral sialidases.

INTRODUCTION

For the past several years our laboratory has been involved in the development and application of affinity labeling reagents for identifying and characterizing proteins of sialic acid metabolism. As yet, these techniques have not been applied to the analysis of polysialic acid metabolic processes. It is our hope that during the course of this symposium that avenues for exploitation of these approaches in this area will become evident. We report, here, a summary of our results characterizing mammalian and microbial sialidases with sialic acid-based affinity labeling reagents.

Sialidases (neuraminidase, E.C. 3.2.2.18) are an important family of glycohydrolases which cleave sialic acid residues from the oligosaccharide units of glycoproteins and gangliosides. Their significance was initially recognized by simple, incisive experiments carried out in several laboratories nearly half a century ago. Hirst (Hirst 1941, 1942) was the first to discover the influenza sialidase while investigating the agglutination of chicken red blood cells in the presence of influenza virus. He observed that once cells were agglutinated by the virus they did not readsorb the eluted virus or freshly added virus. He concluded that a receptor at the cell surface was modified by the virus during the initial agglutination process. Unknown at the time, the modifying agent was the sialidase of the influenza viral envelope and the cellular receptor was sialic acid. Shortly thereafter, bacterial sialidases were discovered by Burnet and coworkers (Burnet et al., 1946, Burnet and Stone 1947) who carried out agglutination studies of similar experimental design to those of Hirst, except that Vibrio cholerae cell culture fluid was used to alter the surface of human red blood cells. This fluid contained the "receptor destroying enzyme" which, like the virus enzyme, also prevented subsequent agglutination by influenza virus. Stone (1948) extended these experiments to whole animals and demonstrated that infusion of V. cholerae extracts into the lungs of mice prevented influenza infection. Identification of the "receptor-destroying enzyme" as an N-acetyl-neuraminosyl-glycohydrolase was made nearly ten years after these pioneering discoveries, when Gottschalk (1956, 1957) showed that N-acetylneuraminic acid was the cleavage product that resulted from the action of the bacterial and viral extracts on submaxillary mucins and glycoprotein substrates. Thus, he suggested the name neuraminidase for the receptor destroying enzyme.

Together, these early observations not only established the existence of the sialidase enzyme, but they also provided exciting hints about its possible role in the viral infective process. As a result, efforts to elucidate the structure, function, and mechanism of action of the influenza sialidases have remained steadfast for the past fifty years. Presently the viral enzymes are the best characterized of all sialidases (for review see Air and Laver 1989); even so, many interesting questions about these molecules remain to be resolved.

Although they are of no lesser importance, the bacterial sialidases have not been analyzed in similar detail, probably because of the focused effort on the influenza system. Many pathogenic bacteria secrete large amounts of sialidase into the surrounding milieu during their growth. The functional role of this extracellular enzyme is not clear, but it has often been speculated that the sialidase may be an important virulence factor which provides an adaptive advantage for infection for these organisms. Detailed kinetic and substrate studies have been carried out with many highly purified bacterial enzymes. However, their mechanism of action is not understood and tertiary structure analysis of the protein is lacking. Only recently have bacterial sialidase structural genes been isolated (Table I), revealing primary sequence identity between them of up to 35 %, with an overall similarity of about 50 % (Hoyer et al., 1992).

In contrast to the microbial enzymes, mammalian sialidases are, without question, the most poorly characterized proteins of the sialidase family. To date only a single mammalian sialidase has been convincingly purified to homogeneity (Miyagi and Tsuiki 1985) and no sialidase gene has been cloned. This is unfortunate because the enzyme has been implicated in several critical metabolic processes including the regulation of cell proliferation (Usuki et al., 1988a), clearance of plasma proteins (Ashwell and Morell 1974), catabolism of gangliosides and glycoproteins (Usuki et al., 1976b), and the developmental modeling of myelin (Saito and Yu 1992). With such functional diversity it is not surprising that the enzyme has been identified in a number of cellular organelles such as, the plasma membrane (Schengrund et al., 1976), lysosomes (Tulsiani and Carubelli 1970) and cytosol (Miaygi and Tsuiki 1985).

Table I: Sialidase Genes Characterized

Source	Date Reported
Viral	
Influenza A and B	Fields et al. (1981); Shaw et al. (1982); Elleman et al. (1982); Davis et al. (1983)
Sendai	Blumberg et al. (1985)
Newcastle HN	Jorgensen et al. (1987)
Avian A	Harley et al. (1989)
Parainfluenza	Bando et al. (1990)
Bacterial	
Vibrio cholerae	Vimr et al.(1988)
Streptococcus pneumoniae	Berry et al. (1988); Camara et al.(1991)
Clostridium-perfringens	Roggentin et al.(1988)
sordellii	Rothe et al. (1989)
septicum	Rothe et al. (1991)
Bacteroides fragilis	Yeung and Fernandez (1991)
Salmonella typhimurium LT2	Hoyer et al. (1992)
Protozoan	
Trypanosoma cruzi	Kahn et al. (1991)
Mammalian	None

The interrelationship between these various cellular enzyme forms is not known since their primary or secondary structures have not been determined. Thus, it has not been possible to identify similar sequence or structural domains, compare mechanisms of action, or establish a common genetic origin. All of these intriguing aspects provide a pressing impetus for obtaining mammalian sialidases in a highly purified state. However, progress at purifying the mammalian enzymes has been hampered because, in general, these proteins are extremely thermally labile. For example, the lysosomal sialidase loses activity within several hours upon tissue disruption or cell homogenization (Warner and O'Brien 1979) and it does not survive freezing even in intact tissues. In addition, many mammalian sialidases are in relatively low abundance and particulate in nature, additional factors which have also impeded detailed studies.

In order to characterize mammalian and microbial sialidases, and other proteins involved in sialic acid metabolism, our laboratory has developed novel affinity labeling reagents which are derivatives of sialic acid (Warner and Lee 1986, 1988, Warner et al., 1987,1989). Our goals are to utilize these compounds to: (i) identify, characterize and facilitate purification of proteins involved in sialic acid metabolism, especially those which are difficult to purify by conventional means (e.g. sialidases); (ii) characterize the substrate interactive domains of mammalian and microbial sialidases in order to better understand their mechanisms of action.

For these purposes, photoreactive affinity labeling reagents of sialic acid were prepared containing arylazide moieties. This functional group is attractive since it is relatively chemically inert until it is exposed to ultra-violet light. Thus the probe can easily be manipulated during preparation and it is stable for storage. When irradiated, an arylnitrene is generated which is extremely chemically reactive and can readily form covalent bonds by a variety of mechanisms under moderate conditions with molecules in close proximity (Bayley and Staros 1984). If the probe is also radioactive, then the specific protein labeled with the sialic acid derivative can be identified and purified by following its radioactivity without concern for maintaining the catalytic integrity of the enzyme. This strategy has been utilized extensively to identify and isolate labile enzymes, binding proteins, membrane transport proteins, and cell surface receptors. The technique is especially useful for characterizing the latter class of proteins because they do not possess catalytic function and, consequently, their purification cannot be monitored by enzyme assays (for a general review on similar applications see Bayley and Staros 1984).

Some specific proteins that we have identified as good target candidates for investigation with photoaffinity probes are summarized in Table II. These include the lysosomal sialidase which is one of the most poorly characterized of the lysosomal glycohydrolases because of its thermal lability. This enzyme is important because deficiencies of the enzyme lead to the inherited metabolic disorder, sialidosis (Lowden and O'Brien 1979). Similarly, the lysosomal membrane sialic acid transport protein has no enzymatic function; but is important because defects in this molecule result in sialic acid storage disease (Renlund et al., 1986). Also included here are ganglioside specific sialidases and endosialidases which are not presently associated with any metabolic disease but are important nonetheless because they may be involved in developmental modeling of glycolipids (Saito and Yu 1992) and protein adhesion molecules of the nervous system (Livingston et al., 1988), respectively.

RESULTS AND DISCUSSION

IDENTIFYING A SIALIDASE IN A COMPLEX MIXTURE OF PROTEINS: One of the primary goals of our research has been utilization of photoaffinity reagents to identify mammalian sialidases in partially purified protein preparations. In order to test the feasibility of this approach and to evaluate the selectivity and sensitivity of the probes for this purpose, we carried out affinity labeling of two different, sialidase preparations. The first was a partially purified sialidase isolated from human placenta (Verheijen et al., 1987) and the second was a model system composed of the highly purified bacterial sialidase from C. perfringens added to a crude homogenate of cultured skin fibroblasts. In all cases, each of the probe molecules were also evaluated by kinetic analysis to determine the inhibitor potency. For these studies, the bacterial enzyme or the lysosomal enzyme in fibroblast homogenates was used as an enzyme source.

Table II. Proteins of Sialic Acid Metabolism Potentially Investigated With Photoaffinity Labeling Reagents

Protein	Function	Disease
Lysosomal sialidase	Catabolism of glyco-proteins & glycolipids	Sialidosis
Sialic acid lysosomal membrane transporter	Exports sialic acid from the lysosome	Sialic acid storage disease
Ganglioside specific sialidase	Catabolism of gang-liosides in lysosomes, plasma membrane, mylein	?
Endosialidases	Catabolism/modeling of polysialic acid	?

<u>Preparation of arylazido thioglycosides of sialic acid:</u> Initially, thioglycosides of sialic acids containing nitrophenylazides were prepared because the thioether linkage was presumed to be resistant to enzymatic cleavage. This was a desired feature of the probe since it was important that the structural integrity of the molecule be maintained during photolysis and during the subsequent isolation of the labeled protein. Coupling of the azide unit with the appropriate stereochemical orientation (alpha) was achieved by preparing the 2- thioacetate ester of sialic acid from peracetyl sialyl chloride methyl ester (Warner and Lee 1986,1988). Selective hydrolysis of the thio ester, followed by condensation with either 4-fluoro-3- nitrophenyl azide or 3-fluoro-4,6 dinitrophenyl azide gave (4-azido-2-nitrophenyl)-5-acetamido-2,3,5-trideoxy-2-thio-α-D-*glycero*-D-*galacto*-2-nonulopyranosidonic acid (2-thio-PANP-Neu5Ac) or (3-azido-4,6 dinitro)-5-acetamido-2,3,5-tridoexy-2-thio-α-D-*glycero*-D-*galacto*-2-nonulopyranosidonic acid (2-thio-MADNP-Neu5Ac) in good yield, respectively (Figure 1).

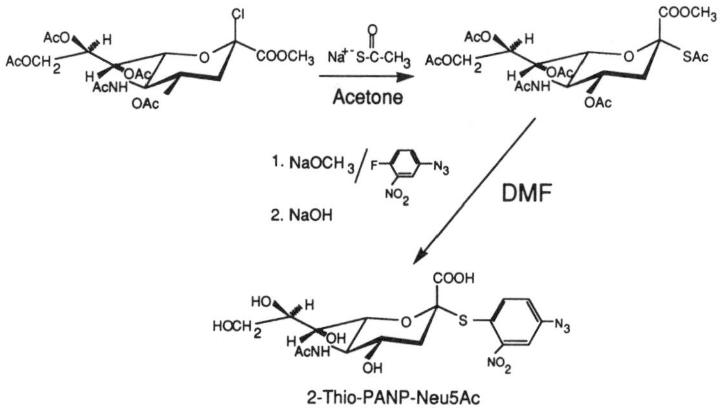

Acetone

1. $NaOCH_3$ / F–C₆H₃(NO₂)–N_3

2. NaOH

DMF

2-Thio-PANP-Neu5Ac

Figure 1: Synthetic scheme for the preparation of thioglycosidases of sialic acid containing photoreactive arylazides (Warner et al., 1988).

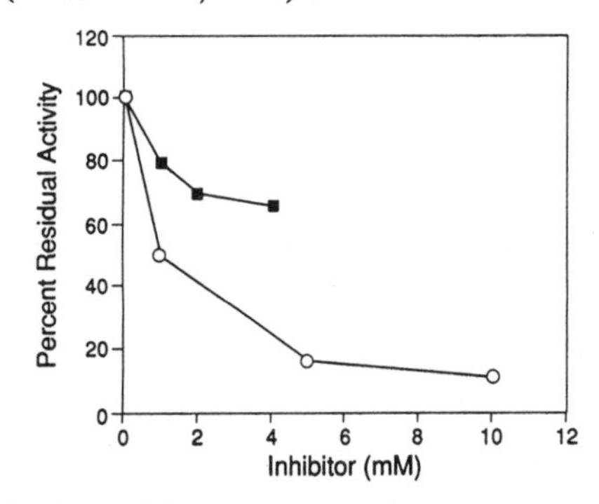

Figure 2: Inhibition of the <u>C. perfringens</u> sialidase with arylazido thioglycosides of sialic acid. Enzyme assays were carried out using 4-MU-Neu5Ac substrate at the Km (Warner and O'Brien 1979) with varying concentrations of inhibitors as shown: n = 2-thio-PANP-Neu5Ac, ○= 2-thio-MADNP-Neu5Ac.

Surprisingly, both compounds were weak inhibitors of the bacterial enzyme (Figure 2) with inhibitory constants, Ki, of about 1 mM. This was unexpected since the Km for O-linked aryl glycoside substrates (e.g. 4-methylumbelliferyl-N-acetyl neuraminic acid, 4-MU-Neu5Ac) is in the micromolar range (Warner and O'Brien 1979). Apparently, substitution of the less electronegative sulfur in place of oxygen at the anomeric center resulted in a lower affinity by the enzyme. Although these compounds were not effective inhibitors of the sialidase, they are included here because they may be useful for identifying and characterizing other proteins which recognize sialic acid such as binding proteins, transport proteins or lectins.

Preparation of derivatives of Neu5Ac2en with linker groups at C-9: Since the arylazido thioglycosides were weak enzyme inhibitors they were not expected to be suitable for labeling a highly impure preparation of the mammalian enzyme. We anticipated that a high binding affinity

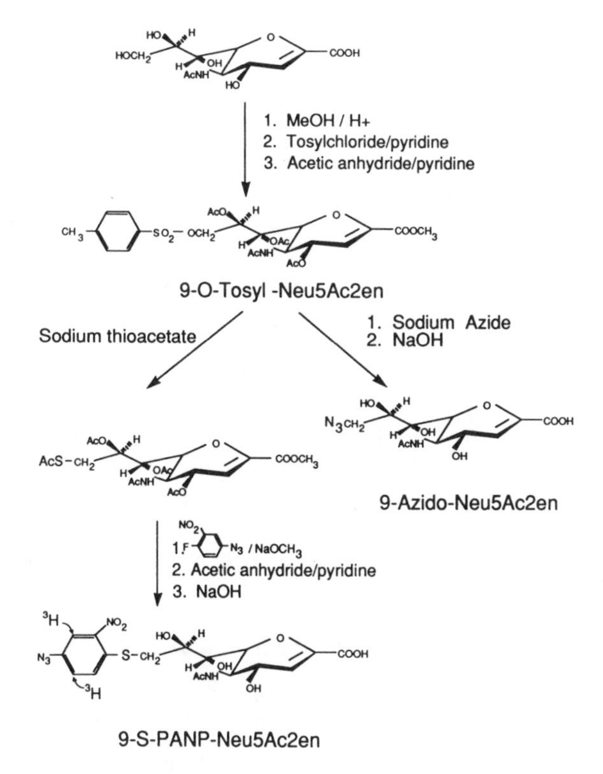

Figure 3: Scheme for modification of Neu5Ac2en at C-9. Shown are preparations of 9-S-PANP-Neu5Ac2en and 9-azido-Neu5Ac2en (Warner et al., 1987, 1991). The probe was made radioactive (400 mCi/mmol) by coupling with 4-fluoro-3-nitro-(2, 6-^{3}H) phenylazide .

for the enzyme by the probe was required in order to increase efficiency of labeling and to reduce non-specific labeling of unrelated proteins in a complex mixture. For this reason, we prepared derivatives of the well known sialidase inhibitor, 5-acetamido-2,6-anhydro-3,5-dideoxy-D-*glycero*-D-*galacto*-non-2-enonic acid (Neu5Ac2en), which has a Ki of about 10 μM for sialidases from many sources including the mammalian enzyme (Figure 3).The primary hydroxyl at C-9 was chosen as the site for modification since it could be selectively derivatized by tosylation over other hydroxyl positions and would allow the introduction of other functional groups such as thioacetate, azides, and nitrophenyl azides .

The relative inhibitor potency of the Neu5Ac2en derivatives with different linker groups at C-9 was evaluated using the lysosomal sialidase in crude homogenates of cultured human skin fibroblasts and the 4MU-Neu5Ac substrate (Figure 4) (Warner et al., 1991).

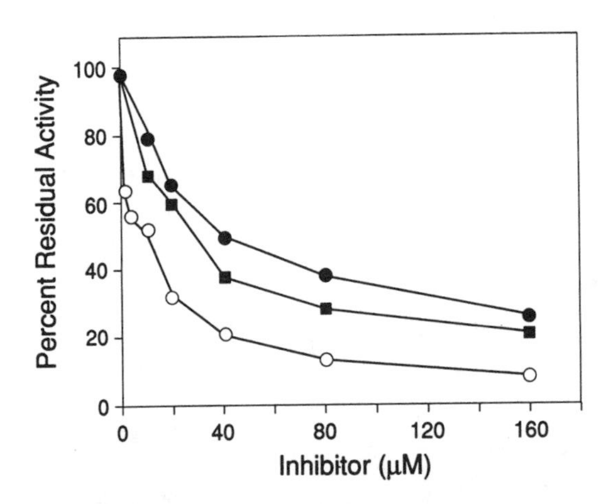

Figure 4: Relative inhibition of the lysosomal sialidase by derivative of Neu5Ac2en modified at C-9. Assays were carried at the Km of 4 MU-Neu5Ac. n= Neu5Ac2en, • = 9-S-PANP-Neu-5Ac2en, ○ = 9-azido-PANP-Neu5Ac2en.

Remarkably, substitution of the C-9 hydroxyl group with azide or arylazide did not adversely affect the inhibitory potency of the molecule and both compounds gave Ki values identical to unmodified Neu5Ac2en of about 10 μM. These results were unexpected because previous studies in a number of laboratories had shown that minor modifications of the glycerol side chain of Neu5Ac2en, such as oxidative cleavage (Meindl et al., 1970), dehydroxylation (Zbrial et al., 1987a) and epimerization (Zbrial et al., 1987b), greatly reduced inhibitory potency. Along these lines, it had also been shown that 9-O-acetyl sialic acid glycosides were very poor sialidase substrates (Kendal 1975). In spite of these anticipated difficulties, our results indicate that the terminal C-9 hydroxyl group is not a major factor in substrate binding, at least with the mammalian enzyme, and that the unsaturated pyran portion of the molecule contains the primary contact points with the protein.

Labeling the placental sialidase: Photolabeling studies using 9-S-[^{3}H]-PANP-Neu5Ac2en were carried out to identify the placental sialidase (Verheijen et al., 1987) and to evaluate the feasibility of using the incorporated radioactive marker to facilitate further purification of the protein from the mixture. In addition to the sialidase, the placental preparation contains a large number of other proteins when analyzed with polyacrylamide gel electrophoresis under denaturing conditions (Figure 5; Panel I lane B and C). When the protein mixture was photolyzed with the radiolabeled probe, two protein bands , 61 and 46 kDa, were visualized with autoradiography (Figure 5: Panel II lane A). When surfactant was included in the photolysis mixture, labeling of the 46 kDa band was greatly reduced with a concomitant increase in labeling of the 61 kDa band. These conditions also coincided with a 25 % increase in sialidase activity (Panel II lane B). No labeling was observed when the complete mixtures were maintained in the dark (Panel II lane C). Labeling of the 61 kDa band was reduced when Neu5Ac2en was included during photolysis (Panel II lane D).

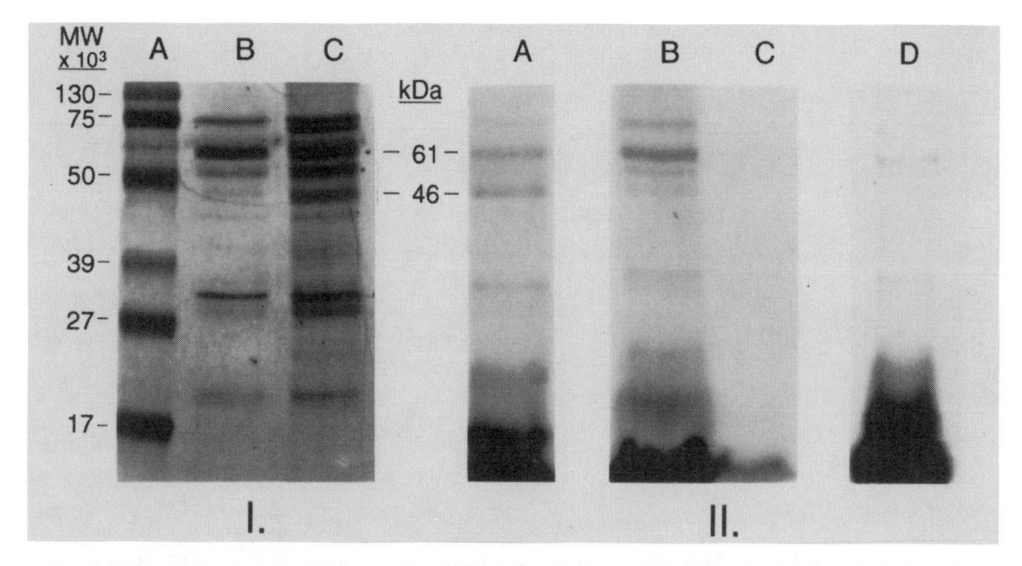

Figure 5: Panel I: Lane B and C, Polyacrylamide gel of two placental sialidase preparations. Gel is stained with silver stain. Shown also are molecular weight markers, Lane A. Panel II: Autoradiogram of polyacrylamide gel of photolysis mixture containing enzyme preparation and radiolabeled probe. Lane A, complete photolysis mixture. Lane B, Photolysis mixture with 1% Triton X-100. Lane C, Dark control. Lane D, Photolysis mixture with 1% Triton X-100 and Neu5Ac2en (1 mM).

These preliminary studies were encouraging since they support the idea that the sialidase lies within the 61 kDa band along with many other proteins. These results corroborated those of Van Der Horst and coworkers (Van Der Horst et al., 1989) who have employed reconstitution experiments and have identified the sialidase in similar placental preparations as a protein which comigrates with β-galactosidase in the 60 kDa range.

Preparation and testing of photoreactive derivatives of Neu5Ac2en containing ^{125}I: Although the tritiated probe was helpful for identifying the sialidase in these preparations, the labeling efficiency of the enzyme was very low and only several hundred counts of radioactivity were incorporated. This low level of labeling precluded further purification of the protein. Other means for incorporating an isotope into the inhibitor which would give higher specific radioactivity were developed (Warner et al., 1992). Ji and coworkers (Ji and Ji 1982) have shown that the attachment of 4-azido salicylic acid (ASA) provides a convenient means for introducing a photoreactive functionality which can be made radioactive with ^{125}I. In order to link ASA to Neu5Ac2en, we incorporated a primary amino group at C-9 of Neu5Ac2en by selective reduction of the 9-azido Neu5Ac2en derivative prepared in earlier studies. Reduction using propane 1,3 dithiol (Bayley et al., 1978) was the preferred route since conventional hydrogenation catalysts such as $Pd/BaSO_4$ (Meindl and Tuppy 1973) did not provide the specificity required and we observed the concomitant reduction of the 2, 3 double bond (Figure 6).

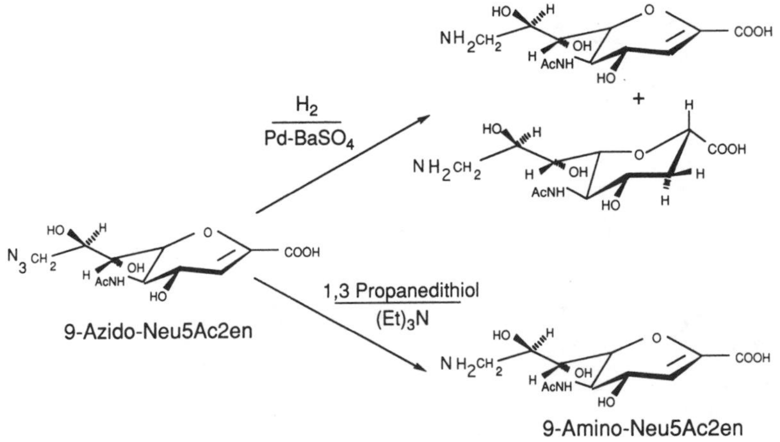

Figure 6: Reaction scheme for the introduction of an amino group at C-9 of Neu5Ac2en by reduction of the 9-azido precursor.

Coupling of the amine was accomplished using ASA as its hydroxysuccinimidyl ester (Pierce Chem. Co.). Iodination of the 9-N-ASA-Neu5Ac2en was carried out using mild oxidative conditions compatible with the presence of the unsaturated double bond. As might be expected, three positional isomers were identified as reaction products when iodine was a limiting reagent. These reaction conditions were similar to those employed when ^{125}I was incorporated (Figure 7) except that the radiolabeling reaction was carried out on a greatly reduced scale. All three isomeric compounds were also detected as radioactively labeled derivatives. The specific activity of the ^{125}I iodinated probe was about 2175 mCi/μ mol.

While this work was in progress, Van Der Horst (Van Der Horst et al., 1990 a and b) reported the use of a similar Neu5Ac2en derivative in a preliminary communication. However, the method employed for the preparation of the 9-amino Neu5Ac2en starting material or the characterization of the resulting iodinated products was not reported. For these reasons, it is not possible to compare their results with ours.

Labeling C. perfringens sialidase added to a mixture of proteins with 9-N-[^{125}I-ASA]-Neu5Ac2en: Photolysis experiments were carried out with 9-N-[^{125}I-ASA]-Neu5Ac2en and *C. perfringens* sialidase in a crude cell homogenate as a test to determine the limits of detection of the iodinated probe (Figure 8). This model system was advantageous over using the placental sialidase preparation because the mammalian enzyme is tedious to isolate in large quantities and

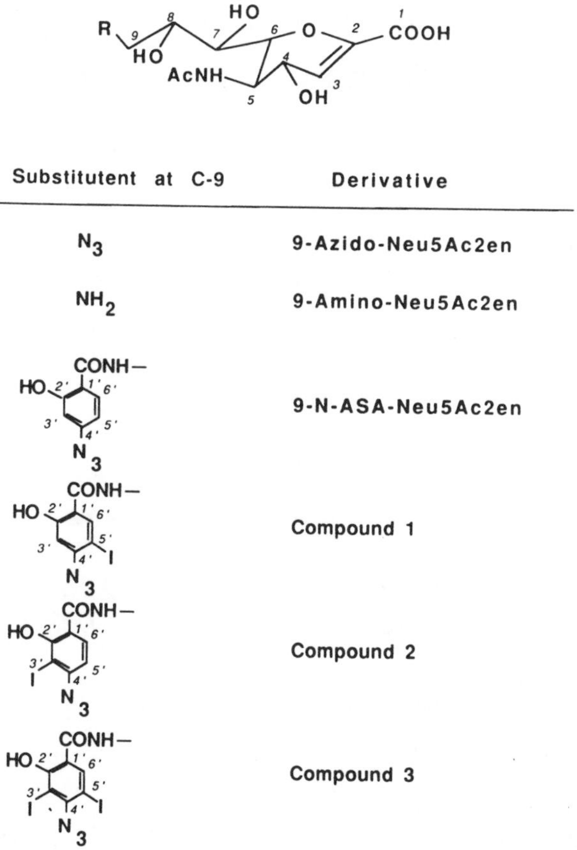

R 9 8 HO 7 HO 6 O 2 1 COOH AcNH 5 4 OH 3

Substitutent at C-9	Derivative
N_3	9-Azido-Neu5Ac2en
NH_2	9-Amino-Neu5Ac2en
CONH—, HO (2'), N_3 (4')	9-N-ASA-Neu5Ac2en
CONH—, HO (2'), I (5'), N_3 (4')	Compound 1
CONH—, HO (2'), I (3'), N_3 (4')	Compound 2
CONH—, HO (2'), I (3'), I (5'), N_3 (4')	Compound 3

Figure 7: Neu5Ac derivatives modified at C-9 to incorporate ASA and ^{125}I (Warner et al., 1992).

it has not been completely characterized. In contrast, the bacterial enzyme has been well studied and it is commercially available in highly purified form (Sigma Chem Co). In these experiments the C. perfringens sialidase was easily detected as an intensely labeled band (about 20,000 dpm) on the gel at the 60 pmol level, when it was about 2% of the total protein present. The detection limit is near the 6 pmol level when the sialidase is between 0.02 and 0.2% of the protein in the photolysis mixture. This level of sensitivity suggests that it is possible to detect a relatively low abundance sialic acid binding protein using an enriched subcellular fraction. For example, most lysosomal proteins are about 0.01 % of the total cellular protein or about 0.6 % of the total protein in a lysosome preparation. This is well within the detection limits of the probe as determined in this model system and it also suggests that it is feasible to use this marker to facilitate further purification of the protein.

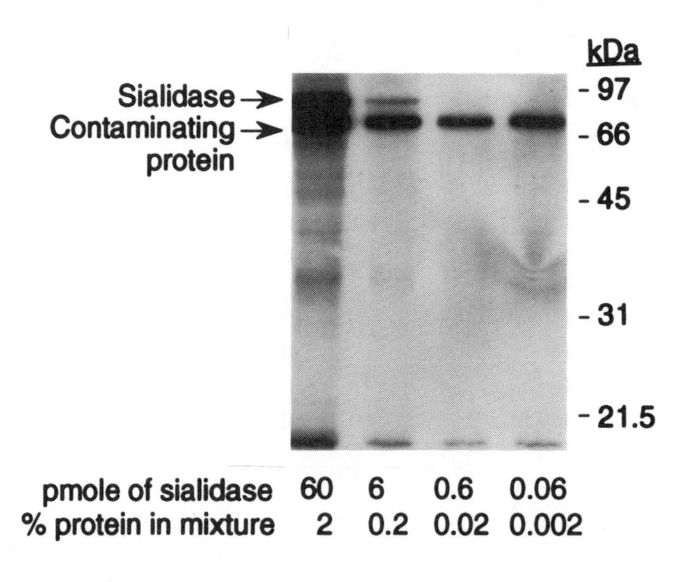

Figure 8: Autoradiogram of polyacrylamide gel containing photolysis mixtures with varying amounts of C. perfringens sialidase added to cell homogenates of skin fibroblasts in the presence of 9-N-[^{125}I-ASA]-Neu5Ac2en. Each mixture contained 25 μg protein of cell extract, 3 μCi probe, 15 mM BME, 25 mM acetate, pH 5.5, and varying amounts of added purified sialidase; lane 1, 0.5 μg, 60 pmol sialidase; lane 2, 0.05 μg, 6 pmol; lane 3, 0.005 μg, 0.6 pmol; lane 4, .0005 μg, 0.06 pmol. The mixtures were irradiated for 60 sec with a hand-held ultraviolet light source and then analyzed with polyacrylamide gel electrophoresis under denaturing conditions. After development, the gel was dried and exposed to x-ray film for 4 hr at -70°C. The protein labeling at 66 kDa is a contaminating protein in the cell extract that labels nonspecifically.

<u>CHARACTERIZATION OF THE SUBSTRATE INTERACTIVE DOMAIN OF A PURIFIED SIALIDASE:</u> The second major focus of our research effort (Warner et al., 1992) has been to use photolabeling techniques to characterize the substrate interactive domains of sialidases. The highly purified sialidase of <u>Salmonella typhimurium</u> LT2 was an attractive enzyme to explore with this technique because the complete protein sequence could be deduced from the cloned structural gene and the enzyme was available in large quantities as a recombinant product overexpressed in <u>E. coli</u> (Hoyer et al., 1992).

<u>Photolabeling Salmonella typhimurium LT2 sialidase:</u> The purified sialidase gave a single major band (41 kDa) when analyzed with polyacrylamide gel electrophoresis under denaturing conditions (Fig.9, lane A). Photolabeling of the protein with 9-N-(^{125}I-ASA)-Neu5Ac2en was determined to be specific for a region of the molecule near the active site by protection experiments using various additives during photolysis. These included a nitrene quenching reagent, a potent competitive inhibitor, and a compound with structural features similar to the probe, but which was not a sialidase inhibitor.

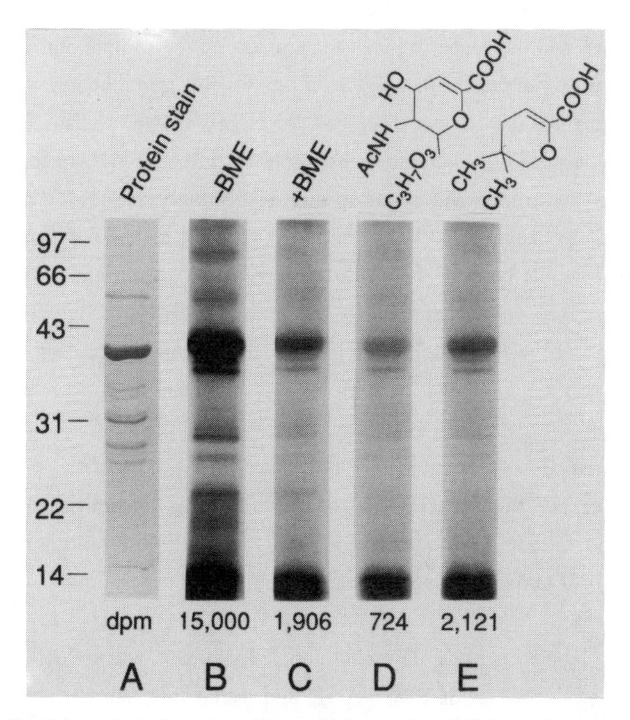

Figure 9: Polyacrylamide gel of the recombinant <u>Salmonella</u> sialidase. Lane A. purified enzyme used in photolabeling studies stained for protein. Lane B. Autoradiogram of gel containing the protein (2 μg) photolyzed with 3-5 μCi of probe. Lane C. Protein photolyzed with probe and 5 mM BME. Lane D. Protein photolyzed with BME and 2 mM Neu5Ac2en. Lane E. Protein photolyzed with BME and 2 mM DOPC. The gel was exposed overnight. Molecular weight markers are indicated.

The thiol, beta-mercaptoethanol (BME), greatly reduced photolabeling of the protein (Fig 9, lanes B and C). This small molecular weight compound is known to decrease non-specific labeling of proteins by scavenging the solution-phase nitrenes which randomly interact with the protein. Presumably, probe molecules bound in the enzyme active site are shielded from the effects of thiol quenching and labeling of this region of the protein is not altered under these conditions. When the sialidase inhibitor Neu5Ac2en was included, the labeling was lowered substantially. (Fig. 9, lane D) although it was not at background levels. The inhibitor competes with the probe for the active site, displacing it from the protein, and reducing the amount of label incorporated. We attribute the residual labeling under these conditions to be due to the incomplete elimination of non-specific labeling of the protein at this concentration of BME. It was not practical to increase the concentration of the thiol above 5 mM since this completely eliminated labeling.

As a further indication for specific labeling of the the protein, photolysis experiments were carried out in the presence of 3,4-dihydro-2,2-dimethyl-4-oxo-2H-pyran-6-carboxylic acid (DOPC) Fig. 9 lane E). Although this compound contains some structural and chemical features similar to Neu5Ac2en and the probe, it is not recognized by the protein and does not bind at the active site. Thus, it does not protect this region of the protein from labeling with the probe. These are important results since they demonstrate that the effects of Neu5Ac2en and the probe cannot be explained on the basis of simple ionic or hydrophobic interactions due to the presence of the carboxylate group or the unsaturated pyranose ring, respectively. It should also be noted, that the reduction in labeling by Neu5Ac2en is not due to quenching by a direct reaction of the nitrene with the unsaturated double bond of the Neu5Ac2en molecule. Such addition reactions might be expected to occur; however, it apparently does not contribute significantly under these conditions, since DOPC also contains an unsaturated pyran moiety but it does not reduce labeling.

Together these results are consistent with highly specific labeling of the protein. We envisage that the probe molecule orients itself with the pyranose ring portion in the substrate binding pocket with the aryl nitrene positioned away from the catalytic amino acids near the protein water interface where it is only partially protected from thiol quenching.

Identification of the labeled peptide: The photolabeled sialidase was treated with CNBr and the resulting peptides separated with SDS-PAGE and electrophoretically transferred to PVDF membrane (Fig. 10). After staining for protein, the membrane was exposed to X-ray film overnight. Two of the fragments were radioactively labeled, 18 kDa and 9.6 kDa (Fig. 10 lane A and B). In order to identify the peptides, N-terminal sequence analysis was conducted on these fragments (Fig. 10, lane B). The fragment at 18 kDa gave no signal in the sequencer, indicating

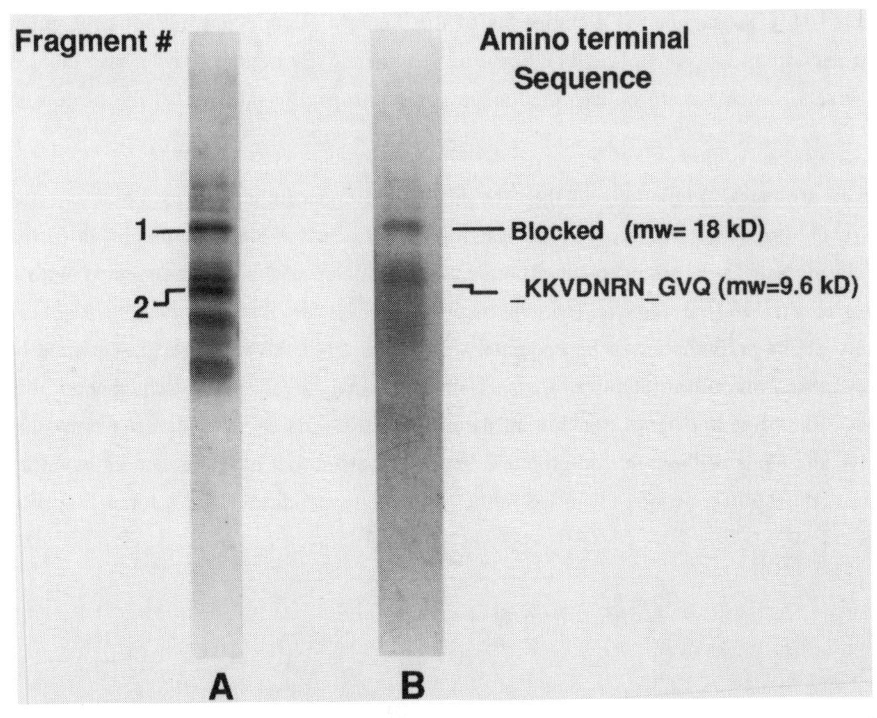

Figure 10: PVDF membrane containing CNBr derived peptide fragments separated by SDS-PAGE. Lane A. peptide fragments stained for protein. Lane B. autoradiogram of the gel. As indicated, the labeled fragments were subjected to N-terminal sequencing for identification. Size estimations were made with marker proteins (not shown).

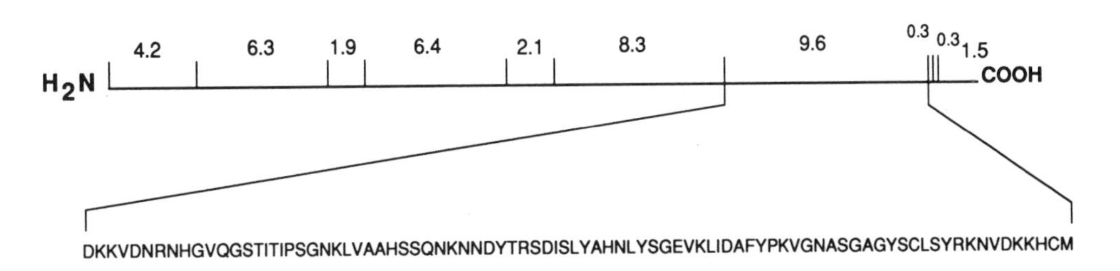

Figure 11: Representation of the Salmonella sialidase protein sequence showing the position of methionine residues and the anticipated size of the resulting fragments. The expanded region shows the sequence of the 9.6 kDa peptide labeled with the probe.

that the N-terminus was blocked. The 9.6 kDa fragment was identified as comprising the C-terminal third of the protein (Fig 11). This is expected to be the largest CNBr derived fragment on the basis of the methionine content and the predicted amino acid sequence. We conclude that bands larger than 9.6 kDa are due to incomplete cleavage of the the peptide. It seems likely that the labeled 18 kDa fragment is a composite of the 9.6 kDa peptide and the adjacent, preceding 8.3 kDa peptide in the protein (Fig 11). Some of the other CNBr fragments were also labeled, but at low levels, which was to be expected since some nonspecific labeling of the protein should occur.

Secondary structural predictions of the labeled peptide and comparison to the active sites of influenza sialidases: Crystallographic analysis of the influenza A sialidase has revealed that the active site domain is an arrangement of alternating segments of β-sheets connected with loops (Colman et al., 1983). Jorgensen (Jorgensen et al., 1987) has observed that the β-sheet-loop segments at the active site can be accurately predicted from the amino acid sequence by the Garnier-Robson algorithm (Garnier, et al.). Using this analysis and other sequence similarities they have identified analogous domains in parainfluenza sialidases which are presumed to form the active site clefts of these related proteins. We have carried out a similar predictive structural analysis of the 9.6 kDa peptide identified in the Salmonella sialidase with photolabeling (Fig 12).

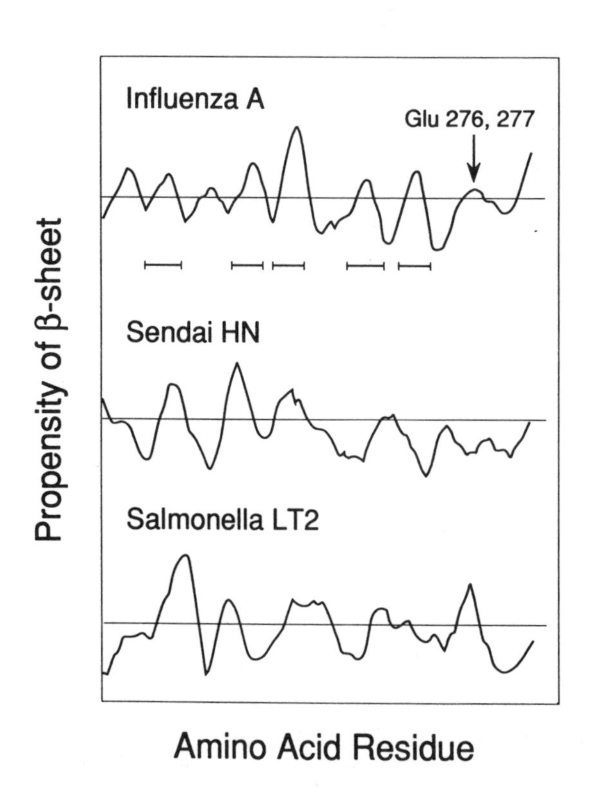

Figure 12: Regions of β-sheets as predicted by the Garnier-Robson algorithm within the active site of influenza A sialidase (residues 210-333), an analogous portion of the Sendai HN sialidase (residues 233-310) and the 9.6 kDa peptide of S. typhimurium sialidase (residues 270-356). Bars show the β-sheet locations within the influenza A enzyme as determined from the crystal structure (Colman et al., 1983).

The results of his treatment suggest that the peptide could be organized in a similar manner. At least four alternating segments with a high probability of β-sheet structure were predicted. With this analysis, the 9.6 kDa peptide bears a general similarity to the active site of influenza A sialidase with an apparently remarkable similarity to the enzyme of Sendai HN virus. Mutagenesis studies of the influenza A sialidase have implicated histidine 274 and glutamic acids 276 and 277 to be directly involved in catalysis (Lentz et al., 1987). The mechanism of the bacterial sialidase may differ from the viral enzyme since a similar juxtaposition of glutamic acid residues was not present in the 9.6 kDa peptide or elsewhere in the sequence of the Salmonella enzyme. In addition, the active sites of other viral sialidases may be organized differently from the influenza A enzyme. Recent crystallographic structural data indicates that the catalytic residues in influenza B sialidase are not situated on the analogous peptide region (residues 210-333) as predicted by Jorgensen (Jorgensen et al., 1987) or Lenz (Lenz et al., 1987; Burmeister et al., 1992). However, segments of this peptide comprise portions of the active site of the influenza B sialidase and serve an important role in substrate binding. For example, Burmeister (Burmeister et al., 1992) positions glutamic acid residue 274 of the protein in close proximity to the glycerol side chain of the sialic acid; hydrogen bonding with the C-8 and C-9 hydroxyl group of the substrate molecule. If the Salmonella enzyme has a related arrangement of its active site such that the 9.6 kDa peptide interacts with the C-8 and C-9 hydroxyl groups of the substrate, then the aryl nitrene group of the probe molecule, which substitutes for the C-9 hydroxyl moiety, would be in an ideal orientation to react with the amino acids in this peptide.

Molecular modeling of 9-N-ASA-Neu5Ac2en- Finally, our data does not verify that the catalytic amino acids or the residues directly involved in binding are located on the peptide identified with the probe, although, this possibility cannot be ruled out since the protection experiments with Neu5Ac2en and DOPC are consistent with this notion. In order to gain some insight into the relative orientation between the labeled peptide and the catalytic amino acids, we have estimated the optimal conformation states of the probe molecule and calculated the distance between the anomeric center of the Neu5Ac2en moiety and the aryl nitrene.

Three conformation models of the probe were developed based on: (i) the solution conformation of sialic acid as determined from magnetic-resonance studies (Czarniecki and Thornton 1976), (ii) the crystal structure coordinates of Neu5Ac2en (Furuhata et al., 1988) and (iii) a general examination of all feasible orientations of the probe. The models were assembled with Version 5.41 of the SYBYL software system (Tripos Associates) and subjected to simple energy minimizations using the Tripos force field.

With either of these models, the distance between the anomeric center at C-2, which is presumed to be situated very near the catalytic residues, and the aryl nitrene was about 12 Angstroms. On this basis, regardless of the conformation assumed by the substrate analog when bound to the enzyme, its covalent structure imposes about a 12 Angstrom limit on the distance between the anomeric center and the aryl nitrene nitrogen. If it is assumed that the catalytic residues were located on the labeled peptide at a site contiguous to the position of attachment of the probe, and if the secondary structure of this peptide were β-sheet, then the catalytic residues would be about 3 to 4 amino acids distant from those labeled with the probe. However, it should be noted that the probe, as its reactive nitrene form, may diffuse away from the catalytic residues before it encounters a suitable nucleophilic amino acid and attaches to the peptide. Thus, although the conformational constraints of the probe limit the distance between its reactive center and the site of bond cleavage at the anomeric carbon, the actual site of attachment to the protein may be farther from the catalytic residues than that expected by this analysis. Due to the low level of labeling, it was not possible to further characterize the labeled peptide and identify the specific amino acid residues which were derivatized by the probe.

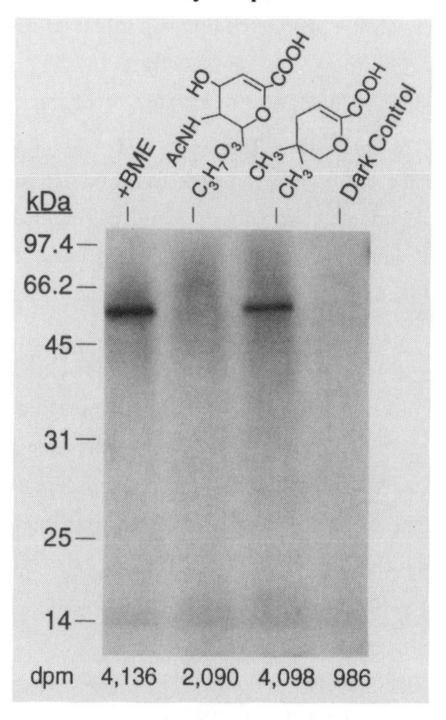

Figure 13: Autoradiogram of polyacrylamide gel containing photolysis mixtures of highly purified Sendai virus sialidase (0.4 μg , 7 pmol). Shown are the effects of various additives on labeling

In summary, both active-site models for the influenza sialidases suggest an important role in sialyl conjugate bond cleavage for the analogous Salmonella peptide. In influenza A, the peptide may contain the catalytic residues. In influenza B, the peptide may be involved in substrate binding and recognition. The functional role of the peptide in the Salmonella enzyme will hopefully be clarified when the crystal structure of the protein becomes available. However, the phtotolabeling data and comparative analysis with the active sites of the viral sialidases strongly indicates that the identified peptide comprises an intregal part of the substrate interactive domain of the bacterial enzyme.

Work in progress-photolabeling Sendai HN sialidase: Given the predicted structural similarity between the 9.6 kDa Salmonella peptide and the presumptive active site region of Sendai HN sialidase, we are currently carrying out photolabeling studies using the highly purified sialidase dimers from Sendai virus. It is of interest to determine if, indeed, the segment of the Sendai protein which is labeled with the probe corresponds with that region predicted by Jorgensen (Jorgensen et al., 1987) to form the active site of the viral enzyme. Thus far, the labeling studies have been encouraging and the enzyme is labeled specifically at high levels (Fig.13). Isolation and characterization of the labeled peptide is currently in progress.

ACKNOWLEDGMENTS

We wish to thank our collaborators; Dr. Michel Potier, University of Montreal, Dr. Eric Vimr, University of Illinois, and Dr.Alan Porter, St. Judes Childrens Research Hospital, who have contributed significantly to this work by supplying their valuable purified proteins and their insightful suggestions and ideas. Also, we thank our Genentech collaborators, Dr. Robert McDowell for his help in molecular modeling and Reed Harris for amino acid sequencing. We also appreciate D. A. Warner and Dr. Miçhael Spellman for their helpful criticisms of this manuscript. This work was supported by Genentech, Inc. and by NIH grant NS 22323.

REFERENCES

Air, G.M. and Laver, W.G. (1989) Proteins: Struct. Func. Genet. 6: 341-356.
Ashwell, G. and Morell, A. (1974) Adv. Enzymol. 41: 99-128.
Bayley, H. and Staros, J.V. (1984) in Azides and Nitrenes (Scriven E. F. V. ed.) pp.433-490. Academic Press, New York.
Bayley, H., Standring, D.N., and Knowels, J. R. (1978) Tetrahedron Lett. 39: 3633-3634.
Bando, H., Kondo, K., Kawano, M., Komada, H., Tsurudome, M., Nishio, M., Ito, Y. (1990) Virology 175: 307-312.
Berry, A.M., Paton, J. C., Glare, E.M., Hansman, D., and Catcheside, D.E.A. (1988) Gene 71:299-305
Blumberg, B., Giorgi, C., Roux, L. Raju, R. Dowling,P., Chollet, A. and Kolakofsky, D. (1985) Cell 41: 269-278.
Burmeister, W.P. Ruigrok, R.W.H. and Cusack, S. (1992) EMBO J. 11: 49-56.
Burnet, F.M., McCrea, J.F. and Stone, J.D. (1946) Brit. J. Exp. Path. 27: 228-236.
Burnet, F.M., Stone, J.D. (1947) Austral. J. Exp. Bio. Med. Sci.25: 227-232.
Camara, M., Mitchell, T.J., Andrew, P.W., and Boulnois, G.J. (1991) Infec.and Immun. 59: 2856-2858.
Colman, P.M., Varghese, J.N., and Laver, W.G. (1983) Nature 303 :41-44.
Czarniecki, M.F., and Thornton, E.R. (1976) J. Am. Chem. Soc. 98: 1023-1025.
Davis, A.R., Bos, T.J. , and Nayak, D. (1983) Proc. Natl. Acad. Sci. USA. 80:3976-3980.

Elleman, T. C. , Azad, A.A., and Ward, C.W. (1982) Nucleic Acids Res. 10: 7005-7015.
Fields, S. , Winter, G. and Brownlee, G.G. (1981) Nature 290: 213-217.
Furuhata, K., Sato, S. Goto, M., Tatayanagi, H. and Ogura, H. (1988) Chem. Pharm. Bull 36: 1872-1876.
Garnier, J., Osguthorpe, D., and Robson, B. (1978) J. Mol. Biol. 120: 97-120.
Gottschalk, A. (1956) Biochem. Biophys. Acta 20: 560-561.
Gottschalk, A. (1957) Biochem. Biophys. Acta 23: 645-636.
Harley, V.R., Ward, C.W., and Hudson, P.J. (1989) 169: 239-243.
Hirst, G.K. (1941) Science 94: 22.
Hirst, G.K. (1942) J. of Exp. Med. 76: 195-209.
Hoyer, L.L., Hamilton, A.C., Steenbergen, S.M., and Vimr, E.R., (1992) Molec. Microbiol. 6: 873-884.
Ji, I. and Ji, T.H. (1982) Anal. Biochem 121: 286-289.
Jorgensen, E. D., Collins, P.L. and Lomedico, P.T. (1987) Virology 156: 12-24.
Kahn, S., Colbert, T.G., Wallace, J.C., Hoagland, N.A., and Eisen H. (1991) Proc. Natl. Acad. Sci. USA. 88: 4481-4485.
Kendal, A.P. (1975) Virology 65: 87-99.
Lentz, M.R., Webster, R.G., and Air, G.M. (1987) Biochemistry 26: 5351-5358.
Livingston, B.D., Jacobs, J.L., Glick, M.C. and Troy, F.A. (1988) J. Biol. Chem. 263: 9443-9448.
Lowden, J.A. and O'Brien, J.S. (1979) Am J. Hum. Genet. 31: 1-18.
Meindl, P., Bodo, G. Palese, P, Schulman, J., and Tuppy, H. (1970) Monatsh. Chem. 101: 639-647.
Meindl, P. and Tuppy, H. (1973) Monatsch. Chem. 104: 402-414.
Miyagi, T. and Tsuiki, S. (1985) J. Biol. Chem. 260: 6710-6714.
Renlund, M., Tietze, T. and Gahl, W.A. (1986) Science 232: 759-762.
Roggentin, P., Rothe, B., Lottspeich, F. and Schauer, R. (1988) FEBS Lett. 238: 31-34.
Rothe, B., Roggentin, P. Frank, R.,Blocker, H., and Schauer, R. (1989) J. Gen. Microbiol. 135: 3087-3096.
Rothe, B., Rothe, B., Roggentin, P. and Schauer, R. (1991) Mol. Gen. Genet. 226: 190-197.
Saito, M. and Yu, R.K. (1992) J. Neurochem.58: 83-87.
Schengrund, C., Rosenberg, A., and Repman, M.A. (1976) J.Cell Biol. 79: 555-561.
Shaw, M.W., Lamb, R.A., Erickson B.W., Briedies, D.J., and Choppin, P.W. (1982) Proc. Natl. Acad. Sci, USA. 79: 6817-6821.
Stone, J.D. (1948) Austral. J. Exp. Bio. Med. Sci. 26: 281-298.
Tulsiani, D.R.P. and Carubelli, R. (1970) J. Biol. Chem. 245: 1821-1827.
Usuki, S., Hoops, P., and Sweeley, C.C. (1988 a) J. Biol.Chem. 10: 595-610.
Usuki, S., Lyu, S., and Sweeley, C.C. (1988 b) J. Biol. Chem. 263: 6874-6853.
Van Der Horst, T.J.G., Galjart, N.J., Azzo, A. Galjaard, H. and Verheijen, F.W. (1989) J. Biol. Chem. 264: 1317-1322.
Van Der Horst, G.T.J., Mancini, G.M.S., Brossmer, R., Rose, U and Verheijen, F.W. (1990a) J.Biol. Chem. 265: 10801-10804.
Van Der Horst, G.T.J., Mancinin, G.M.S. Brossmer, R. Rose, U., and Verheijen, F.W. (1990b) J. Biol. Chem. 265: 10801-10804.
Verheijen, F.W., Palmeri, S. and Galjaard, H.(1987) Eur. J. Biochem. 162: 63-67.
Vimr, E.R., Lawrisuk, L. Galen, J., and Kaper, J.B. (1988) J. Bacteriol. 170: 1495-1504.
Warner, T.G. and O'Brien, J.S. (1979) Biochemistry 18: 2783-2787.
Warner, T.G. and Lee, L.A. (1986) Fed. Proc.45: 1816.
Warner, T.G. (1987) Biochem. Biophys. Res. Commun. 148: 1323-1329.
Warner, T.G. and Lee, L.A. (1988) Carbohyr. Res. 176: 211-218.
Warner, T.G. and Loftin, S. K. (1989) Enzyme 42: 103-109.
Warner, T.G., Louie, A., Potier, M. and Ribeiro, A. (1991) Carbohyr. Res. 215: 315-321.
Warner, T.G., Harris, R. McDowell, R. and Vimr, E.R. (1992) Biochem J. 285: 957-964
Yeung, M.K. and Fernandez, S.R. (1991) Applied Environ. Microbiol. 57: 3062-3069.
Zbiral, E. Brandstetter, H.H., Christian, R., and Schauer, R. (1987 a) Justus Liebigs Ann. Chem. 159-165.
Zbiral, E., Bandstetter, H.H., Christian, R., and Schauer, R. (1987 b) Justus Liebigs Ann. Chem. 781-786.

Polysialic Acid
J. Roth, U. Rutishauser and F. A. Troy II (eds.)

STUDIES OF THE O-ACETYLATION AND (IN)STABILITY OF POLYSIALIC ACID

Ajit Varki* and Herman Higa**

University of California, San Diego, La Jolla, CA, USA* and Glycomed Inc., Alameda, CA, USA**

SUMMARY:
Bacterial polysaccharides that consist of polysialic acid (PSA) can be O-acetylated. Expression of this O-acetylation in K1+ E. coli correlates with the expression of a specific O-acetyltransferase activity. This enzyme uses acetyl-Coenzyme A as a donor and shows a marked preference for polymers of length >12-14. Although the enzyme is very stable in the crude state, it becomes extremely labile upon further purification. This destabilization can be reproduced by exposing the crude enzyme to a bacteriophage endosialidase specific for PSA. It appears that the enzyme requires the continuous presence of PSA for its stability.
Another interesting property of PSA is its propensity to undergo depolymerization under mildly acidic conditions. However, a search of the literature indicates that the primary linkage unit of PSA (Siaα2-8Sia, which is found also in gangliosides) must be quite stable under such conditions. This paradox is being investigated to determine if it is due to selective destabilization of the internal linkages of the polymer. Regardless of the mechanism(s) involved, this instability must be taken into account in exploring the chemistry and biology of PSA in various systems.

BACTERIAL POLYSIALIC ACIDS CAN BE O-ACETYLATED

Polysialic acid (PSA) is an unusual homopolymer of sialic acid (Sia) found in certain animal glycoproteins e.g. the Neural Cell Adhesion Molecule (N-CAM), and in the capsular polysaccharides of certain pathogenic bacteria (Troy, 1992; Rutishauser et al., 1988). In the latter case, the polymer is known to be subject to O-acetylation at the 7- or 9-positions. This occurs in some, but not all, strains of E. coli and meningococcus (Orskov et al., 1979; Arakere et al., 1991; Vann et al., 1978). O-acetylation is also found in the PSA of fish egg glycoproteins (Kitajima et al., 1988; Iwasaki et al., 1990). In the case of N-CAM, the possibility of O-acetylation has not been investigated.

Pathogenic K1+ E. coli bearing the Siaα2-8Sia PSA homopolymer can be fixed in an O-acetyl negative or an O-acetyl-positive state. Alternatively, some strains undergo a form variation between the O-acetyl-positive and O-acetyl-negative states. This form variation has been noted to be rather rapid, occurring as frequently as every 1/30 doublings (Orskov et al., 1979). We have investigated the enzymatic basis of the O-acetyl-positive state and of the O-acetyl form variation.

THE O-ACETYL POSITIVE STATE CORRELATES WITH THE EXPRESSION OF A SPECIFIC O-ACETYLTRANSFERASE ACTIVITY

A simple assay for the O-acetyltransferase was developed, that uses PSA as an acceptor, and [^{3}H]acetyl CoA as the radioactive donor molecule. Following incubation, the macromolecular product is separated from the radioactive donor molecule by gel filtration (Higa et al., 1988; Vann et al., 1978). The radioactive product was confirmed to be O-acetylated PSA by sequential digestion with endo- and exosialidases. The monomeric product that resulted proved to be a mixture of [^{3}H-acetyl]7- and 9-O-acetyl-N-acetyl-neuraminic acid (the latter being the predominant product). Significant activity of this enzyme was found in E. coli strains known to be O-acetyl positive (C375:K1,D698:K1), and not in strains known to be O-acetyl negative (016:K1,C940:K1) or K1 negative (HB101:K12,JM103:K1) (Higa et al., 1988).

THE O-ACETYL FORM VARIATION IS CORRELATED WITH FLUCTUATIONS IN THE LEVEL OF EXPRESSION OF THE O-ACETYLTRANSFERASE

By picking random clones from D698:K1(known to show the O-acetyl form variation), we followed the expression of the O-acetyl-transferase through several generations. Progeny of a colony positive for enzyme expression were mostly positive, but the occasional negative one (~1/30-40) gave mostly negative progeny, with an occasional positive one. This "flip-flop" was followed through four generations. Thus, the pattern of expression of the O-acetyltransferase indicates that it is the determinant of O-acetyl form variation. The genetic mechanism by which the form switch occurs is currently unknown.

PARTIAL PURIFICATION AND CHARACTERIZATION OF THE O-ACETYLTRANSFERASE

The enzyme activity was solubilized by Triton X-100 (0.05%), and was found to be quite stable at 4°C. The crude preparation showed significant endogenous acceptor activity, presumably arising from PSA present in the bacteria that were extracted. The activity was partially purified by chromatography on DEAE-cellulose, giving ~100-fold purification over total cellular protein. This step also freed the enzyme from almost all detectable endogenous acceptor. Although the enzyme was very stable at this step, all attempts to purify it further failed (the probable reasons for this phenomenon are discussed below). Further characterization of the enzyme was therefore done with material from the DEAE step. We found that the enzyme had a pH optimum of 7.0 - 7.5, was not dependent on divalent cations, and showed an apparent Km of 300 μM for AcCoA. Coenzyme A was inhibitory, giving half-maximal inhibition at 100 μM (Higa et al., 1988).

THE O-ACETYLTRANSFERASE PREFERENTIALLY O-ACETYLATES HIGH-MOLECULAR WEIGHT POLYSIALIC ACIDS

The apparent Km of the transferase for PSA was 3.7 mM (expressed as sialic acid concentration). However, this substrate (commercial "colominic acid") is actually a mixture of PSA fragments of varying sizes, with an average of ~15 residues. Gel filtration analysis of the radiolabeled reaction product indicated that it was rather large. To explore this further, we studied the radioactive reaction product by MONO-Q HPLC, which separates fragments of different sizes. We found that the enzyme preferentially acetylates the polymers with a DP>14. Thus, the enzyme appears to recognize some feature of PSA that is confined to very large polymers (Higa et al., 1988). This may be related to the secondary structure of PSA discussed by others in this book.

THE O-ACETYLTRANSFERASE IS STABLE IN THE CRUDE STATE, BUT EXTREMELY LABILE TO FURTHER PURIFICATION

After the DEAE purification step, the enzyme is very stable at 4°C (>5 years) and even at 60°C (>1 hour). However, all further attempts to purify the enzyme failed because it became unstable. It was puzzling that this instability became manifest upon using many different purification methods that rely on entirely different principles (e.g. cation exchange, dye-matrix affinity, Coenzyme-A affinity and gel filtration). All attempts to add-back fractions from various steps failed to regain the activity. This change in stability is best explained by the findings presented below, that indicate that the stability of the transferase is dependent upon continous association with PSA (Higa et al., 1988).

THE O-ACETYLTRANSFERASE APPEARS TO BE STABILIZED BY CONSTANT ASSOCIATION WITH POLYSIALIC ACID

We reasoned that if the O-acetyltransferase was stabilized by constant association with PSA, the activity should be destabilized by treatment with the phage endosialidase specific for PSA. This proved to be the case. If the transferase was first treated with endosialidase at 57°C, and then cooled 37°C, most of the activity was lost. This was clearly due to the activity of the endosialidase itself (and not a contaminating protease), because the effect could be completely protected against by the presence of competing PSA in the initial incubation (Higa et al., 1988). These data indicate that the transferase may be stabilized by constant association with PSA of endogenous origin. This might explain why the enzyme is stable to the DEAE purification step, in which PSA would be expected to co-elute from the column. Thus, this step would presumably result in co-purification of endogenous PSA, which can bind to this anion exchange column. On the other hand, the fact that the endosialidase could not inactivate the transferase at 37°C of this, it is somewhat surprising that activity is so easily lost during attempts at further purification at 4°C. This paradox remains unresolved, and suggests that other unknown factors may also contribute to the destabilization of the enzyme.

THE TRANSFERASE CAN O-ACETYLATE POLYSIALIC ACIDS OF EMBRYONIC N-CAM

The E. coli O-acetyltransferase was found to be able to O-acetylate the PSA chains of chicken embryonic N-CAM. All of the radioactively labeled products were found to have a molecular weight of >210 Kda, indicating that the enzyme specifically recognized the molecules with the longest length of PSA (Higa et al., 1988). This suggests that the enzyme could be used as a probe to detect such structures in a variety of systems. Since the other available probes (e.g. antibodies, endosialidase) recognize PSA of shorter lengths (8-10 residues) (Finne et al., 1985; Hallenbeck et al., 1987; Häyrinen et al., 1989; Hekmat et al., 1990; Lackie et al., 1990), the O-acetyltransferase could theoretically be used to further differentiate adult from embryonic N-CAM.

PARADOXICAL FINDINGS CONCERNING THE STABILITY OF THE Siaα2-8Sia LINKAGE OF POLYSIALIC ACIDS

During our studies of the O-acetyltransferase, we noted that PSA has a tendency to break down gradually into smaller fragments even when stored in the freezer for prolonged periods of time. This problem was particularly evident when the samples were slightly acidic. Other observations mentioned in the literature indicate that PSA is a very labile molecule. The PSA in N-CAM is easily degraded when preparing samples for SDS-PAGE gels (i.e. boiling in pH 6.5 Laemli buffer). Thus samples are usually heated at 60°C to minimize this problem. On the other hand, PSA is sometimes deliberately degraded into smaller fragments for analysis by incubation at moderately low pH (Kitazume et al., 1992). In some instances, even boiling at close to neutral pH is used to obtain partial breakdown of PSA (Troy, 1979). Paradoxically, the primary α2-8 linked Sia linkage unit of PSA is reported to be more stable than other Sia linkages, being difficult to degrade in hot sulfuric acid (Troy, 1979; Nadano et al., 1986). Similarly, the α2-8 linked Sia residues of gangliosides are notoriously difficult to release with acid hydrolysis.

Is the long PSA intrinsically more unstable than short PSA? Taken together, the evidence from the literature would suggest that this is indeed the case. We have carried out some preliminary comparisons that confirm and extend this phenomenon. Further studies are required to define the exact extent of difference, and the point at which increasing length results in decreased stability.

POSSIBLE EXPLANATIONS FOR THE INSTABILITY OF POLYSIALIC ACID

The following possibilities must be considered:

1. Polysialic acids are well known to develop secondary structure as the polymer size increases (Kabat et al., 1988; Jennings et al., 1989; Häyrinen et al., 1989; Yamasaki et al., 1991; Brisson et al., 1992) (see also other chapters in this book). It is possible that some unknown feature of this secondary structure results in destabilization of the glycosidic bond.
2. Polysialic acids are well known to develop internal lactones (Lifely et al., 1981). It is possible that formation of these five-membered rings results in destabilization of the glycosidic bond.
3. Acidic polymers can undergo changes in the pKa of internal acidic groups as they increase in length (Katchalsky, 1947). It is possible that this may favor an intra-molecular general acid catalysis involving a protonated carboxyl group and the immediately adjacent glycosidic bond (Karkas et al., 1964). If this is so, the mechanism would be similar to that of glycosidases, in which glutamic or aspartic acid residues with a high pKa are involved in catalysis (Tull et al., 1991; Rouvinen et al., 1990 ; Reddy and Maley, 1990). It is also possible that some combination of these mechanisms is involved. We are currently investigating each of these possibilities.

IS THE INSTABILITY OF PSA SUFFICIENT TO BE BIOLOGICALLY RELEVANT?

From the practical point of view, the instability of PSA at mildly acidic pH should be taken into account by all investigators designing *in* vitro studies involving these molecules. The question arises whether the instability is sufficient to be physiologically relevant in vivo. The precise pH values at which long-lasting stability is achieved need to be defined. It is quite possible that, in certain specialized situations, the extracellular pH might be low enough to be significant (e.g. in renal tissues, in infections, in hypoxic states, and in tumors). Likewise, at the pH values encountered in certain intracellular compartments (endosomes and lysosomes), PSA might be destabilized. In this regard, it is worthy of note that the following questions about N-CAM remain unanswered:

1. Is N-CAM normally internalized, or is it confined to the cell surface?
2. Is the length of the PSA on N-CAM molecules stable after synthesis?
3. Does the PSA on N-CAM turn over faster than the protein itself?
4. What is the mechanism for the switch from the embryonic to adult form of N-CAM seen in many systems? Does this always involve new synthesis of N-CAM molecules?
5. What is the mechanism for terminal degradation of the PSA of N-CAM?

The answers to these questions may require a better understanding of the instability of PSA. Our current efforts are directed at elucidating this fundamental issue.

REFERENCES

Arakere, G. and Frasch, C.E. (1991) Infect. Immun. 59: 4349-4356.

Brisson, J.-R., Baumann, H., Imberty, A., Pérez, S. and Jennings, H.J. (1992) Biochem. 31: 4996-5004.

Finne, J. and Makela, P.H. (1985) J. Biol. Chem. 260:1265-1270.

Hallenbeck, P.C., Vimr, E.R., Yu, F., Bassler, B. and Troy, F.A. (1987) J. Biol. Chem. 262: 3553-3561.

Hekmat, A., Bitter-Suermann, D. and Schachner, M. (1990) J. Comp. Neurol. 291:457-467.

Higa, H. and Varki, A. (1988) J. Biol. Chem. 263: 8872-8878.

Häyrinen, J., Bitter-Suermann, D. and Finne, J. (1989) Mol. Immunol. 26: 523-529.

Iwasaki, M., Inoue, S. and Troy, F.A. (1990) J. Biol. Chem. 265: 2596-2602.

Jennings, H.J., Gamian, A., Michon, F. and Ashton, F.E. (1989) J. Immunol. 142: 3585-3591.

Kabat, E.A., Liao, J., Osserman, E.F., Gamian, A., Michon, F. and Jennings, H.J. (1988) J. Exp. Med. 168: 699-711.

Karkas, J.D. and Chargaff, E. (1964) J. Biol. Chem. 239:949-957.

Katchalsky, A. and Spitnik, P. (1947) J. Polym. Sci. 2: 432-446.

Kitajima, K., Inoue, S., Inoue, Y. and Troy, F.A. (1988) J. Biol. Chem. 263: 18269-18276.

Kitazume, S., Kitajima, K., Inoue, S. and Inoue, Y. (1992) Anal. Biochem. 202: 25-34.

Lackie, P.M., Zuber, C. and Roth, J. (1990) Development 110: 933-947.

Lifely, M.R., Gilbert, A.S. and Moreno, C. (1981) Carbohydr. Res. 94: 193-203.

Nadano, D., Iwasaki, M., Endo, S., Kitajima, K., Inoue, S. and Inoue, Y. (1986) J. Biol. Chem. 261: 11550-11557.

Orskov, F., Orskov, I., Sutton, A., Schneerson, R., Lin, W., Egan, W., Hoff, G.E. and Robbins, J.B. (1979) J. Exp. Med. 149: 669-685.

Reddy V.A., Maley F. (1990) J. Biol. Chem. 265: 10817-20.

Rouvinen, J., Bergfors, T., Teeri, T., Knowles J.K., Jones, T.A. (1990) Science 249: 380-386.

Rutishauser, U., Acheson, A., Hall, A.K., Mann, D.M. and Sunshine, J. (1988) Science 240: 53-57.

Troy, F.A., II (1992) Glycobiology 2: 5-23.

Troy-F. A., II (1979) Annu-Rev-Microbiol. 33:519-60.

Tull, D., Withers, S.G., Gilkes, N.R., Kilburn, D.G., Warren, R.A., Aebersold, R. (1991) J. Biol. Chem. 266: 15621-5.

Vann, W.F., Liu,T.Y., Robbins, J.B. (1978) J. Bacteriol. 133: 1300-1306.

Yamasaki, R. and Bacon, B. (1991) Biochem. 30: 851-857.

Polysialic Acid
J. Roth, U. Rutishauser and F. A. Troy II (eds.)

Glycobiology of Fish Egg Polysialoglycoproteins (PSGP) and Deaminated Neuraminic Acid-Rich Glycoproteins (KDN-gp)

Yasuo Inoue

Department of Biophysics and Biochemistry, Faculty of Science
University of Tokyo, Hongo-7, Tokyo 113, Japan

Fish egg polysialoglycoproteins (PSGP), a new class of glycoproteins, were first isolated from the unfertilized eggs of rainbow trout by Inoue and Iwasaki (Inoue and Iwasaki, 1978; 1980). PSGPs were ubiquitously found in *Salmonidae* fish eggs (Shimamura *et al.*, 1983; Iwasaki and Inoue, 1985; Iwasaki *et al.*, 1985) and were the major glycoprotein components of cortical vesicles (alveoli) of these fishes (Inoue and Inoue, 1986; Inoue *et al.*, 1987).

In 1988, we showed the first example of oligo/poly(KDN)-containing glycoprotein (KDN-gp) isolated from the vitelline envelope (VE) of rainbow trout eggs (Inoue *et al.*, 1988; Kanamori *et al.*, 1990). In KDN-gp, KDN (deaminated neuraminic acid) was the only ulosonic acid found in this glycoprotein and occurred both at the nonreducing termini as KDNα2→3Galβ1→ and in α-2→8-linked oligo/poly(KDN) chains. A similar KDN-gp was also isolated from the ovarian fluid of ovulating rainbow trout (Kanamori *et al.*, 1989). In this review article, I describe the results of the structural and functional studies on these PSGPs and KDN-gp carried out in our research group and perspective of glycobiology of poly(Sia) and related biopolymers.

Occurrence and carbohydrate structure of fish egg PSGP

Fish egg PSGP contains α-2→8-linked oligo/poly(Sia) residues that are attached to *O*-linked carbohydrate chains (Nomoto *et al.*, 1982; Shimamura *et al.*, 1983; Iwasaki and Inoue, 1985; Iwasaki *et al.*, 1984a, 1984b; Kitajima *et al.*, 1984; Shimamura *et al.*, 1984):

Structures of the five different types of oligo/polysialylglycan units present in *Salmonidae* egg PSGPs (Sia = Neu5Ac or Neu5Gc)

(a) Short-core unit

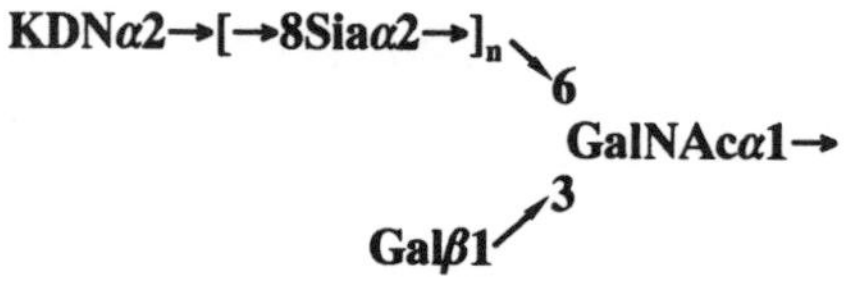

```
KDNα2→[→8Siaα2→]n↘
                   6
                   GalNAcα1→
                   3
           Galβ1↗
```

(b) Trisaccharide-core unit

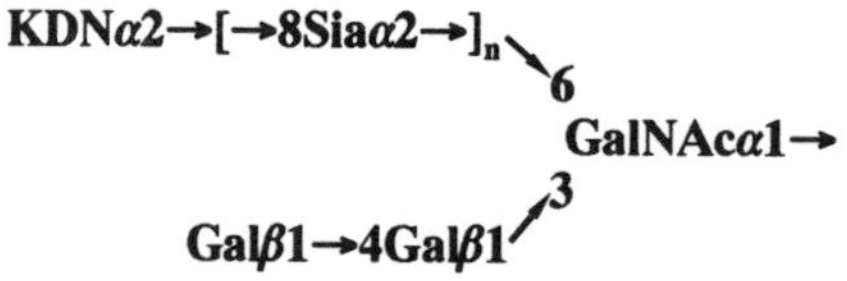

```
KDNα2→[→8Siaα2→]n↘
                   6
                   GalNAcα1→
                   3
  Galβ1→4Galβ1↗
```

(c) Tetrasaccharide-core unit

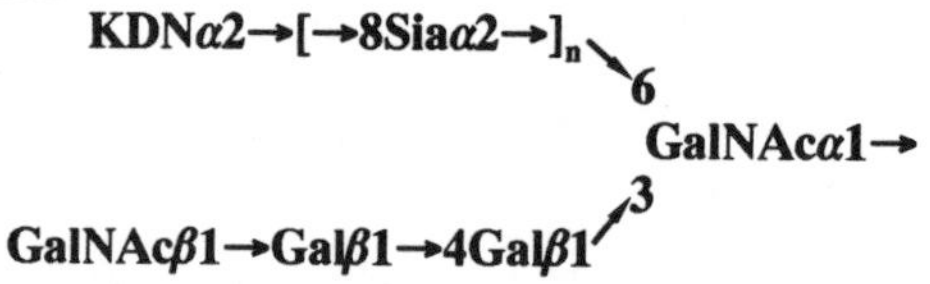

```
           KDNα2→[→8Siaα2→]n↘
                            6
                            GalNAcα1→
                            3
GalNAcβ1→Galβ1→4Galβ1↗
```

(d) Fucose-containing core unit

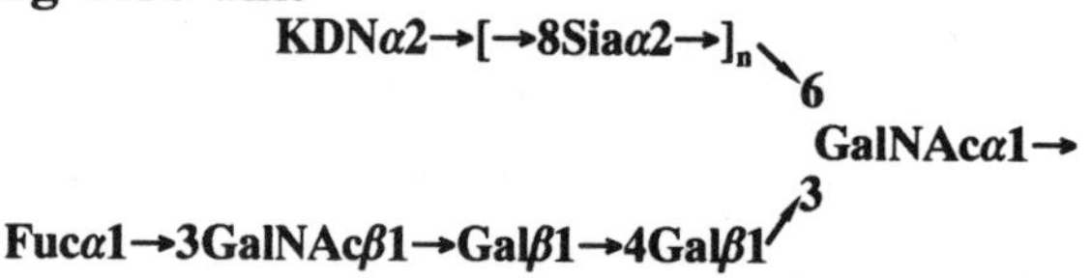

```
                   KDNα2→[→8Siaα2→]n↘
                                    6
                                    GalNAcα1→
                                    3
Fucα1→3GalNAcβ1→Galβ1→4Galβ1↗
```

(e) Long-core unit

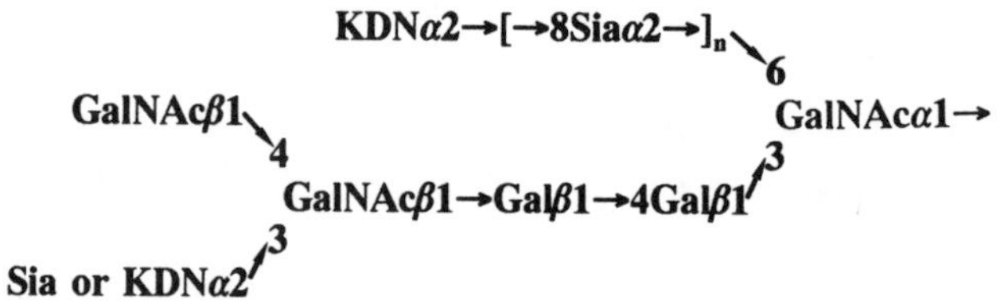

```
                     KDNα2→[→8Siaα2→]n↘
                                      6
GalNAcβ1↘                             GalNAcα1→
         4                            3
         GalNAcβ1→Galβ1→4Galβ1↗
         3
Sia or KDNα2↗
```

Sialic acid accounts for 50~60% of the mass of the glycoprotein. The presence of α-2→8-linked oligo/poly(Sia) structures was identified by methylation analysis (Inoue and Matsumura, 1979,1980), anion-exchange chromatography (Inoue and Iwasaki, 1980; Nomoto *et al.*, 1982), and ^{1}H-nmr studies (Nomoto *et al.*, 1982; Kitajima *et al.*, 1984; Shimamura *et al.*, 1984). These PSGP molecules are the major glycoprotein components in cortical vesicles (alveoli) of the eggs of 8 species (from 3 genera) of *Salmonidae* so far examined [*Oncorhynchus mykiss* (rainbow trout), *Oncorhynchus keta* (chum salmon), *Oncorhynchus masou ishikawai* (land-locked cherry salmon), *Oncorhynchus nerka adonis* (kokanee salmon), *Salmo trutta fario* (brown trout), *Salvelinus leucomaenis pluvius* (Japanese common char), *Salvelinus namaycush* (lake trout), and *Salvelinus fontinalis* (brook trout)]. A species-specific structural diversity has been revealed in poly(Sia) chains of *Salmonidae* PSGPs, while no distinct difference was found in the structures of the asialo-oligosaccharide moieties among *Salmonidae* species (Iwasaki and Inoue, 1985). For example, the PSGPs from *Oncorhynchus* (rainbow trout, chum salmon, cherry salmon, and kokanee salmon) eggs are all composed exclusively of Neu5Gc residues. In brown trout and *Salvelinus* species egg PSGPs contain both α-2→8-linked poly(Neu5Ac) and poly(Neu5Gc) or a hybrid type of oligo/polysialyl chains having both Neu5Ac and Neu5Gc residues (Sato *et al.* to be published). *O*-Acetyl substitution at C-4, C-7, and C-9 was extensive in some species (Iwasaki *et al.*, 1990) while it was not significant in other species. The most extensive modification of the sialic acid residue in fish egg PSGP was found in deaminated neuraminic acid (KDN) that occurred at the nonreducing termini of α-2→8-linked oligo/poly(Neu5Gc) chains (Iwasaki *et al.*, 1987; Nadano *et al.*, 1986). *O*-Acetylation also occurred at C-9 of KDN (Iwasaki *et al.*, 1990).

Localization and fertilization-induced depolymerization of fish egg PSGP

Four-fold evidence has been presented to establish that PSGP is a soluble component of cortical alveoli (Inoue and Inoue, 1986; Inoue *et al.*, 1987; Kitajima *et al.*, 1988b): **(i)** dramatic change in the molecular weight of PSGP (200 K → 9 K) was observed following egg activation, and the kinetics of the transformation temporally coincides with that of morphological cortical reaction; **(ii)** a fraction rich in cortical alveoli was isolated and 200-kDa PSGP was shown to be a major water-soluble component of this fraction; **(iii)** the perivitelline space fluid was isolated from the activated eggs and found to contain 9-kDa PSGP; **(iv)** cortical alveoli isolated from lake trout eggs whose PSGP has α-2→8-linked poly(Neu5Ac) chains were shown to bind with anti-poly(Neu5Ac) H.46 antiserum by the indirect immunofluorescence staining method.

Cortical vesicles are specialized secretory organelles found in the peripheral cytoplasm of mature eggs of almost all animal species including human (for review see Guraya (1982)). Upon fertilization (activation) of the egg, the cortical vesicles fuse with the plasma membrane of the egg and release their contents into the perivitelline space (cortical reaction), this process being a prerequisite for normal development of the embryos under natural conditions. In fish eggs, these vesicles are 10 times larger (2~40 μm) than those in other animals and frequently called cortical alveoli instead of cortical granules. Only limited information is available on the chemical nature of cortical vesicle contents and their functional role in fish and higher vertebrates until now. Unique structural characterization of both the carbohydrate and protein moieties of PSGP suggested that it possibly performs important roles during embryogenesis. The focal point was the dramatic depolymerization of 200–kDa PSGP into 9–kDa PSGP on fertilization or egg activation and this point was further studied.

Polyprotein nature of PSGP and the molecular mechanism of fertilization–induced depolymerization

First, we showed that 9–kDa PSGP (L–PSGP) isolated from fertilized rainbow trout eggs had the amino acid and carbohydrate compositions identical to those of 200–kDa PSGP. The peptide moiety of L–PSGP comprised single tridecapeptide whose sequence was determined as Asp–Asp–Ala–Thr*–Ser*–Glu–Ala–Ala–Thr*–Gly–Pro–Ser–Gly (* indicates glycosylation site) (Inoue and Inoue, 1986). Second, we showed that 200–kDa PSGP (H–PSGP) isolated from unfertilized rainbow trout eggs was made up of a tandem repeat of the glycotridecapeptide (Kitajima *et al.*, 1986). Thirdly, we showed the rapid depolymerization of H–PSGP following egg activation (by fertilization or parthenogenetic method) was a result of proteolysis catalyzed by a specific proteinase designated as PSGPase into the repeating unit. We studied some properties of PSGPase in an *in vitro* system and found that the enzyme is active only at concentrations of NaCl below 40 mM and the optimal temperature is about 16°C (Kitajima and Inoue, 1988). These conditions are also most suitable for egg activation and we concluded that the salt concentration may regulate the enzyme activity also *in vivo*. PSGP is most likely present together with its physiological substrate, H–PSGP, in cortical alveoli and its activity must be inhibited in the dormant egg by the high ionic strength inside the alveoli. Upon fertilization salmonid fish eggs are forced to be in an environment of low osmolarity, and water will tend to flow into the perivitelline space causing it to swell. This resulted in the estimated salt concentration of 10~20 m*M* within the perivitelline space where PSGPase translocated from the cortical alveoli can exhibit activity.

Structures of the apoPSGP from unfertilized and fertilized eggs of 4 *Salmonidae* species were compared (Kitajima *et al.*, 1988a; Yu Song *et al.*, 1990). The occurrence of tandem repeats of dodeca– or tridecapeptide was found in apo H–PSGP from all species. Their amino acid sequences were highly homologous with that of rainbow trout (see below). In all species, H–PSGP was depolymerized into the repeating unit, L–PSGP, after fertilization. Sequence analysis of L–PSGP from each species revealed the unique sequence specifity of PSGPase: proteolytic cleavages occur at the position two residues C–terminal to the Pro residue, i.e., –Pro–Ser–Xaa–Asp (Xaa = either Gly, Ser, or Asp).

Amino acid sequences of H– and L–PSGP isolated from unfertilized and fertilized eggs of salmonid fishes (* indicates O–glycosylation site)

(A) H–PSGP (<N> = about 20)

Rainbow trout: (Asp–Asp–Ala–Thr*–Ser*–*Glu*–Ala–Ala–Thr*–Gly–Pro–Ser–*Gly*)$_N$

Chum salmon: (Asp–Asp–Ala–Thr*–Ser*–z*Glu*–Ala–Ala–Thr*–Gly–Pro–Ser–*Ser*)$_N$

Cherry salmon: (Asp–Asp–Ala–Thr*–Ser*–*Glu*–Ala–Ala–Thr*–Gly–Pro–Ser–*Ser*)$_N$

Kokanee salmon: (Asp–Asp–Ala–Thr*–Ser*–*Asp*–Ala–Ala–Thr*–Gly–Pro–Ser–*Gly*)$_N$

(Asp–Asp–Ala–Thr*–Ser*–*Asp*–Ala–Ala–Thr*–Gly–Pro–*Ser*)$_N$

Iwana or Japanese common char:
(Asp–Asp–Ala–Thr*–Ser*–*Glu*–Ala–Ala–Thr*–Gly–Pro–*Ser*)$_N$

(B) L–PSGP

Rainbow trout: Asp–Asp–Ala–Thr*–Ser*–*Glu*–Ala–Ala–Thr*–Gly–Pro–Ser–*Gly*

Chum salmon: Asp–Asp–Ala–Thr*–Ser*–*Glu*–Ala–Ala–Thr*–Gly–Pro–Ser–*Ser*

Cherry salmon: Asp–Asp–Ala–Thr*–Ser*–*Glu*–Ala–Ala–Thr*–Gly–Pro–Ser–*Ser*

Kokanee salmon: Asp–Asp–Ala–Thr*–Ser*–*Asp*–Ala–Ala–Thr*–Gly–Pro–Ser–*Gly*

Asp–Ala–Thr*–Ser*–*Asp*–Ala–Ala–Thr*–Gly–Pro–Ser–Asp

Iwana or Japanese common char:
Asp–Ala–Thr*–Ser*–*Glu*–Ala–Ala–Thr*–Gly–Pro–Ser–Asp

Possible function of fish egg PSGP

General consensus on the function of cortical vesicles of animal eggs is that they are a devise to transfer the enzymes and materials prepared by maternal cells to the extracellular space (the perivitelline space) and possibly to the vitelline envelope (zona pellucida) for the protective, regulative, and other yet unknown functions necessary for the normal development of the embryo. A few of speculative functions are osmoregulation and protection of polyspermy. Increase in polymer concentration in the perivitelline space may contribute to introduce water from the environment. In the case of *Salmonidae* the eggs are spawn in fresh water and the hydrated polyanionic PSGP together with its counter ions may serve to keep ionic concentrations in the perivitelline space similar to those inside the cells. H–PSGP has a property of agglutinating sperm as do other polyanions in general. However, in fish the sperm can enter into the fertilizing egg only through the special opening called micropyle, so that they may not have any chance to contact with PSGP that is in the perivitelline space. Our careful examination failed to detect PSGP outside of the perivitelline space. We were unable to obtain any evidence to show that PSGP molecules are incorporated in the fertilized envelope.

It should be noted that in none of these speculative functions, if any, of PSGP, we can find any reasoning for such a dramatic depolymerization of H–PSGP to L–PSGP after fertilization. For polyanions such as PSGP, increase in terms of molar concentration by 20 times does not result in increase in calculated osmotic pressure because of the large contribution from the Donnan term. We again emphasize that though H–PSGP has sperm agglutinating activity L–PSGP does not. At the present stage we must consider that fish egg PSGPs are multifunctional molecules. Osmoregulation described above may be one of the important function. We should also appreciate calcium binding properties of PSGP. We have observed that substrate quantity of calcium is required in hardening of the vitelline envelope, that is transformation into the fertilization envelope. It is probable that these calcium ions are stored in cortical vesicles and transported into the perivitelline space at the time when fertilization and development occur. We have recently studied calcium binding of PSGP by equilibrium dialysis method and found that PSGP has moderate affinity and high capacity of calcium binding (Shimoda *et al.*, to be published). Apart from these protective functions classically discussed for the cortical vesicle exocytosis, recently increasing interest in regulatory function of poly(Sia) residues in NCAM–mediated cell–cell interactions makes us to expect such function of regulating cell–cell interaction and cell migration for L–PSGP, the molecule that retains in the extra embryonic space during embryogenesis.

Biosynthesis of fish egg PSGP

cDNAs coding for the apoPSGPs of rainbow trout eggs have been cloned (Sorimachi *et al.*, 1988). The expression of apoPSGP mRNA was organ- and stage-specific: it was detected only in immature ovary (6 months before ovulation), and not in the mature eggs, nor in any other organs. Nucleotide sequence analysis showed that apoPSGP mRNA contains tandem repeats, each composed of 39-base encoding a tridecapeptide and that the sequences of the repeating units are completely conserved at the nucleotide level. Multiple mRNA species are present that are transcribed from multiple genes for apoPSGP having diverged numbers of the repetitive sequence.

Recently, we initiated biosynthetic studies of poly(Sia) residues of rainbow trout PSGP. Two distinct sialyltransferases, CMP-Neu5Ac (or Neu5Gc):α-*N*-acetylgalactoside α-2→6-sialyltransferase (α2,6-ST) that catalyzes the initial transfer of Sia to the C-6 of the proximal GalNAc residues on asialo PSGP and CMP-Neu5Ac (or Neu5Gc):α-2→8-sialosyl sialyltransferase (α2,8-ST) that is responsible for polyα-2→8-sialylation have been identified. Interestingly, both α2,6- and α2,8-ST activities are found in cortical alveoli suggesting that extensive sialylation occurs in these vesicles (Kitazume *et al.*, 1991).

Fish egg PSGP is a member of hyosophorin--cortical alveolar glycopolyproteins

Since cortical vesicles and their exocytosis at fertilization are generally found in wide range of animal eggs, our study was directed toward the search for PSGP in eggs of animals other than *Salmonidae*. Initial attempts were focussed on the eggs of various fish species and we found that in fish other than *Salmonidae*, PSGP was not the major cortical alveolus glycoprotein component. Each species of fish had homologous cortical alveolus glycopolyprotein which we designated as "hyosophorin", hyosoho (Japanese) means cortical alveoli (Kitajima *et al.*, 1989; Seko *et al.*, 1989). Hyosophorin is a family of glycoproteins that would satisfy the following requirements: **(i)** it is a component of cortical vesicles; **(ii)** it has a high carbohydrate content (80~90%, w/w); **(iii)** apo-hyosophorin comprises tandem-repetitions of the identical peptide sequence; **(iv)** it is completely cleaved into repeating units when a resting egg cell begins to develop in response to stimulus (sperm fusion or parthenogenetic activation). We have isolated hyosophorins from the eggs of *Cyprinus carpio* (common carp), *Tribolodon hakonensis* (a dace), *Plecoglossus altivelis* (a kind of fresh-water trout), *Oryzias latipes* (medaka), and *Paralichthys olivaceus* (flounder); the first four species are fresh water species while flounder is a species spawning in marine water.

Strikingly, in none of the major form of these hyosophorins isolated from non-salmonid fishes were present poly(Sia) residues. The glycan chains of these hyosophorins are

N–glycosidically linked. Single glycan chain is attached to the repeating unit in contrast to PSGP in which three *O*–glycosidic glycan chains are linked to each repeating peptide unit. Despite such great difference in carbohydrate structure, the tandem repeat structures made up of 7~9 kDa glycopeptide units and the fertilization–induced rapid depolymerization into the repeating unit may provide firm evidence that these hyosophorins and PSGP share common biological functions. For example, the amino acid sequence of medaka L–hyosophorin isolated from the fertilized eggs has been determined to be Asp–Ala–Ala–Ser–Asn*–Gln–Thr–Val–Ser (* indicates the glycosylation site), and the structure of H–hyosophorin isolated from the unfertilized eggs of this fish species has been established as (Asp–Ala–Ala–Ser–Asn*–Gln–Thr–Val–Ser)$_N$, where N is a diverged number of repeat units and 12 for the major molecular species (Kitajima *et al.*, 1989). The *N*–glycan chains attached to these hyosophorins are large (>7,000) and multiantennary, and the determination of their complete structures is difficult in general (papers in preparation). Structures of peripheral region of such multiantennary glycans of hyosophorins show species–specific features. One of the interesting findings in this regard is that hyosophorins isolated from fishes that spawn in fresh water are highly acidic molecules and their anionic properties are ascribed to the presence of sialic acid residues and/or sulfate groups linked to the carbohydrate units. In anion exchange chromatography, these hyosophorins behave similarly to PSGP, i.e, a hyosophorin from *Salmonidae*. By sharp contrast, the major hyosophorin molecules from flounder, fish spawning in marine water, had only neutral carbohydrate chains containing no sialic acid nor sulfate and the molecules show nonpolyanionic nature (Seko *et al.*, 1989). It is thus quite possible that hyosophorins may be involved in such functions as transport of calcium from the egg cell to the perivitelline space or maintenance of the salt–water balance for developing embryos.

Occurrence of KDN–rich glycoproteins containing oligo/poly(KDN) in fish

An unusual family of glycoproteins containing ~50%(w/w) KDN and no *N*–acylneuraminic acid was first isolated in 1988 by Inoue *et al.* from the vitelline envelope of rainbow trout eggs and designated as KDN–gp. KDN–gp contained ~15% protein and ~85% carbohydrate which was *O*–glycosidically linked to Thr/Ser. The molar ratio of carbohydrate component was Gal/GalNAc/KDN = 1:2:~5. The amino acid composition of KDN–gp was also unusual. Thr and Ala residues accounted for 40 and 27 residue %, respectively, of apo–KDN–gp. The percentages of aromatic amino acid residues were less than 1. KDN–gp was polydisperse glycoproteins whose molecular weights were estimated to be 700~4,000 K (the peak ~3,000 K) by gel chromatography.

Similar KDN–gp was also found to be a major glycoprotein component of the ovarian fluid (coelomic fluid of ovulating female) of rainbow trout and referred to as KDN–gp–OF. Although small differences were noted in amino acid compositions and the range of molecular weights, KDN–gp–VE (from the vitelline envelope) and KDN–gp–OF are considered to belong to the same family. The biosynthetic and functional relationships between KDN–gp–VE and KDN–gp–OF are yet unknown. The amounts of KDN–gp–VE and KDN–gp–OF were estimated to be ~10 mg/100 g eggs and ~1.2 mg/ml of ovarian fluid, respectively.

Structures of oligo/poly(KDN)–linked carbohydrate chains of KDN–gp

A 3,000 kDa molecule of KDN–gp consists of 500 kDa polypeptide to which 1,300 carbohydrate chains of 2 kDa (in average) are *O*–glycosidically linked. A series of acidic oligosaccharide alditols were released by alkaline borohydride treatment of KDN–gp. By analyzing these oligosaccharide alditols, we have shown that the carbohydrate chains of KDN–gp have a common core trisaccharide Galβ1→3GalNAcα1→3GalNAc in which the terminal Gal residue is blocked by a single residue of KDN and the proximal GalNAc residue is linked to α–2→8–linked oligo/poly(KDN) chains with different degree of polymerization (Kanamori *et al.*, 1990). Several lines of evidence for the presence of oligo(→8KDNα2→) chains (DP 2~6) was obtained through **(a)** methylation analysis (Inoue, S., unpublished), **(b)** analysis of the products of periodate oxidation, **(c)** the formation of KDNα2→8KDN and KDNα2→8KDNα2→8KDN after treatment of KDN–gp with mild acid, and **(d)** 400–MHz ^{1}H–nmr spectral analysis (Kanamori *et al.*, 1990).

Oligo/poly(KDN)–containing glycan unit

KDNα2→[→8KDNα2→]$_n$↘6
GalNAcα1→Thr/Ser
KDNα2→3Galβ1→4GalNAcα1↗3

Biosynthesis and possible function of KDN–gps

KDN–gp was detected in rainbow trout ovary 3 months prior to ovulation but not in the ovaries of earlier stages (Inoue, S., unpublished). Thus KDN–gp appeared to be synthesized in the ovary during relatively late stages of oogenesis. Although the cell types that synthesize KDN–gp have not been defined, it is most likely synthesized under hormonal control in some cells surrounding the oocytes, secreted and partly incorporated in the outer layer (2nd layer from outside) of the vitelline envelope just before and at the time of ovulation. KDN–gp may thus be considered as the molecule homologous to oviduct glycoproteins of mammals that are reported to be secreted and partly incorporated in the egg (Gandolfi *et al.*, 1989; Oikawa *et al.*, 1988; Wegner and Killian, 1991). The function of KDN–gp is yet to be clarified. Their presence in the ovarian fluid, viscous fluid surrounding the ovulated eggs before and during spawning, and more directly associated with the vitelline envelope may suggest a role of facilitating fertilization. Our pilot experiment showed that KDN–gp has specific interaction with homologous sperm (Inoue, S., to be published).

REFERENCES

Gandolfi, F., Brevini, Th., Richardson, L., Brown, C.R., and Moor, R.M.(1989) *Development* **106**, 303–312

Guraya, S.S.(1982) *Int. Rev. Cytol.* **78**, 257–358

Inoue, S. and Inoue, Y.(1986) *J. Biol. Chem.* **261**, 5256–5261

Inoue, S. and Iwasaki, M.(1978) *Biochem. Biophys. Res. Commun.* **83**, 1018–1023

Inoue, S. and Iwasaki, M.(1980) *Biochem. Biophys. Res. Commun.* **93**, 162–165

Inoue, S., Kanamori, A., Kitajima, K., and Inoue, Y.(1988) *Biochem. Biophys. Res. Commun.* **153**, 172–176

Inoue, S., Kitajima, K., and Inoue, Y.(1987) *Dev. Biol.* **123**, 442–454

Inoue, S. and Matsumura, G.(1979) *Carbohydr. Res.* **74**, 361–368

Inoue, S. and Matsumura, G.(1980) *FEBS Lett.* **121**, 33–36

Iwasaki, M. and Inoue, S.(1985) *Glycoconjugate J.* **2**, 209–228

Iwasaki, M., Inoue, S., Kitajima, K., Nomoto, H., and Inoue, Y.(1984a) *Biochemistry* **23**, 305–310

Iwasaki, M., Inoue, S., Nadano, D., and Inoue, Y.(1987) *Biochemistry* **26**, 1452–1457

Iwasaki, M., Inoue, S., Tazawa, I., and Inoue, Y.(1985) *Glycoconjugates* 51–52

Iwasaki, M., Inoue, S., and Troy, F.A.(1990) *J. Biol. Chem.* **265**, 2596–2602

Iwasaki, M., Nomoto, H., Kitajima, K., Inoue, S., and Inoue, Y.(1984b) *Biochem. Int.* **8**, 573–579

Kanamori, A., Inoue, S., Iwasaki, M., Kitajima, K., Kawai, G., Yokoyama, S., and Inoue, Y.(1990) *J. Biol. Chem.* **265**, 21811–21819

Kanamori, A., Kitajima, K., Inoue, S., and Inoue, Y.(1989) *Biochem. Biophys. Res. Commun.* **164**, 744–749

Kitajima, K., and Inoue, S.(1988) *Dev. Biol.* **129**, 270–274

Kitajima, K., Inoue, S., and Inoue, Y.(1989) *Dev. Biol.* **132**, 544–553

Kitajima, K., Inoue, S., Inoue, Y., and Troy, F.A.(1988b) *J. Biol. Chem.* **263**, 18269–18276

Kitajima, K., Inoue, Y., and Inoue, S.(1986) *J. Biol. Chem.* **261**, 5262–5269

Kitajima, K., Nomoto, H., Inoue, Y., Iwasaki, M., and Inoue, S.(1984) *Biochemistry* **23**, 310–316

Kitajima, K., Sorimachi, H., Inoue, S., and Inoue, Y.(1988a) *Biochemistry* **27**, 7141–7145

Kitazume, S., Kitajima, K., Inoue, S., and Inoue, Y.(1991) *Glycoconjugate J.* **8**, 265

Nadano, D., Iwasaki, M., Endo, S., Kitajima, K., Inoue, S., and Inoue, Y.(1986) *J. Biol. Chem.* **261**, 11550–11557

Nomoto, H., Iwasaki, M., Endo, T., Inoue, S., Inoue, Y., and Matsumura, G.(1982) *Arch. Biochem. Biophys.* **218**, 335–341

Seko, A., Kitajima, K., Iwasaki, M., Inoue, S., and Inoue, Y.(1989) *J. Biol. Chem.* **264**, 15922–15929

Shimamura, M., Endo, T., Inoue, Y., and Inoue, S.(1983) *Biochemistry* **22**, 959–963

Shimamura, M., Endo, T., Inoue, Y., Inoue, S., and Kambara, H.(1984) *Biochemistry* **23**, 317–322

Sorimachi, H., Emori, Y., Kawasaki, H., Kitajima, K., Inoue, S., Suzuki, K., and Inoue, Y.(1988) *J. Biol. Chem.* **263**, 17678–17684

Oikawa, T., Sendai, Y., Kurata, S.J., and Yanagimachi, R.(1988) *Gamete Res.* **19**, 113–122

Wegner, C.C. and Killian, G.J.(1991) *Molec. Reproduc. Develop.* **29**, 77–84

Yu Song, Kitajima, K., and Inoue, Y.(1990) *Arch. Biochem. Biophys.* **283**, 167–172

Polysialic Acid
J. Roth, U. Rutishauser and F. A. Troy II (eds.)

Analytical Methods for Identifying and Quantitating Poly(α2→8Sia) Structures Containing Neu5Ac, Neu5Gc, and KDN

Sadako Inoue¶, Mariko Iwasaki¶, Ken Kitajima§, Akiko Kanamori§, Shinobu Kitazume§, and Yasuo Inoue§

¶School of Pharmaceutical Sciences, Showa University, Hatanodai-1, Tokyo 142 and the §Department of Biophysics and Biochemistry, Faculty of Science, University of Tokyo, Hongo-7, Tokyo 113, Japan

In 1978 Inoue and Iwasaki reported isolation from rainbow trout eggs of a novel highly sialylated (>50% w/w) *O*-glycan type glycoprotein. The sialic acid residues were exclusively Neu5Gc and the presence of Neu5Gcα2→8Neu5Gc structure was identified by methylation analysis and exosialidase digestion (Inoue and Iwasaki, 1978). The conclusive evidence for the poly(α2→8Neu5Gc) structure having DP 2 to 24 was presented by the same authors in 1980 by anion-exchange chromatographic resolution on DEAE-Sephadex A-25 of the (8Neu5Gcα 2→)$_n$-bearing oligosaccharide alditols released by alkali-borohydride treatment of the parent polysialylated glycoprotein (Inoue and Iwasaki, 1980). Ubiquitous occurrence of polysialoglycoproteins (PSGP) as cortical vesicular components of salmonid fish eggs (and other fish species in much lesser amounts) has since been reported (Shimamura *et al.*, 1984; Iwasaki *et al.*, 1985; Iwasaki and Inoue, 1985). Our studies based mainly on chemical methods have revealed extensive structural variation in the type of sialic acid residues that constitute poly(α2→8Sia) chains of fish egg PSGP. These include the difference in *N*-acyl groups (Neu5Ac or Neu5Gc) (Kitajima *et al.*, 1988) and the presence of *O*-acetyl substitution (Iwasaki *et al.*, 1990).

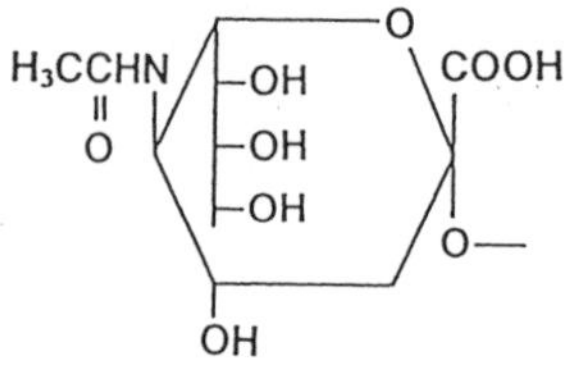

Neu5Ac 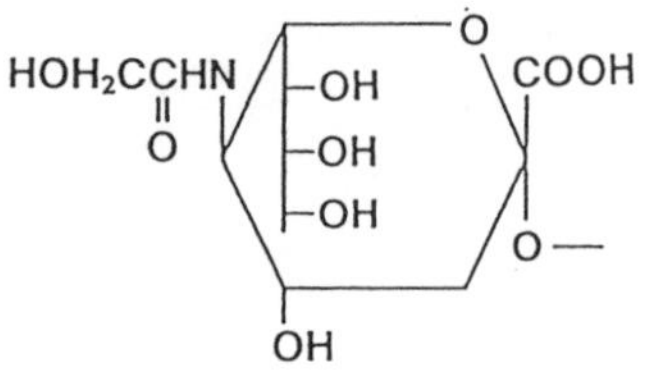 *Neu5Gc*

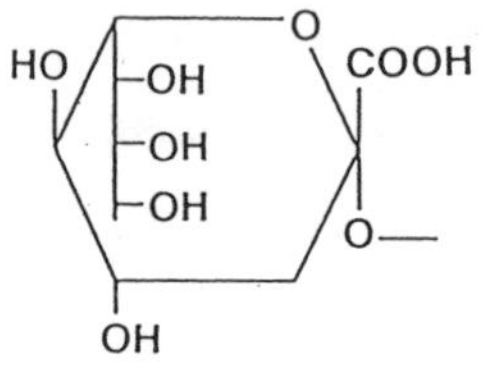

KDN

Fig. 1. Structures of Neu5Ac, Neu5Gc, and KDN residues

The most extensive modification of the sialic acid residue in poly(Sia) chains may be the deaminated form of neuraminic acid (3-deoxy-D-*glycero*-D-*galacto*-nonulosonic acid, KDN; see Fig. 1) that we found in poly(α2→8Neu5Gc) chains of rainbow trout PSGP in 1986 (Nadano *et al.*, 1986). Furthermore, a new type of KDN-rich glycoprotein containing poly(α2→8KDN) chains was isolated first from the vitelline envelope of the eggs (Inoue *et al.*, 1988) and later in the ovarian fluid of rainbow trout (Kanamori *et al.*, 1989). Since KDN is structurally and biosynthetically considered as a sialic acid analogue, poly(α2→8KDN) can be regarded as a type of poly(Sia) chains.

In mammalian systems, only one type of poly(Sia) structure, i.e. poly(α2→8Neu5Ac) has so far been reported (Troy, 1992) since the first report by Finne (Finne, *et al.*, 1982). This is a logical consequence of the methodology. Except for the earlier work by Finne, detection of poly(Sia) residues in mammalian systems are relied primarily on monoclonal or polyclonal antibodies that specifically recognize α2→8-linked homopolymers of Neu5Ac. Moreover, these antibodies recognize higher oligomers (DP>8) of poly(α2→8Neu5Ac). Therefore, studies using only antibodies as the probe fail to detect other types of poly(oligo)sialylated molecules which might have biological functions. There may be no reason to consider that structural variation in poly(Sia) chains is a phenomenon observed only in fish.

Development of highly sensitive and specific analytical methods may reveal structural diversity of poly(Sia) chains in mammalian systems, particularly in transformed tissues or cells. Preparation of monoclonal antibodies specific for each different type of poly(Sia) chains may be the goal of this project. As the fundamental approach to such goal, major purpose of this report is to summarize the analytical methods we used in identifying and quantitating a variety of nonulosonate residues that occur in glycoconjugates with special emphasis on the newly found sialic acid analogue, KDN. Some preliminary accounts on poly- and monoclonal antibodies that recognize oligo(α2→8KDN) chains are also presented.

Detection and determination of KDN in polynonulosonates by color reaction (Kitajima et al., 1992).- Both free and bound KDN exhibit color in the TBA reaction without prior hydrolysis (hereafter referred to as the direct TBA reaction). The chromophore formed from KDN has the same absorption maximum at 552 nm in the mixed solvent of water and 2-methoxyethanol as exhibited by Neu5Ac and Neu5Gc with the molar extinction coefficients ±SD, 9.1 ± 0.4 (n=7) (× 10^{-7} cm^2/mol) [cf. Neu5Ac, 8.5 ± 0.2 (n=11), Neu5Gc, 4.9 ± 0.3(n=5)].

In contrast to Neu5Ac or Neu5Gc which does not show the TBA reaction in bound forms, bound KDN was found to produce the TBA chromogen by periodate oxidation. The color yield obtained from oligo(α2→8KDN) chains by the direct TBA reaction was not less than 80% of the color yield obtained after mild acid hydrolysis of oligo(α2→8KDN). Since no other method is available to estimate the concentration of oligo(α2→8KDN) without prior hydrolysis, estimation by the direct TBA method is sufficient for ordinary purpose. A 50 μg quantity of KDN monomer gives no color in the resorcinol reaction or the orcinol reaction, thus it is clearly differentiated from Neu5Ac or Neu5Gc.

Gas chromatographic analysis of polynonulosonates containing Neu5Ac, Neu5Gc, and KDN (Kitajima et al., 1992).- For GLC analysis, each residue of Neu5Ac, Neu5Gc, and KDN is first converted to the respective methyl ester methyl ketosides (α and β forms) by methanolysis, and derivatized to the trimethylsilyl (TMS) ethers. The TMS derivatives are analyzed using either 1.5% OV-17 or 2% OV-101 on Chromosorb W or Gas Chrom Q (3 mm × 1 m) with temperature gradient from 130 to 230°C. The peak of the KDN derivative appeared before the peak of the Neu5Ac derivative. We studied the conditions to obtain the maximum yield of nonulosonate monomers from the polynonulosonates without cleaving *N*-acyl groups. The conditions recommended previously (Yu and Ledeen, 1970) for the differential quantitation of terminal sialic acid (0.05 *N* methanolic HCl, 80°C, 1 h) were found to cleave only 3 and 16% of the ketosidic linkages from colominic acid and rainbow trout PSGP, respectively. Elongation of the methanolysis time up to 4 h increased these values to 7 and 30%, respectively, but still insufficient for quantitative analysis. Our recommending conditions for cleaving the ketosidic linkages in poly(α2→8Sia) residues are to subject them first to mild hydrolysis (0.1 *N* TFA, 80°C, 3h) and then to mild methanolysis (0.05 *N* methanolic HCl, 80°C, 1 h). In this procedure, mild hydrolysis is effective in cleaving ketosidic linkages while keeping *N*-acyl groups uncleaved, and subsequent mild methanolysis is necessary to convert liberated sialic acid monomers to the methyl ester methyl ketosides. We found under these conditions of mild hydrolysis–mild methanolysis (MH–MM) 80% of sialic acid residues in fish egg PSGP was converted to the monomers. In the differential determination of individual sialic acid residues in poly(α2→8Sia) containing both Neu5Ac and Neu5Gc, MH–MM method is obviously superior to conventional MM method.

In contrast to Neu5Ac and Neu5Gc, the KDN residues in poly/oligo(α2→8KDN) are liberated in high yield (98%) by mild methanolysis (0.05 *N* HCl–methanol, 80°C) if a longer methanolysis time (3 h) is employed, whereas the value obtained by 1 h methanolysis was 60% of that obtained by MH–MM method. High yield (94%) of KDN monomer was also obtained under the standard conditions of methanolysis (0.5 *N* HCl–methanol, 65°C, 16 h) generally used in carbohydrate composition analysis of glycoproteins. These results are expected from the differences between *N*-acylneuraminic acids and KDN, namely the difference in the hydrolysis rate of ketosidic bonds and the difference in the stability of the liberated monomers in the acid: KDN (devoid of *N*-acyl group) is relatively stable.

Partial acid hydrolysis of poly(α2→8Sia) chains and detection of the liberated oligo(Sia) chains (Kitajima et al., 1988; Kitazume et al., 1992).– The presence of poly/oligo(α2→8Sia) chains in glycoconjugates may be easily detected by analyzing oligo(α2→8Sia) chains formed by partial acid hydrolysis, if sufficient amount of material is available. We studied conditions of partial hydrolysis of poly(α2→8Sia) chains that liberate sufficient amount of oligomers and negligible monomers, and found that incubation of fish PSGP for 3 days in sodium acetate buffer (pH 4.8) at 37°C resulted in 60% release of sialic acid from the core oligosaccharide chains and 90% of the released sialic acid was recovered as oligosialic acids. Milder conditions (pH 5.6, 37°C, 24 h or pH 5.6, 80°C, 1–3 h) increased the recovery of higher oligomers both from fish PSGP and colominic acid. Even milder conditions (pH 6.0) were suitable for oligo(α2→8)KDN. Thin-layer chromatography (TLC) and high performance liquid chromatography (HPLC) have been used for the analysis of the products. While TLC on silica gel plate in solvent, 1-propanol/25% aq.ammonia/water (6 : 1 : 2.5) was suited for differential separation of $(8\text{Neu5Ac}\alpha2\rightarrow)_n$, $(8\text{Neu5Gc}\alpha2\rightarrow)_n$, and $(8\text{KDN}\alpha2\rightarrow)_n$ according to DP in each series as well as according to the difference in substituent at C–5, this method requires much amount of the material because of the detection limit of the spray reagents: 1 μg of Neu5Ac or Neu5Gc by the resorcinol reaction and 2.5 μg of KDN by 5% H_2SO_4 in ethanol.

We have found that adsorption–partition chromatography with HPLC system was effective in separating (8Neu5Acα2→)$_n$, (8Neu5Gcα2→)$_n$, and (8KDNα2→)$_n$ ($n<7$) according not only to the chain length but also to the nature of the substituent at C–5. We used Shodex RS pak (150 × 6 mm i.d.) DC–613 (polystyrene cation–exchange resin gel) and eluted isocratically with the 1:2 to 1:3 mixture of 0.02–0.025 *M* sodium phosphate buffer (pH 7.4) and acetonitrile at 35°C. The elution pattern was monitored by UV absorption below 220 nm. 10 ~ 100 ng of Neu5Ac and Neu5Gc and 100 ng ~ 1 μg of KDN were detectable. The *O*–acetyl derivatives were separated from the parent oligo(α2→8Sia) also by this HPLC system. The HPLC method is superior to TLC by the lower limit of detection but the detection lacks specificity.

Anion–exchange chromatographic separation of oligo(α2→8Sia) chains have been used both in preparative and microanalytical scales. Anion–exchange HPLC of ^{3}H–labeled alditols derived from oligo(α2→8Sia) may be the most sensitive chemical method for analysis of oligo(α2→8Sia) formed by mild acid or endosialidase treatment of poly(α2→8Sia) chains.

Table I Reactivity and Susceptibility of Different Types of PSA toward Available Immunochemical and Enzymatic Probes

Polysialic Acids	Endo–N substrate	H.46 antigen	Anti–(KDN–gp), RM antigen
Poly(α2→8Neu5Ac)	+	+	–
Poly(α2→8Neu5Ac) +OAc	–(?)	–	n.d.
Poly(α2→8Neu5Gc)	+	–	–
Poly(α2→8Neu5Gc) +OAc	–	–	n.d.
Poly(α2→8Neu5Ac, Neu5Gc)	+	–	n.d.
Poly(α2→8KDN)	–	–	+
Poly(α2→8KDN) +OAc	–	–	?

n.d. = not determined.

Though anion–exchange chromatography on various types of exchangers is widely used in the research of poly(α2→8Sia), the separation is only based on the chain length and no information on the type of *N*–acylneuraminic acid or the presence of *O*–acetyl groups is available. In this regard, we found the elution of oligo(α2→8KDN) is retarded as compared with oligo(α2→8Neu5Ac) or oligo(α2→8Neu5Gc) of the same DP's.

Specificities of equine antipolysialyl antibody (H.46) and bacteriophage-derived endosialidase. – H.46 is equine antiserum containing polyclonal antibodies against α2→8-linked poly(Neu5Ac) prepared by Dr. J.B. Robbins (NIH, U.S.A.). We examined immunospecificity of H.46 toward fish egg PSGPs (a) that contain exclusively poly(α2→8Neu5Gc) chains, (b) that constitute a mixture of molecules having only poly(α2→8Neu5Ac) and poly(α2→8Neu5Gc) chains, and (c) that contain poly(α2→8Neu5Ac, Neu5Gc) chains. H.46 was found to react only with poly(α2→8Neu5Ac) chains but not with poly(α2→8Neu5Gc).

In contrast to this immunospecificity, we found endo-α2→8-sialidase (Endo-N) derived from bacteriophage K1F (kindly donated by Dr. F.A. Troy, UC Davis, U.S.A.) can depolymerize the poly(α2→8Neu5Gc) chains of fish egg PSGP. The major product of Endo-N digestion was $(8Sia\alpha2\rightarrow)_3$ from both poly(α2→8Neu5Gc) and poly(α2→8Neu5Ac). Endo-N failed to depolymerize oligo(α2→8KDN) chains.

These results indicate that while H.46 is a probe specific for poly(α2→8Neu5Ac), Endo-N can be used as a reagent for detecting or degrading both types of poly(Sia) chains (**Table I**).

Preparation and properties of antibodies recognizing poly(α2→8KDN). – We prepared rabbit antiserum (RM) raised against KDN-gp. The antibodies belonging to IgG were found to recognize specifically oligo(α2→8KDN) but not oligo(α2→8Neu5Ac) or oligo(α2→ 8Neu5Gc) (Kanamori *et al.*, 1991) (**Table I**). Since RM also recognizes KDNα2→3Gal, present in KDN-gp and KDN-ganglioside, e.g. (KDN)G_{M3}, preparation of monoclonal antibody specific to poly/oligo(α2→8KDN) is desired, and the work is underway. RM was used in immunochemical localization of KDN-gp in the second layer of rainbow trout vitelline envelope.

ACKNOWLEDGEMENTS

This research was supported in part by Grants-in-Aid for Developmental Scientific Research (02558014), and for International Scientific Research Program: Joint Research (04044055) from the Ministry of Education, Science, and Culture of Japan, by a fund from the Mitsubishi Foundation to YI.

Abbreviations used: KDN, 2-keto-3-deoxy-D-*glycero*-D-*galacto*-nononic acid; KDN-gp, poly/oligo(α2→8KDN)-containing glycoprotein; PSGP, polysialoglycoprotein; Neu5Ac, *N*-acetylneuraminic acid; Neu5Gc, *N*-glycolylneuraminic acid; GLC, gas-liquid chromatography; TLC, thin-layer chromatography; HPLC, high performance liquid chromatography; TBA, thiobarbituric acid.

REFERENCES

Finne, J.(1982) *J. Biol. Chem.* **257**, 11966-11970

Inoue, S., and Iwasaki, M.(1978) *Biochem. Biophys. Res. Commun.* **83**, 1018-1023

Inoue, S., and Iwasaki, M.(1980) *Biochem. Biophys. Res. Commun.* **93**, 162-165

Inoue, S., Kanamori, A., Kitajima, K., and Inoue, Y. (1988) *Biochem. Biophys. Res. Commun.* **153**, 172-176

Iwasaki, M., Inoue, S., Tazawa, I., and Inoue, Y.(1985) in *Glycoconjugates* (Davidson, E.A., Williams, J.C., and Di-Ferrante, N.M., eds.), pp. 51-52, Praeger, New York

Iwasaki, M., and Inoue, S.(1985) *Glycoconjugate J.* **2**, 209-228

Iwasaki, M., Inoue, S., and Troy, F.A.(1990) *J. Biol. Chem.* **265**, 2569-2602

Kanamori, A., Kitajima, K., Inoue, S., and Inoue, Y. (1989) *Biochem. Biophys. Res. Commun.* **164**, 744-749

Kanamori, A., Kitajima, K., Inoue, Y., and Inoue, S.(1991) *Glycoconjugate J.* **8**, 222

Kitajima, K., Inoue, S., Inoue, Y., and Troy, F.A.(1988) *J. Biol. Chem.* **263**, 18269-18276

Kitajima, K., Inoue, S., Kitazume, S., and Inoue, Y.(1992) *Analytical Biochem.* **205**, in press

Kitazume, S., Kitajima, K., Inoue, S., and Inoue, Y.(1992) *Analytical Biochem.* **202**, 25-34

Nadano, D., Iwasaki, M., Endo, S., Kitajima, K., Inoue, S., and Inoue, Y.(1986) *J. Biol. Chem.* **261**, 11550-11557

Shimamura, M., Endo, T., Inoue, Y., Inoue, S., and Kambara,H.(1984) *Biochemistry* **23**, 317-322

Troy, F.A.(1992) *Glycobiology* **2**, 5-23

Yu, R.K., and Ledeen, R.W.(1970) *J. Lipid Res.* **11**, 506-516

Polysialic Acid
J. Roth, U. Rutishauser and F. A. Troy II (eds.)

CMP-3-Deoxynonulosonate Synthetase Which Catalyzes the Transfer of CMP to KDN from CTP

Partial Purification and Characterization of the Enzyme from Rainbow Trout Testis.

Takaho Terada¶, Shinobu Kitazume¶, Ken Kitajima¶, Sadako Inoue§, Fumio Ito‖, Frederic A. Troy‖, and Yasuo Inoue¶

¶Department of Biophysics and Biochemistry, Faculty of Science, University of Tokyo, Hongo-7, Tokyo 113, Japan, the §School of Pharmaceutical Sciences, Showa University, Hatanodai-1, Tokyo 142, Japan, and the ‖Department of Biological Chemistry, University of California, School of Medicine, Davis, California 95616

SUMMARY

The sugar nucleotide, cytidine 5'-(3-deoxy-D-*glycero*-D-*galacto*-2-nonulosonic phosphate) (CMP-KDN) is expected to serve as a donor of KDN residues in the synthesis of KDN-containing glycoconjugates. We report here the identification and characterization of CMP-KDN synthetase, a novel enzyme responsible for synthesis of CMP-KDN from KDN and CTP. The enzyme was partially purified from the testis of rainbow trout (*Oncorhynchus mykiss*), where KDN-gangliosides were first discovered (Yu Song, Kitajima, K., Inoue, S., and Inoue, Y.(1991) *J.Biol.Chem.* **266**, 21929-21935), and used to synthesize CMP-[^{14}C]KDN, which was characterized by ^{1}H NMR. V_{max}/K_m studies showed that KDN was a preferred nonulosonic acid substrate compared to Neu5Ac or Neu5Gc. In contrast, Neu5Ac and Neu5Gc were the preferred nonulosonic acid substrates for the calf brain CMP-sialic acid synthetase. The presence of either Mg^{2+} or Mn^{2+} is essential for CMP-KDN synthetase activity. Kinetic and substrate specificity studies also showed that the trout testis enzymes could synthesize activated sugar nucleotides required for synthesis of both (KDN)G_{M3} and (Neu5Ac)G_{M3}. The expression of CMP-KDN synthetase was shown to be temporally correlated with development of sperm.

INTRODUCTION

KDN (3-deoxy-D-*glycero*-D-*galacto*-2-nonulosonic acid), a unique deaminated analogue of sialic acid was first reported as naturally occurring sialic acid in fish egg polysialoglycoproteins (PSGPs) in 1986 (Nadano *et al.*, 1986). Since then, an increasing number of KDN-containing glycoconjugates have been reported (Iwasaki *et al.*, 1987, 1990; Inoue *et al.*, 1988; Kanamori *et al.*, 1989, 1990; Yu Song *et al.*, 1991; Knierl *et al.*, 1989; Strecker *et al.*, 1989). Possible physiological roles of these KDN-glycoconjugates have also been reported recently (Kanamori *et al.*, 1990; Yu Song *et al.*, 1991; Troy, 1992). A logical step in the investigation of KDN-glycoconjugates is to determine how the KDN residues are incorporated into KDN-glycan chains. Presumably, the penultimate step in synthesis of KDN-glycoconjugates would be synthesis of the activated KDN nucleotide, CMP-KDN, catalyzed by CTP:CMP-KDN cytidylyltransferase (CMP-KDN synthetase). A related enzyme, CTP:CMP-sialic acid cytidylyltransferase (CMP-sialic acid synthetase) has been identified in cells and tissues of animals (van den Eijnden and van Dijk, 1972; Kean and Roseman, 1966a; Kean, 1970; Coates *et al.*, 1980) and bacteria (Warren and Blacklow, 1962; Kean and Roseman, 1966b; Vann *et al.*, 1987; Shames *et al.*, 1991). Since CMP-KDN synthetase has not been previously reported, we cannot rule out *a priori* the possibility that Neu5Ac or Neu5Gc residues are deaminated at the level of the sugar nucleotide or after incorporation into polymeric products. To investigate these unresolved questions, studies were initiated to determine: 1) if CMP-KDN synthetase activity exists in rainbow trout testis, a tissue known to synthesize KDN-containing glycoconjugates; 2) to partially purify and characterize such an enzyme activity, if found; and 3) to use the enzyme to synthesize CMP-[^{14}C]KDN, thereby providing a key substrate for future biosynthetic studies to determine how expression of KDN-glycoconjugates are regulated.

EXPERIMENTAL PROCEDURES

Materials. Rainbow trout testes were obtained from 1-year old fish collected monthly between May and November. Fish were provided through the courtesy of Gunma Prefectural Fisheries Experimental Station at Kawaba, and stored at -80°C until use. Mature sperm obtained in November were centrifuged at 5,000 rpm for 10 min to remove the seminal fluid and kept separately at -80°C until use. Calf brain was obtained from 4-month old calf through Tokyo Shibaura Zouki. [U-^{14}C]Neu5Ac (10 GBq/mmol), D-[U-^{14}C]Man (10 GBq/mmol), and sodium [1-^{14}C]pyruvate (1.18 GBq/mmol) were purchased from Amersham (England). Neu5Ac was obtained from Nacalai (Kyoto), and Neu5Gc was isolated and purified from rainbow trout PSGP, as previously described (Nomoto *et al.*, 1982; Kitazume *et al.*, 1992). CMP–Neu5Ac was purchased from Wako Chemicals (Osaka), and CMP–Neu5Gc was kindly provided from Mect. 5'–CTP (trisodium salt) was obtained from Seikagaku Kogyo, Co. (Tokyo). KDN was isolated from rainbow trout ovarian fluid KDN–rich glycoprotein (Kitazume *et al.*, 1992) or enzymatically synthesized using the *Escherichia coli* acylneuraminate pyruvate lyase (Wako Chemicals, Osaka) according to the procedure of Augé and Gautheron (Augé and Gautheron, 1987).

Synthesis and purification of [^{14}C]KDN and [^{14}C]Neu5Gc. [^{14}C]KDN was prepared from D–[U–^{14}C]mannose and sodium pyruvate by the reverse action of acylneuraminate pyruvate lyase by the procedure similar to that described by Augé and Gautheron (Augé and Gautheron, 1987). Neu5Gc was first converted by acylneuraminate pyruvate lyase to *N*–glycolylmannosamine and pyruvate, according to the method of Comb and Roseman (Comb and Roseman, 1960). [^{14}C]Neu5Gc was then synthesized from ManNGc and [^{14}C]pyruvate by an aldol condensation, catalyzed by the same enzyme.

CMP–nonulosonate synthetase assay. Reaction mixtures contained 4 μl of the enzyme fraction in 20 μl of 100 m*M* Tris–acetic acid buffer (pH 9.0) with 0.18 m*M* [^{14}C]KDN (400 Bq) or [^{14}C]Neu5Ac or [^{14}C]Neu5Gc, 9.6 m*M* 5'–CTP, and 25 m*M* $Mg(OAc)_2$. After incubation for 3 h at 25°C, the reaction was stopped by the addition of 2 volumes (40 μl) of cold ethanol. After standing for 30 min in an ice–bath, the mixture was centrifuged at 5,000 rpm for 10 min. A 10 μl aliquot of the supernatant was applied on a cellulose sheet (20 × 20 cm) with fluorescent indicator (Kodak), which was developed in 95% ethanol/1 *M* ammonium acetate (7:3, v/v) for 2 h. After air drying, the CMP–[^{14}C]KDN formed was detected by a Bio–imaging analyzer (Fujix BAS 2000). One unit of the enzyme was defined as a quantity of the enzyme required to synthesize 0.01 μmol of CMP–nonulosonate per 1 h at 25°C.

Preparation of rainbow trout testis CMP–nonulosonate synthetase. The procedures of van den Eijnden and van Dijk (van den Eijnden and van Dijk, 1972) and Higa and Paulson (Higa and Paulson, 1985) were followed in extraction of the CMP–nonulosonate synthetase from rainbow trout testis. Testis (44 g) from rainbow trout collected on August 19, 1991 (average 36 g/fish) were sliced after removal of connective tissue and blood–vessels was homogenized in 90 ml of 0.01 *M* sodium pyrophosphate (pH 10.2) containing 9 mg of soybean trypsin inhibitor (SBTI). The homogenate was centrifuged at 15,000 rpm for 20 min and the resulting supernatant (94 ml) was ultracentrifuged at 100,000 × g for 1 h, resulting in 81 ml of the supernatant. Saturated ammonium sulfate was gradually added to this supernatant (up to 30%) and the mixture was gently stirred at 4°C for 15 min and then left at 4°C for 1 h. Precipitated proteins were removed by centrifugation at 12,000 rpm for 20 min, and 148 ml of the supernatant fraction was obtained. Saturated ammonium sulfate was again added with stirring to a final concentration of 60%. The mixture was stirred for 15 min and left overnight at 4°C. Precipitated proteins were collected after centrifugation at 12,000 rpm for 20 min, dissolved in 35 ml of 5 m*M* Tris–acetate buffer (pH 7.5) containing 0.05% 2-mercaptoethanol, and dialyzed against 5 m*M* Tris–HCl buffer (pH 7.6) containing 0.1% 2-mercaptoethanol and 1 m*M* $MgCl_2$ for 4 to 12 h. This enzyme fraction thus obtained was used in experiments to characterize the CMP–nonulosonate synthetase activity and in preparation of CMP–nonulosonates.

RESULTS

Partial purification of CMP–nonulosonate synthetase from rainbow trout testes. A nucleotide phosphatase activity present in the crude homogenate, and which hydrolyzed CTP, was mostly eliminated by precipitation with 30% saturated $(NH_4)_2SO_4$.

Effects of temperature and incubation time on the formation of CMP–nonulosonates. The time course of the CMP–nonulosonates at 15, 25, and 37°C was followed, and the data suggested that all three sialic acids were converted to their CMP–derivatives. The formation of CMP–[^{14}C]KDN was significantly greater than either CMP–[^{14}C]Neu5Ac or CMP–[^{14}C]Neu5Gc. In contrast, the calf brain CMP–Neu5Ac synthetase was highly active on Neu5Ac and Neu5Gc, but was only slightly active on KDN (data not shown). Thus, the rainbow trout testis enzyme can be designated a CMP–KDN synthetase, while the calf brain enzyme should be referred to as a CMP–sialate synthetase. The optimum temperature for the rainbow trout testis CMP–KDN synthetase activity was approximately 25°C.

The effects of Mg^{2+} ion on formation of CMP–nonulosonates. The presence of Mg^{2+} is essential for the enzyme activity. Optimum formation of CMP–nonulosonates occurred between 25 and 100 m*M* of Mg^{2+} (data not shown).

Effects of pH on the formation of CMP–nonulosonates. With three different nonulosonates as substrates, the enzyme showed maximal activity between pH 9.0 and 10.0 in the presence of 25 m*M* Mg^{2+} (**Fig. 1**).

Kinetic studies of the rainbow trout testis CMP–nonulosonate synthetase. The Michaelis constants of these nonulosonate at a fixed saturated concentration of CTP(9.6 m*M*) were determined. The apparent Michaelis constants were estimated by Lineweaver–Burk plots (**Fig. 2**). The V_{max}/K_m value for KDN is nearly two times greater than that of Neu5Ac and Neu5Gc.

This leads us to conclude that the partially purified CMP–KDN synthetase prefers KDN as its nonulosonate substrate, and that this enzyme can effectively provide CMP–KDN for synthesis of KDN-containing glycoconjugates, such as (KDN)G_{M3} in trout testis.

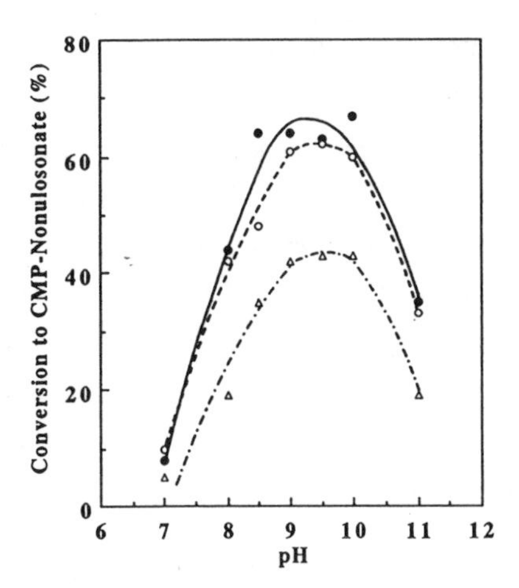

Fig. 1. Effect of pH on the rainbow trout testis CMP–nonulosonate synthetase activity. The pH activity profiles were carried out using KDN, Neu5Ac, and Neu5Gc as substrates. [14C]KDN (●), [14C]Neu5Ac (○), and [14C]Neu5Gc (△).

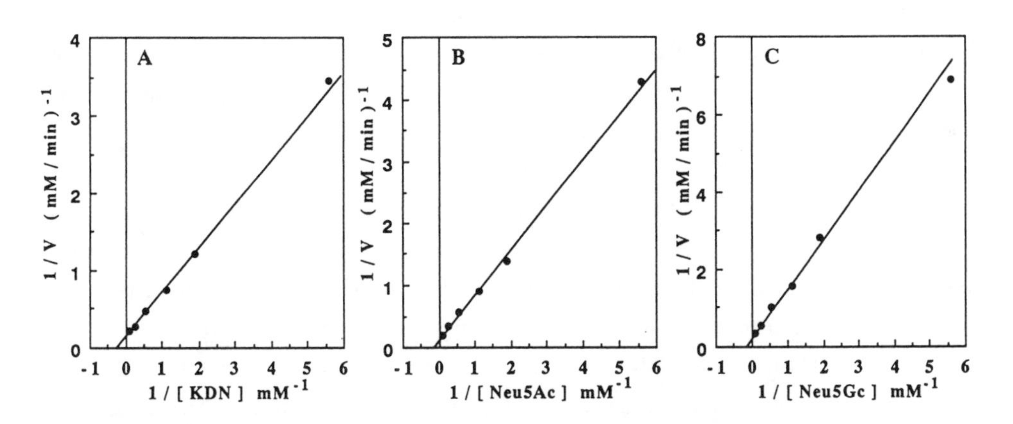

Fig. 2. Lineweaver–Burk analysis for the CMP–nonulosonate synthetase–catalyzed synthesis of (A) CMP–KDN, (B) CMP–Neu5Ac, and (C) CMP–Neu5Gc. In each of these two-substrate systems, the concentration of free nonulosonate was varied while that of 5'-CTP remained constant at 9.5 m*M*. Reaction conditions were identical to those described in the text except that 75 m*M* $Mg(OAc)_2$ was used.

Synthesis and characterization of CMP–nonulosonates. The product of the synthetase with KDN was identified to be CMP–[^{14}C]KDN by TLC analysis and ^{1}H NMR measurement (data not shown).

Developmental changes of CMP–nonulosonate synthetase activity. The developmental expression of CMP–KDN and CMP–Neu5Ac synthetase activities was determined in freshly isolated testis at different stages of spermatogenesis. The level of enzyme activity in the testis per single fish rose rapidly with maturation of testis, assessed by an increase in weight, until the middle of August (data not shown). The developmental expression of CMP–KDN synthetase activity paralleled the expression of CMP–Neu5Ac synthetase activity (data not shown). To determine the fate of CMP–nonulosonate synthetase, the enzyme activity in the mature sperm (spermatozoa) after spermiation was measured, and found to be undetectable (data not shown). No activity was detected in the seminal plasma, suggesting that the regulated expression of the CMP–nonulosonate synthetase activity may be restricted to a relatively narrow stage of development.

DISCUSSION

Recent finding of the unique nonulosonic acid residue, KDN, in glycoconjugates (Nadano *et al.*, 1986; Iwasaki *et al.*, 1987, 1990; Inoue *et al.*, 1988; Kanamori *et al.*, 1989, 1990; Yu Song *et al.*, 1991; Knierl *et al.*, 1989; Strecker *et al.*, 1989) has raised several interesting questions regarding the biosynthetic pathway, and how KDN may be activated and transferred to pre–existing endogenous acceptors. We have considered it particularly important to determine how these new classes of glycoconjugates are synthesized. The first question we addressed was if KDN was activated with CTP to form CMP–KDN, prior to being transferred to endogenous oligosaccharide acceptors. Nothing is known about the mechanism of KDN activation, nor formation of KDN–glycan units. An alternative pathway would be that Neu5Ac or Neu5Gc residues were first transferred from CMP–Neu5Ac or CMP–Neu5Gc into glycan chains, then deacylated and deaminated. These two alternative hypotheses could be tested because the first would predict the existence of a CMP–nonulosonate synthetase with a preference for KDN, while the absence of such an activating enzyme would support the latter hypothesis. The results of our studies show that rainbow trout testis contains a CMP–KDN synthetase that catalyzes the formation of CMP–KDN from CTP and KDN, as shown in reaction 1:

$$\textbf{(1)}\qquad \textbf{CTP + KDN} \xrightarrow{Mg^{2+}} \textbf{CMP–KDN + PPi}$$

On the basis of these results, we hypothesize that the subsequent reaction leading to formation of KDN–glycans will be catalyzed by a CMP–KDN:sialyltransferase that transfers KDN from CMP–KDN to acceptor glycans (reaction 2):

(2) CMP–KDN + acceptor glycan → KDN–glycan + CMP

The CMP–KDN synthetase was purified 116–fold from trout testis and found to catalyze the preference activation of KDN, although Neu5Ac and Neu5Gc could also be activated. CMP–KDN synthetase showed optimal activity between pH 9–10, and at 25°C. The presence of either Mg^{2+} is essentially required for the enzyme activity. We haven't yet determined if the different nonulosonate activating activities are catalyzed by the same or different enzymes. This point does not distract, however, from the importance of our major finding of a CMP–KDN synthetase activity that has not been previously described. Also of major importance is our finding that the developmental expression of CMP–KDN activity is temporally correlated with development during spermato–genesis and with the developmental expression of (KDN)G_{M3}. The level of enzyme activity exhibits characteristic developmental changes, increasingly rapidly from June and reaching a maximum level in mid–August. Thereafter, spermiation begins (Billard, 1983) and this is correlated with a dramatic decrease in enzyme activity that then tapers off slowly until November. No CMP–KDN synthetase activity was found in spermiated mature sperm.

Although we have yet to ascertain where the synthetase is localized within spermatozoa, we presume that like CMP–Neu5Ac, CMP–KDN is also synthesized in the cell nucleus, diffuses into the cytosol (Coates *et al.*, 1980), and is imported into Golgi to be available for the KDN–transferases (Sommers and Hirschberg, 1982). These aspects of how KDN–glycoconjugates are synthesized await further studies that are now possible with the availability of CMP–[^{14}C]KDN. Since this sugar nucleotide is more labile than CMP–[^{14}C]Neu5Ac, added precautions will have to be taken in carrying out biosynthetic studies.

Acknowledgments– We also thank Mr. Kensaku Satsumi, Gunma Prefectural Fisheries Experimental Station at Kawaba, for generous provision of trout testes used in this study. This research was supported in part by Grants–in–Aid for Developmental Scientific Research (02558014), and for International Scientific Research Program: Joint Research (04044055), and by a fund from the Mitsubishi Foundation to YI.

REFERENCES

Augé, C., and Gautheron, C. (1987) *J. Chem. Soc. Chem. Commun.* 859–860

Billard, R. (1983) *Cell Tissue Res.* **230**, 495–502

Coates, S. W., Gurney, T., Jr., Sommers, L. W., Yeh, M., and Hirschberg, L. (1980) *J. Biol. Chem.* **255**, 9225–9229

Comb, D. G., and Roseman, S. (1960) *J. Biol. Chem.* **235**, 2529–2537

Higa, H. H., and Paulson, J. C. (1985) *J. Biol. Chem.* **260**, 8838–8849

Inoue, S., Kanamori, A., Kitajima, K., and Inoue, Y. (1988) *Biochem. Biophys. Res. Commun.* **153**, 172–176

Iwasaki, M., Inoue, S., Nadano, D., and Inoue, Y. (1987) *Biochemistry* **26**, 1452–1457

Iwasaki, M., Inoue, S., and Troy, F. A. (1990) *J. Biol. Chem.* **265**, 2596–2602

Kanamori, A., Kitajima, K., Inoue, S., and Inoue, Y. (1989) *Biochem. Biophys. Res. Commun.* **164**, 744–749

Kanamori, A., Inoue, S., Iwasaki, M., Kitajima, K., Kawai, G., Yokoyama, S., and Inoue, Y. (1990) *J. Biol. Chem.* **265**, 21811–21819

Kean, E. L. (1970) *J. Biol. Chem.* **245**, 2301–2308

Kean, E. L., and Roseman, S. (1966a) *Methods in Enzymol.* **8**, 208–215

Kean, E. L., and Roseman, S. (1966b) *J. Biol. Chem.* **241**, 5643–5650

Kitazume, S., Kitajima, K., Inoue, S., and Inoue, Y. (1992) *Anal. Biochem.* **202**, 25–34

Knierl, Yu. A., Kocharova, N. A., Shashkov, A. S., and Kochetkov, N. K. (1989) *Carbohydr. Res.* **188**, 145–155

Nadano, D., Iwasaki, M., Endo, S., Kitajima, K., Inoue, S., and Inoue, Y. (1986) *J. Biol. Chem.* **261**, 11550–11557

Nomoto, H., Iwasaki, M., Endo, T., Inoue, S., Inoue, Y., and Matsumura, G. (1982) *Arch. Biochem. Biophys.* **218**, 335–341

Shames, S. L., Simon, E. S., Christopher, C. W., Schmid, W., Whitesides, G. M., and Yang, L. L. (1991) *Glycobiology* **1**, 187–191

Sommers, L. W., and Hirschberg, C. B. (1982) *J. Biol. Chem.* **257**, 10811–10817

Strecker, G., Wieruszeski, J. M., Michalski, J. C., Alonso, C., Boilly, B., and Montreuil, J. (1992) *FEBS Lett.* **298**, 39–43

Troy, F. A. (1992) *Glycobiology* **2**, 5–23

Yu Song, Kitajima, K., Inoue, S., and Inoue, Y. (1991) *J. Biol. Chem.* **266**, 21929–21935

van den Eijnden, D.H., and van Dijk, W. (1972) *Hoppe-Seyler's Z. Physiol. Chem.* **353**, 1817–1820

Vann, W. F., Silver, R. P., Abeijon, C., Chang, K., Aaronson, W., Sutton, A., Finn, C. W., Lindner, W., and Kotsatos, M. (1987) *J. Biol. Chem.* **262**, 17556–17562

Warren, L., and Blacklow, R. S. (1962) *J. Biol. Chem.* **237**, 3527–3534

Polysialic Acid
J. Roth, U. Rutishauser and F. A. Troy II (eds.)

STRUCTURAL AND BIOSYNTHETIC ASPECTS OF *N*-GLYCOLYL-8-*O*-METHYLNEURAMINIC-ACID-OLIGOMERS, LINKED THROUGH THEIR *N*-GLYCOLYL GROUPS, ISOLATED FROM THE STARFISH *ASTERIAS RUBENS*

Aldert A. Bergwerff[1], Stephan H.D. Hulleman[1], Johannis P. Kamerling[1], Johannes F.G. Vliegenthart[1], Lee Shaw[2], Gerd Reuter[2] and Roland Schauer[2]

[1] Bijvoet Center, Department of Bio-Organic Chemistry, Utrecht University, P.O. Box 80.075, NL-3508 TB Utrecht, The Netherlands, and [2] Biochemisches Institut, Christian-Albrechts-Universität Kiel, Olshausenstrasse 40-60, D-2300 Kiel 1, Germany

SUMMARY: Analyses of hydrolysates of homogenates from the starfish *Asterias rubens* by mass spectrometry and NMR spectroscopy showed the occurrence of *N*-acetylneuraminic acid, *N*-acetyl-8-*O*-methylneuraminic acid, *N*-acetyl-9-*O*-acetyl-8-*O*-methylneuraminic acid, *N*-glycolylneuraminic acid, *N*-glycolyl-8-*O*-methylneuraminic acid, and *N*-glycolyl-9-*O*-acetyl-8-*O*-methylneuraminic acid. In addition, a di- and a trimer of *N*-glycolyl-8-*O*-methylneuraminic acid (Neu5Gc8Me), linked through the anomeric centre to the *N*-glycolyl moiety, namely, Neu5Gc8Meα(2-O5)Neu5Gc8Me and Neu5Gc8Meα(2-O5)Neu5Gc8Meα(2-O5)Neu5Gc8Me were found. Investigations on the biosynthesis of *N*-acyl-8-*O*-methylneuraminic acid in *A. rubens*, using *S*-adenosyl-L-[methyl-^{14}C]methionine, demonstrated that *N*-acylneuraminate 8-*O*-methyltransferase activity was present predominantly in the membrane fraction. CMP-*N*-acetylneuraminic acid monooxygenase activity was detected in the soluble protein fraction, in agreement with studies on the corresponding vertebrate enzyme. On basis of these results, a pathway for the biosynthesis of Neu5Gc8Me is proposed.

INTRODUCTION

The echinodermata are thought to be the most primitive animals containing sialic acids (Corfield and Schauer, 1982). In addition to the common sialic acids *N*-acetylneuraminic acid (Neu5Ac) and *N*-glycolylneuraminic acid (Neu5Gc), the starfish *A. rubens* contains a rather unusual group of bound sialic acids, namely, *N*-acetyl-8-*O*-methylneuraminic acid (Neu5Ac8Me), *N*-acetyl-9-*O*-acetyl-8-*O*-methylneuraminic acid (Neu5,9Ac$_2$8Me), *N*-glycolyl-8-*O*-methylneuraminic acid (Neu5Gc8Me), and *N*-glycolyl-9-*O*-acetyl-8-*O*-methylneuraminic acid (Neu9Ac5Gc8Me) (Schauer *et al.*, 1983; Schauer *et al.*, 1984; Smirnova *et al.*, 1988). The biochemical reactions leading to such a variety of sialic acids are poorly understood. Preliminary data have suggested that the 8-*O*-methylation of bound Neu5Gc is catalysed by an *S*-adenosyl-methionine-dependent *O*-methyltransferase (Schauer and Wember, 1985) :

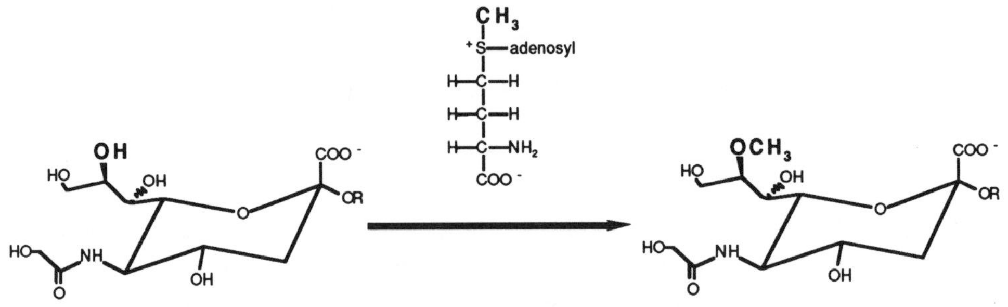

N-acylneuraminate 8-*O*-methyltransferase
R, glycoconjugate

Here, in addition to the series of sialic acids present in bound form in the starfish *A. rubens*, structural data are presented, demonstrating the existence of Neu5Gc8Me-oligomers. Furthermore, investigations on the properties of *N*-acylneuraminate 8-*O*-methyltransferase and the demonstration of CMP-Neu5Ac monooxygenase activity allowed us to propose a pathway for the biosynthesis of Neu5Gc8Me.

MATERIALS AND METHODS

Starfish were collected from the Baltic Sea, and were prepared for sialic acid analysis and enzyme assays, as described by Bergwerff *et al.* (1992). Briefly, the bulk of the digestive system was removed from the starfish, and the remaining material was homogenised in water (sialic acid analysis) or in an appropriate buffer (enzyme assays).

In the case of sialic acid analysis, the homogenates were subjected to mild acid hydrolysis. The released sialic acids were isolated via dialysis and purified by column chromatography. Sialic acids were analysed by GLC-EIMS, and sialic-acid-oligomers by FAB-MS and/or NMR-spectroscopy as previously reported (Bergwerff *et al.*, 1992).

For testing the enzyme activities, various particulate and soluble fractions were collected by differential centrifugation of a homogenate supernatant obtained at 700 *g*. In the case of experiments focused on *O*-methyltransferase activity, membrane fraction **P1** was obtained by centrifugation at 12,800 *g*. The remaining supernatant was centrifuged at 120,000 *g*, giving rise to a membrane fraction, designated **P2**, and a soluble protein fraction, designated **S**. For experiments directed towards monooxygenase activity, the soluble protein fraction **S'** and the total membrane fraction **P'** were collected in one centrifugation step at 120,000 *g*.

Fractions **P1**, **P2** and **S** were tested for *O*-methyltransferase activity by incubation for 3 h at 37°C using defined glycoconjugate-bound sialic acid or free sialic acid as methyl-acceptor probes, and *S*-adenosyl-L-[methyl-^{14}C]methionine as a methyl-donor. Radioactive products were quantified by liquid scintillation counting and were identified, after mild acid hydrolysis, by radio-TLC with reference non-radioactive sialic acids.

Fractions **P'** and **S'** were tested for monooxygenase activity by incubation for 3 h at 37°C in the presence of NADH, $FeSO_4$ and cytidine-5'-mono-phospho-*N*-acetyl-[4,5,6,7,8,9-^{14}C]neuraminic acid. Incubations were stopped by addition of HCl and the released [^{14}C]sialic acids were analysed by radio-TLC. Reference non-radioactive Neu5Ac and Neu5Gc were co-chromatographed on the same lane.

RESULTS

Analysis of Neu5Gc8Me-dimer and Neu5Gc8Me-trimer

GLC-EIMS of per-*O*-trimethylsilylated derivatives of sialic acids (Kamerling *et al.*, 1982; Reuter and Schauer, 1986), obtained after mild acid hydrolysis of pools of homogenised starfish *A . rubens*, revealed the presence of six different sialic acids, namely, Neu5Ac (4%), Neu5Ac8Me (12%), Neu5,9Ac$_2$8Me (< 1%), Neu5Gc (19%), Neu5Gc8Me (47%), and Neu9Ac5Gc8Me (18%). The major sialic acid Neu5Gc8Me was isolated and investigated by ^{13}C- (Fig. 1) and ^{1}H- (Table 1) NMR spectroscopy.

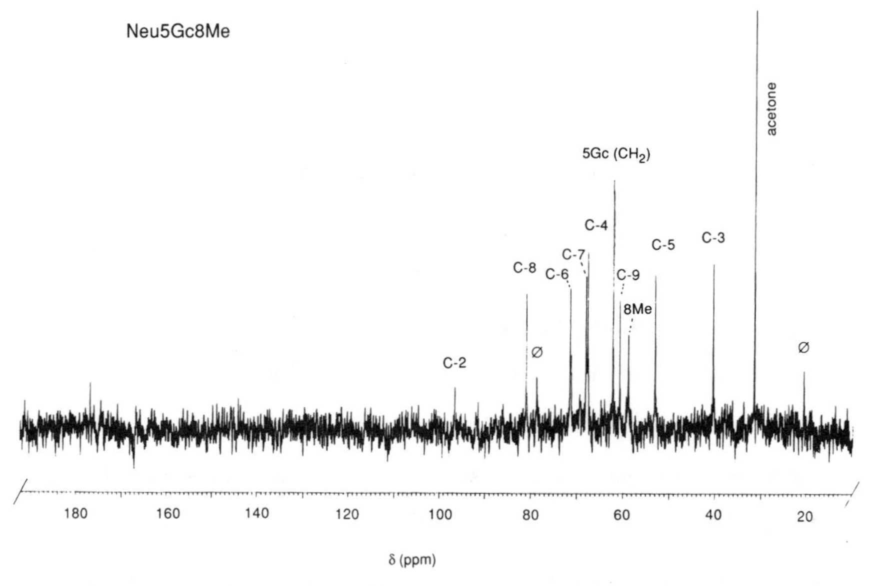

Figure 1. 75-MHz 1D noise ^{1}H-decoupled ^{13}C-NMR spectrum of Neu5Gc8Me recorded in $^{2}H_2O$ at p^{2}H 7 and 300 K.

Besides the free sialic acids, the hydrolysate also contained sialic-acid-positive fractions, which eluted on a Mono Q anion-exchange column at the positions of dimeric and trimeric fragments of colominic acid. These fractions were further fractionated by HPLC on Partisil SAX. Mass spectrometric and NMR spectroscopic analyses of the major subfractions (Table I) showed the presence of di- and tri-sialo-oligomers, linked through the anomeric centre to the N-glycolyl moiety, namely, **Neu5Gc8Meα(2-O5)Neu5Gc8Me** (Figs. 2, 3 and 4) and **Neu5Gc8Meα(2-O5)-Neu5Gc8Meα(2-O5)Neu5Gc8Me**. The analysis of the dimer will be discussed in more detail.

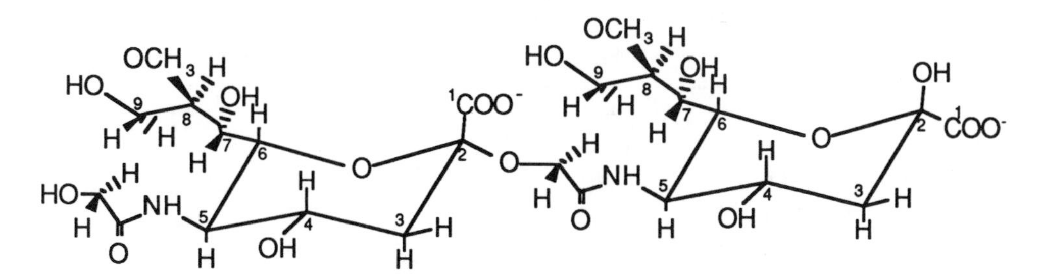

Figure 2. The Neu5Gc8Me-dimer, in which the non-reducing unit is α-glycosidically linked to the hydroxyl group of the N-glycolyl function of the reducing unit. The reducing Neu5Gc8Me-unit is depicted in the β-configuration.

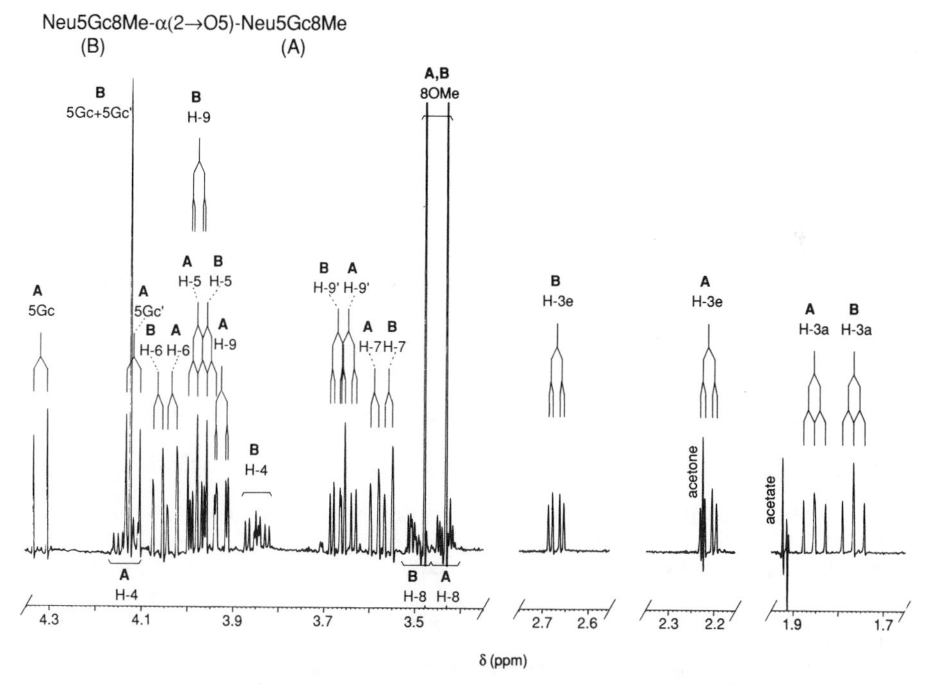

Figure 3. 500 MHz 1D ^{1}H-NMR spectrum recorded in ^2H_2O at p^{2}H 7 and 300 K of Neu5Gc8Meα(2-O5)Neu5Gc8Me, predominantly occurring in the β-anomeric form.

Table I. ^1H-NMR data for Neu5Gc8Me, Neu5Gc8Meα(2-O5)Neu5Gc8Me [(Neu5Gc8Me)$_2$], and Neu5Gc8Meα(2-O5)Neu5Gc8Meα(2-O5)Neu5Gc8Me [(Neu5Gc8Me)$_3$]. Chemical shifts are given in ppm relative to internal acetone (δ 2.225) in ^2H_2O at p^{2}H 7 and 300 K. α and β stand for the anomeric configuration of the reducing residue; nd, not detected.

Protons	Chemical shift (in ppm)					
	Neu5Gc8Me	(Neu5Gc8Me)$_2$		(Neu5Gc8Me)$_3$		
		reducing	terminal	reducing	internal[a]	terminal[a]
H-3a$_α$	1.643	nd	1.767	nd	1.756	1.758
H-3a$_β$	1.863	1.853		1.854		
H-3e$_α$	2.551	nd	2.669	nd	2.655	2.665
H-3e$_β$	2.219	2.212		2.209		
H-4	4.110	4.134	3.846	4.140	3.831	3.868
H-5	3.983	3.999	3.959	3.980	3.950	3.961
H-6	4.017	4.033	4.064	4.036	4.067	4.088
H-7	3.559	3.589	3.558	3.590	3.584	3.543
H-8	3.432	3.433	3.496	3.436	3.504	3.494
H-9	3.932	3.927	3.979	3.926	3.984	3.972
H-9'	3.652	3.647	3.670	3.646	3.664	3.661
5Gc[b]	4.125[c]/4.133[d]	4.320	4.124	4.317	4.309	4.119
5Gc'[b]		4.118		4.113	4.099	
8Me	3.425	3.432	3.479	3.434	3.488	3.476
NH	nd	8.26[e]	8.31[e]	nd	nd	nd

[a] The set of H-3a,3e,4,5,6 of terminal and internal α-Neu5Gc8Me may be interchanged. The same holds for the set of H-7,8,9,9'; [b] 5Gc and 5Gc' represent the protons of the *N*-glycolyl methylene group. In the case of a singlet (2H) only one value is given, in the case of two doublets (2 x 1H) two values are given; [c] The anomeric configuration is α; [d] The anomeric configuration is β; [e] Value obtained at 280K, pH 5.6 and in ^1H_2O, containing 20 mM sodium phosphate.

The compound with the same retention time as Neu5Acα(2-8)Neu5Ac on Mono Q was susceptible to *Vibrio cholerae* sialidase, giving rise to monomeric Neu5Gc8Me. This finding in combination with the FAB mass data [$(M+H)^+$ at *m/z* 661, $(M+NH_4)^+$ at *m/z* 678] and the 1D ^1H-NMR spectrum, showing two sets of signals with equal intensities corresponding to Neu5Gc8Me in α- and β-configuration (Table I), respectively, established the sequence to be Neu5Gc8Meα2→ Neu5Gc8Meβ. The subsets of signals for the glycerol-side-chain- and the ring-protons for each residue, were correlated using 2D Homonuclear Hartmann-Hahn (HOHAHA), and 2D rotating-

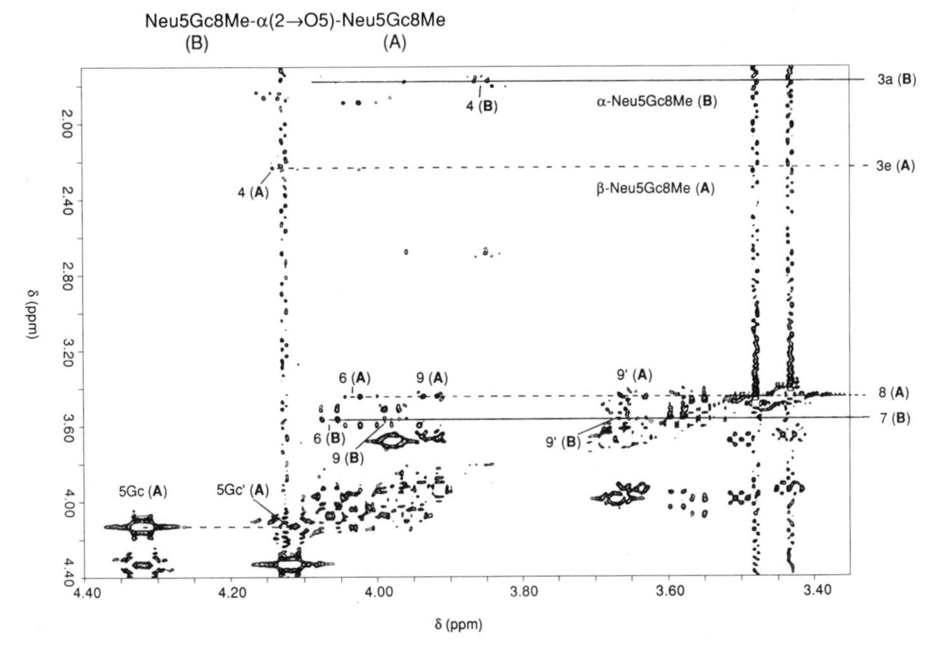

Figure 4. 500-MHz 2D rotating-frame nuclear Overhauser enhancement (ROESY) spectrum of Neu5Gc8Meα(2-O5)Neu5Gc8Me recorded in ^2H_2O at p^{2}H 7 and 300 K, with a spin-lock mixing time of 200 ms. Tracks are drawn for H-3a and H-7 of the non-reducing α-Neu5Gc8Me (**B**) residue (——) and for H-3e, H-8 and 5Gc' of the reducing β-Neu5Gc8Me (**A**) residue (- - - -). Protons are represented by their corresponding position number (cf. Fig. 2).

frame nuclear Overhauser enhancement spectroscopy (ROESY) in ^2H_2O. The 1D ^{1}H-NMR spectrum (^2H_2O) showed a singlet stemming from one *N*-glycolyl methylene group, which gave rise to a r.O.e. cross-peak with an amide proton in a 2D-ROESY spectrum (^1H_2O). This amide proton could be correlated with the ring-protons of the non-reducing α-Neu5Gc8Me residue in a 2D-HOHAHA spectrum (^1H_2O), thereby assigning the *N*-glycolyl methylene singlet as belonging to the latter residue. Following a similar reasoning, the two *N*-glycolyl methylene doublets, $^2J_{H,H}$ -15 Hz, could be assigned to the geminal protons of the *N*-glycolyl group of the reducing β-Neu5Gc8Me residue. The presence of two doublets instead of one singlet indicates a distinct rotamer distribution for the *N*-glycolyl methylene group of β-Neu5Gc8Me. The large downfield shift (Δδ +0.187) of one of these *N*-glycolyl methylene protons (δ 4.320) as compared with the protons of the *N*-glycolyl methylene group (singlet) of the non-reducing unit, together with a small inter-residual r.O.e. cross-peak between α-Neu5Gc8Me H-3a and β-Neu5Gc8Me 5Gc, indicate that the non-reducing Neu5Gc8Me residue is α-glycosidically linked to the hydroxyl group of the *N*-glycolyl substituent of the reducing β-Neu5Gc8Me residue (Fig. 2).

S-*Adenosyl*-L-*methionine*:N-*acyl*-D-*neuraminate 8*-O-*methyltransferase activity*

Experiments focused on 8-*O*-methyltransferase activity were carried out with *S*-adenosyl-L-[methyl-^{14}C]methionine, using defined sialoglycoconjugates (Table II), free sialic acids (Table II) or immobilised porcine submandibular gland mucin (PSM) (Fig. 5) as methyl-acceptors. Using the immobilised substrate, the pH optimum of 8-*O*-methyltransferase was determined to be 8.9. The highest specific *O*-methyltransferase activity was found in the membrane fractions **P1** and **P2** using soluble PSM as substrate, being 0.51 ± 0.16 and 0.59 ± 0.06 pmol radioactive product per mg protein in 3 h, respectively, whereas this value was 0.10 ± 0.05 pmol per mg protein in 3 h for

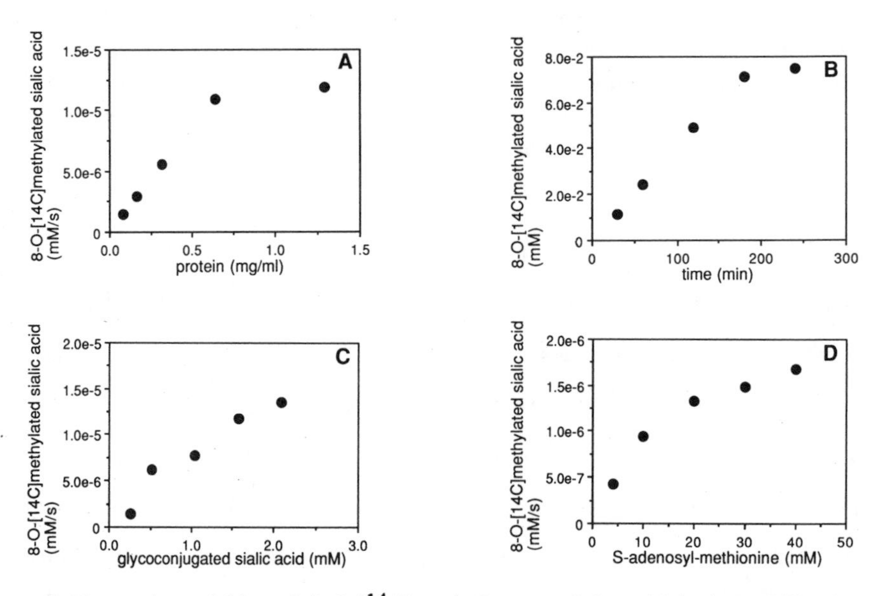

Figure 5. Formation of *N*-acyl-8-*O*-[^{14}C]methyl-neuraminic acid in immobilised porcine submandibular gland mucin by *N*-acylneuraminate 8-*O*-methyltransferase: dependence on protein concentration (A), time (B), sialic acid concentration (C) and *S*-adenosyl-L-methionine concentration (D).

the high-speed fraction **S**. In the presence of 4 mM *S*-adenosyl-L-homocysteine, or 1 mM EDTA, the activity was reduced by 80% and 60%, respectively, whereas the addition of 10 mM EDTA gave rise to almost complete inhibition.

Table II. Substrate specificity of *S*-adenosyl-L-methionine:*N*-acyl-D-neuraminate 8-*O*-methyltransferase. Conjugated sialic acids were released by mild acid hydrolysis. Sialic acids were obtained after dialysis and Dowex 1X8 anion-exchange chromatography. Radioactive sialic-acid products were analysed by radio-TLC on cellulose and on silica plates, which were run in several solvent systems, and were co-chromatographed with reference non-radioactive sialic acids.

Substrate		Sialic acid products
endogenous	→	Neu5Gc8Me
porcine submandibular gland mucin	→	Neu5Gc8Me, Neu5Ac8Me
bovine submandibular gland mucin	→	Neu5Gc8Me, Neu5Ac8Me
collocalia mucin	→	Neu5Ac8Me
Neu5Acα(2-3)Galβ(1-4)Glc	→	Neu5Ac8Me
Neu5Acα(2-6)Galβ(1-4)Glc	→	Neu5Ac8Me
Neu5Ac	→	no detectable reaction
Neu5Gc	→	no detectable reaction

*CMP-*N*-acetyl-D-neuraminate monooxygenase activity*

Experiments focused on the monooxygenase activity were carried out with CMP-[^{14}C]Neu5Ac. The formation of radioactive sialic acids was followed and quantified by radio-TLC. About 98% of the total CMP-Neu5Ac hydroxylase activity was detected in the supernatant fraction **S'**. From the specific activities determined at different pH, the highest CMP-Neu5Ac monooxygenase activity was detected at about pH 7.

DISCUSSION

Evaluating the findings, a biosynthetic pathway is proposed for starfish sialic acids (Scheme 1). The detection of Neu5Ac among the total bound sialic acids by GLC-MS, suggests that the biosynthesis starts with the production of Neu5Ac, as reported for the vertebrate system (Kundig *et al.*, 1966; Comb *et al.*, 1966). The occurrence of 84% of the sialic acids being *N*-glycolylated, and detection of monooxygenase activity, prevalently present in the 120,000 *g* supernatant fraction, indicate that after activation of Neu5Ac to its CMP-glycoside, it is predominantly hydroxylated in the cytoplasm to CMP-Neu5Gc, as previously observed in vertebrate tissue (Shaw and Schauer, 1988). The cytoplasmic CMP-glycosides may then be transported into the Golgi-apparatus by a transport system analogous to that in mouse liver (Lepers *et al.*, 1989), where sialic acids are transferred onto growing carbohydrate chains by sialyltransferases, including the possible involvement of (poly/oligo)sialyltransferases. The presence of sialylate-8-*O*-methyltransferase in the membrane fractions, and its specificity for glycosidically bound sialic acid substrates, suggest that the methylation reaction occurs after transfer of sialic acid to the carbohydrate chains. The stage at which the *O*-acetyltransferase reaction occurs still remains to be clarified.

Scheme 1. Possible biosynthetic pathway for the formation of sialo-glycoconjugates in the starfish *Asterias rubens* starting from *N*-acetylneuraminic acid.

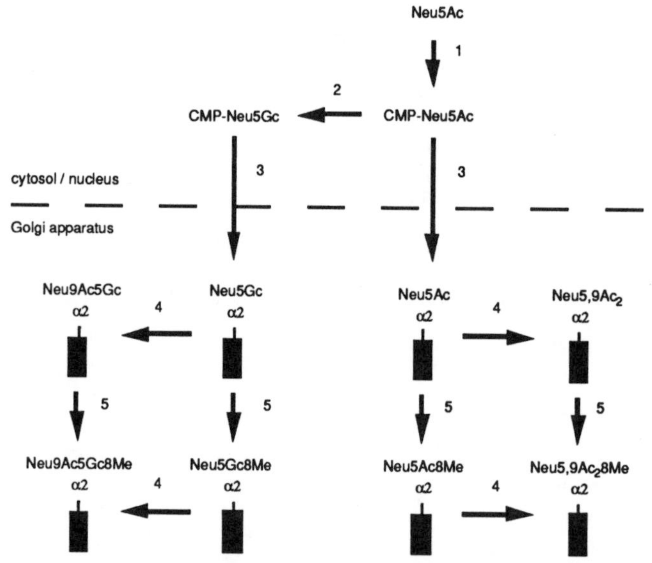

1, *N*-acetyl-neuraminate cytidyltransferase; 2, soluble cytidine monophospho-*N*-acetylneuraminate monooxygenase (**S'**); 3, sialyltransferase; 4, *N*-acylneuraminate-*O*-acetyltransferase; 5, membrane bound *N*-acylneuraminate-8-*O*-methyltransferase (**P2**). The black squares symbolize glycoconjugates.

The oligomers of Neu5Gc8Me, unusually linked through their anomeric centre to the *N*-glycolyl group, were obtained after mild acid hydrolysis. So far, no information is available about either the biological source of these uncommon compounds or the degree of polymerisation of the native sialo-poly/oligomers. An oligosaccharide with the terminal sequence of Neu5Gc8Meα(2-O5)-Neu5Gc8Meα(2-3) has been identified in a ganglioside from the starfish *Aphelaterias japonica* (Smirnova *et al.*, 1987), and oligosaccharides with up to five repeating Neu5Gc8Meα(2-O5) units were detected in gangliosides from the starfish *Asterias amurensis* (Irie *et al.*, 1990). A methyl group at C-8 may not only stimulate elongation of a sialic-acid-containing carbohydrate chain via the *N*-glycolyl group, but also via the hydroxyl group at C-4, as has been found for a ganglioside from the starfish *Asterina pectinifera*, namely, Araβ(1-6)Galβ(1-4)Neu5Gc8Meα(2-3)Galβ(1-4)Glcβ(1-1)-ceramide (Sugita, 1979).

ACKNOWLEDGEMENTS

We are indebted to Uwe Schwenk and the Institut für Meereskunde in Kiel for supplying the starfish. We thank Cees Versluis (Bijvoet Center, Department of Mass Spectrometry, Utrecht University) for recording the FAB mass spectra. The financial supports of EMBO (A.A.B.) and of ERASMUS Program (S.H.D.H.) are gratefully acknowledged. This work was also financially supported by the Fonds der Chemischen Industrie and the Sialic Acids Society (Kiel) and by the Netherlands Foundation for Chemical Research (NWO/SON).

REFERENCES

Bergwerff A.A., Hulleman S.H.D., Kamerling J.P., Vliegenthart J.F.G., Shaw L., Reuter G. & Schauer R. (1992) Biochimie 74, 25-38

Comb D.G., Watson D.R. & Roseman S. (1966) J. Biol. Chem. 241, 5637-5642

Corfield A.P. & Schauer R. (1982) in: Sialic Acids, Cell Biol. Monogr. (Schauer R., ed.), Vol. 10, Springer, Wien, 5-50

Irie A., Kubo H. & Hoshi M. (1990) Abstr. XVth Int. Carbohydrate Symp., Yokohama, Japan, 200

Kamerling J.P. & Vliegenthart J.F.G. (1982) in: Sialic Acids, Cell Biol. Monogr. (Schauer R., ed.), Vol. 10, Springer, Wien, 95-125

Kundig W., Gosh S. & Roseman S. (1966) J. Biol. Chem. 241, 5619-5626

Lepers A., Shaw L., Cacan R., Schauer R., Montreuil J. & Verbert A. (1989) FEBS Lett. 250, 245-250

Reuter G. & Schauer R. (1986) Anal. Biochem. 157, 39-46
Schauer R., Schröder C. & Shukla A.K. (1984) Adv. Exp. Med. Biol. 174, 75-86
Schauer R. & Wember M. (1985) in: Glycoconjugates-Proc. VIIIth Int. Symp. (Davidson E.A., Williams J.C. & Di Ferrante N.M., eds.), Vol. I, Praeger, New York, 266-267
Schröder C., Nöhle U., Shukla A.K. & Schauer R. (1983) Proc. 7th Int. Symp. Glycoconjugates, Rahms Lund, Sweden, 162
Shaw L. & Schauer R. (1988) Biol. Chem. Hoppe-Seyler 369, 477-486
Smirnova G.P., Glukhoded I.S. & Kochetkov N.K. (1988) Bioorg. Khim. 14, 636-641
Smirnova G.P., Kochetkov N.K. & Sadovskaya V.L. (1988) Biochim. Biophys. Acta 920, 47-55
Sugita M. (1979) J. Biochem. 86, 765-772

PART 2.

DEVELOPMENTAL AND CELL BIOLOGY
DISEASED STATES AND TUMORS

Polysialic Acid
J. Roth, U. Rutishauser and F. A. Troy II (eds.)

REGULATION OF CELL-CELL INTERACTIONS BY NCAM AND ITS POLYSIALIC ACID MOIETY

Urs Rutishauser

Departments of Genetics and Neuroscience, School of Medicine, Case Western Reserve University, Cleveland, OH 44106

SUMMARY: In vertebrate embryos, polysialic acid (PSA) is associated with the cell surface adhesion molecule called NCAM. The amount of PSA on NCAM is biologically regulated and correlates with a variety of events during embryo development. In vitro, the presence of NCAM and its PSA moiety have been found to affect cell interactions involving a variety of different cell surface receptors other than NCAM itself. In view of this broad influence, we have proposed that NCAM and PSA can serve as a "pull-push" mechanism for regulation of overall cell-cell contact. This proposal is consistent with observations that NCAM alone enhances cell-cell contact whereas PSA-containing NCAM increases the amount of space between apposing cell membranes.

NCAM AND PSA COMBINE TO MAKE A VERY UNUSUAL CELL SURFACE GLYCOPROTEIN

NCAM is a member of the immunoglobulin family of membrane receptors and has been shown to participate in cell-cell adhesion via a homophilic (NCAM-NCAM) binding mechanism (Figure 1; for a general review of NCAM, PSA and other adhesion molecules, see Rutishauser and Jessell, 1988). In addition to being one of the most abundant proteins at the cell surface, NCAM is an atypical cell adhesion molecule. Unlike many other adhesion receptors which have a restricted pattern of expression,

NCAM is present on many cells in a variety of tissues throughout development. Moreover, while other adhesion molecules, such as cadherins and integrins, are members of multi-gene families with different adhesive specificities, NCAM is encoded by a single gene. While different forms of NCAM are generated by alterative splicing, these developmentally regulated variations appear to influence NCAM function without a change in its intrinsic adhesive function or specificity.

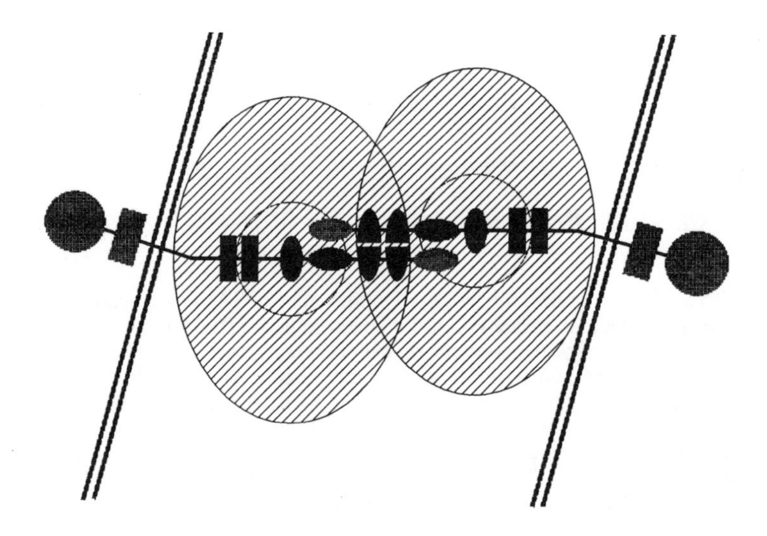

Figure 1. Schematic model of a membrane-membrane bond formed by homophilic binding between immunoglobulin domains of two NCAM molecules. The large hatched ellipses represent the attachment of large amounts of sialic acid to the fifth immunoglobulin domain in the form of linear homopolymers, referred to here as PSA. The smaller circles within the ellipse represents the fact that the amount of PSA on NCAM is independently regulated during development.

Like many other cell surface proteins, NCAM is glycosylated at several sites along its single polypeptide chain. Some of this carbohydrate, such as the HNK-1 moiety, is typical of branched, short chain structures found shared by a broad spectrum of proteins. By contrast, PSA consists of long linear homopolymers of sialic acid which in the vertebrate embryo appear to be confined to the fifth immunoglobulin domain of NCAM (Figure 1).

NCAM AND PSA CAN REGULATE A VARIETY OF CELL INTERACTIONS

Another remarkable feature of NCAM and PSA is that they are capable of affecting a broad spectrum of cell interactions mediated by other receptors. Antibodies that block NCAM function have been shown to result in a decrease in cadherin function, contact-dependent regulation of enzyme synthesis, and gap junctions (Rutishauser et al., 1988). With respect to PSA-dependent attenuation of interactions, it has been found that the presence of PSA on NCAM can inhibit contact-dependent enzyme synthesis, adhesion via the L1 molecule, and attachment of cells to collagen or laminin matrices via integrins (Acheson et al., 1991). In each of these cases, it could be shown that NCAM or PSA themselves did not participate directly in the interaction, but instead appear to play an auxiliary permissive role in regulation of that interaction. While these phenomena were observed in vitro, both PSA and NCAM are highly regulated during development in a variety of spatial and temporal patterns that suggest an important and pleitropic role in embryogenesis.

MECHANISMS BY WHICH NCAM AND PSA COULD REGULATE CELL INTERACTIONS

The precise mechanism by which this broad regulation occurs is not fully understood. For NCAM acting as an adhesion molecule, the situation would seem straightforward: the adhesive action of NCAM could bring the membranes together physically so that other recognition-type receptors, which themselves cannot produce macroscopic adhesion, can interact between cells (Figure 2). However, NCAM is a complex, multidomain protein, and a variety of evidence suggests that it may generate or be influenced by environmental signals either within or outside of the cell (Doherty et al.,

1990; Atashi et al., 1992). Thus the interrelationship between NCAM and other receptors could also involve biochemical pathways.

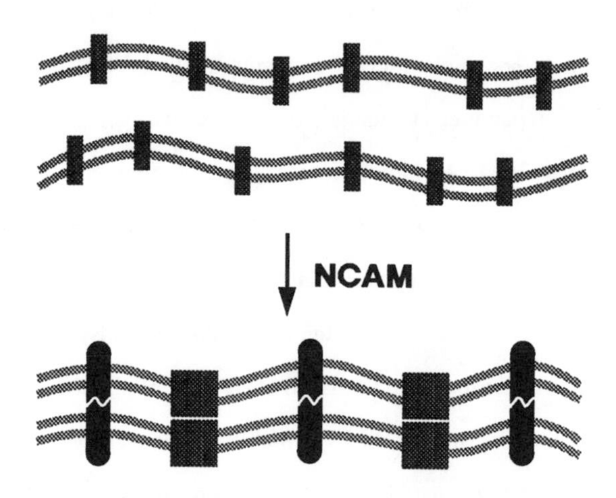

Figure 2. Expression of NCAM without PSA can enhance the probability of effective receptor encounter by promoting membrane-membrane contact. In the example shown, subunits (rectangular receptors) of a specialized junction, which by themselves cannot produce cell-cell adhesion, are able to assemble into a functional unit upon expression of NCAM.

With either mode of action, the negative regulatory role of PSA could be envisioned as a direct inhibition of NCAM function. Remarkably, in situations where NCAM function is inhibited and/or replaced, PSA still maintains its ability to serve as a negative regulator of cell interactions. For example, the ability of plant lectins to promote cell agglutination or contact-dependent enzyme synthesis in the absence of NCAM-mediated adhesion remains can be blocked by the presence of PSA on the inactivated NCAM (Rutishauser et al., 1988). Moreover, the attachment of cells to plastic coated with collagen or laminin, via an integrin but not NCAM-dependent mechanism, is again attenuated by the presence of PSA on the NCAM (Acheson et al., 1991).

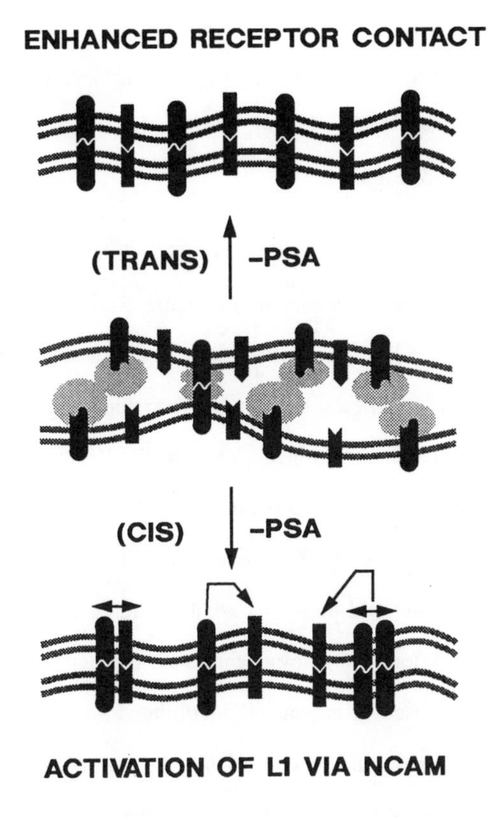

Figure 3. Mechanisms by which NCAM with PSA could regulate cell-cell interactions. Top: *trans* mechanism by which PSA directly affects intercellular space and thus the encounter between receptors on apposing cells. Bottom: *cis*-acting mechanisms in which PSA indirectly affects cell-cell interactions by influencing heterophilic receptor-receptor interactions within the plane of the membrane (left), generating a receptor-activating/inhibiting signal (center), or by perturbing NCAM receptor clustering and thereby affecting other receptors (right).

In view of this regulation by PSA in the absence of NCAM-mediated adhesion, it has been proposed that PSA could affect cell interactions via a physical impedance (via hydrated volume and/or charge repulsion) of overall membrane-membrane contact (Figure 3, top). Simply put, if there were enough space occupied or produced by the presence of PSA, the *trans* interaction of receptors on apposing cells would be impaired. Once again, however, one is confronted by the possibility that PSA on NCAM could alter some signal that affects other receptors on the same cell. Another alternative would be that PSA impedes molecular interactions between molecules in the plane of the membrane (Doherty et al., 1992; Singer, 1992)). These *cis*-type mechanisms are represented at the bottom of Figure 3.

THE SIZE AND DENSITY OF NCAM-PSA ON THE CELL SURFACE.

One requirement of the *trans* mechanism in Figure 3 is that PSA should occupy a substantial volume at the cell surface, producing in effect a thin coating over the cell surface. To examine this parameter, both the surface density of NCAM and the size of hydrated PSA on NCAM was determined (Yang et al., 1992). In Figure 4 is shown a scale representation summarizing these results. The findings suggest that the physical presence of PSA on a cell surface is of sufficient magnitude to influence the intimacy of membrane-membrane contact and thereby *trans* binding between receptors on apposing cells. Of course this fact does not rule out that a *cis* mechanism could operate as well. Another prediction of a space-filling model involving PSA hydration and/or charge is that such on cell interaction should be very sensitive to salt concentration. In fact, it has been found that PSA has almost no effect on cell adhesion at 0.4 M NaCl, whereas the effect at 0.02 M NaCl is several times greater than under physiological conditions (0.15 M) (P. F. Yang and U. Rutishauser, unpublished results).

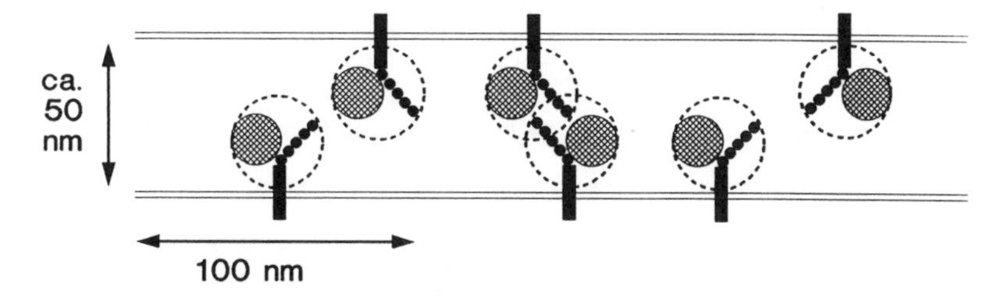

Figure 4. Schematic modeling of the interaction of two cell membranes, each bearing NCAM at the density measured live cells, and PSA with a volume corresponding to that determined by dynamic light scattering (Yang et al., 1992). The dashed circle around each NCAM represents the ability of the molecule to rotate rapidly in the plane of the membrane.

PSA CAN INFLUENCE INTERCELLULAR SPACE

A more definitive test for a *trans* mechanism would be to determine if PSA can increase intercellular space under experimental conditions where *cis* mechanisms are less likely. To provide this test, the influence of PSA on membrane-membrane contact was determined in studies where all adhesion was blocked by specific inhibitors, and cell-cell contact was produced mechanically through gentle centrifugal force (Yang et al., 1992). The results obtained (Table 1), indicate that the presence of PSA increases the volume between cell membranes by about one-quarter.

Table 1. Effect of PSA and other cell surface proteoglycans on the relative amount of space between apposed cell membranes (Yang et al., 1992).

PSA	+	–	+
CS and HS	+	+	–
Space:	100%	76%	99%

Cells were treated with specific enzymes to remove PSA or both chondroitin sulfate and heparan sulfate (CS and HS). Values represent the percentage of trapped ferritin particles per um of apposed membrane, relative to untreated cells.

As depicted in Figure 5, such a change in the distance between membranes would shift the relative positions of apposing receptors by some 10-15 nm. Inasmuch as the protein domains involved in binding are typically about 5nm in length, a change of this magnitude could produce profound changes in receptor-mediated interactions.

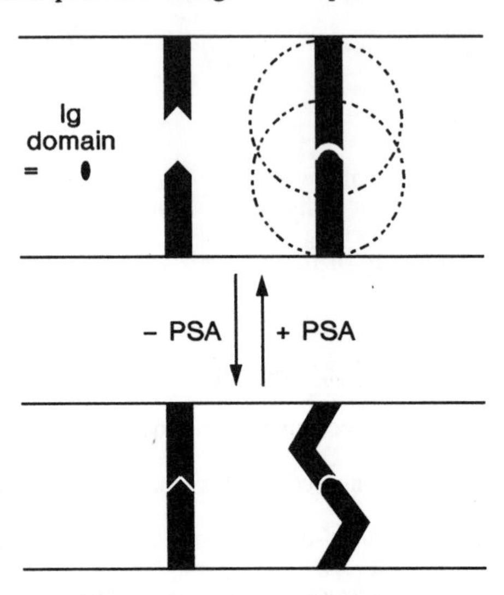

Figure 5. Schematic representation of the effect of PSA (dashed circles) on interactions between receptor pairs. Because the change in intercellular space (about 25%) produced by PSA is large as compared to the size of an immunoglobulin-like domain, it is reasonable to expect that receptor-receptor interactions via those domains would be influenced. Also illustrated is the possibility that a hinged receptor (such as NCAM) might be less susceptible to such differences.

POLYSIALIC ACID PROMOTES AXON OUTGROWTH IN VITRO

The ability of axons to grow on a substrate is dependent in part on the ability of their growth cones to adhere to that substrate. Thus it would seem paradoxical that the presence of PSA on both the axons and the substrate causes an increase rather than a decrease in the rate of axon outgrowth (Zhang et al., 1992). Also remarkable is the fact that this phenomenon appears to involve in particular the effect of PSA on interaction mediated by the L1 adhesion molecule. That is, when axon growth is slowed after specific enzymatic removal of PSA, this effect can be reversed by addition of antibodies that block L1-mediated adhesion, but not by antibodies against other adhesion molecules that promote axon growth. This situation is summarized in Figure 6, which proposes that while some interaction is favorable for growth cone migration, too much involvement of the L1 molecule can slow down the translocation across the substrate.

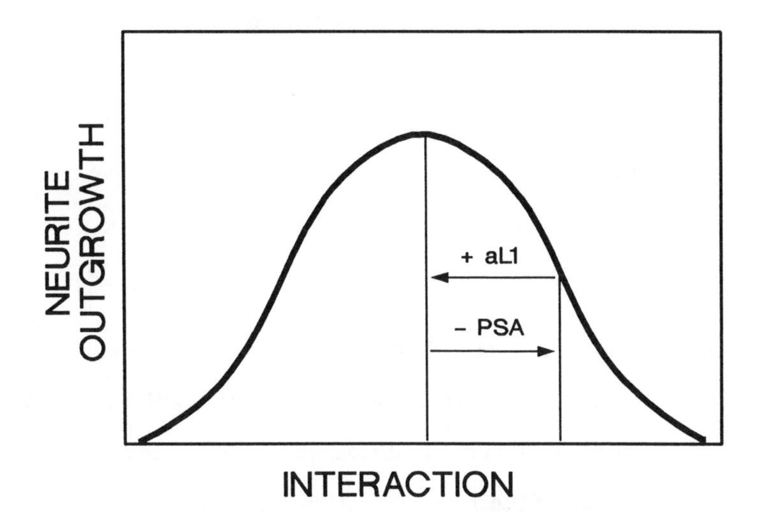

Figure 6. Relationship between cell-cell interaction and the extent of neurite outgrowth on a neuronal substrate (Zhang et al., 1992). The ascending portion of the curve represents the ability of growth cone-substrate interaction adhesion to promote motility, whereas the descending portion indicates that too much interaction can slow down elongation. Experiments suggest that L1 is particularly influential in the latter, and that L1 function is regulated to optimal levels by PSA.

It is curious that in both the above in vitro as well as in vivo studies (see below and chapter by Landmesser), the L1 adhesion molecule appears to be particularly susceptible to regulation by PSA. In this respect it may be relevant that interference reflectance microscopic studies by Drazba et al. (1992) have indicated that growth cones are much more intimately associated with an L1-containing substrate than one composed of other adhesion molecules. Thus L1-mediated outgrowth might be particularly influenced by an increase in the space filled by PSA between the cell membrane and the substrate.

EFFECTS OF POLYSIALIC IN VIVO

Much of the PSA found in vertebrate embryos is associated with large axon bundles. Such bundles are produced by the growth of axons along other axons, that is, through migration of PSA-positive growth cones along PSA-positive axon shafts. While this situation is similar to the in vitro experiment described above, the effects of PSA in vivo appear to be manifested in terms of improved pathway selectivity rather than more rapid outgrowth (Landmesser et al., 1990; Tang et al., 1992; see subsequent chapter by Landmesser). Nevertheless, the cellular mechanism is the same: the presence of PSA appears to attenuate L1-mediated interactions selectively, thus allowing the axon to respond more efficiently to guidance cues in their environment.

This situation is presented schematically in Figure 7, using the example of motor axon innervation of skeletal muscle. The outgrowth of the first PSA^+, $L1^+$ axon over the PSA^-, $L1^-$ muscle substrate appears not be affected by specific enzymatic removal of the axonal PSA or by inhibition of either L1 or NCAM by specific antibodies. However, the subsequent axons, in choosing whether to follow the preceding axon(s) or to branch out over the muscle surface, are highly influenced in vivo by all three perturbations. That is, anti-L1 increases branching by reducing axon-axon interaction, while anti-NCAM decreases branching by having a greater effect on axon-muscle interaction. In terms of mechanism, the system behaves like a competition between growth cone-axon interaction to promote formation of bundles and growth cone-muscle interaction to promote branching. The effect of PSA removal, which is to decrease branching, seems at first glance to be contradictory, since PSA is a negative regulator and its removal might be expected to be the opposite as anti-NCAM. However, as described above, PSA is a particularly potent regulator of interactions when both cells have PSA and L1, which in this situation occurs in axon-axon but not axon-muscle interactions.

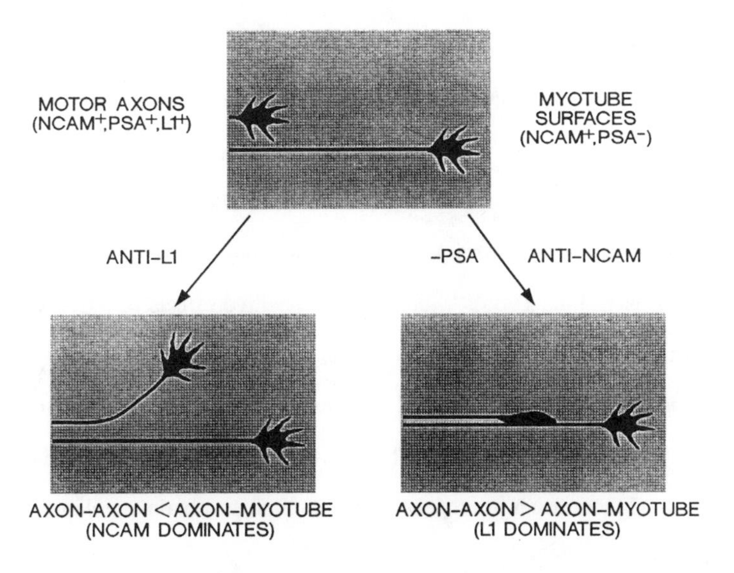

Figure 7. Cellular and molecular interactions in nerve branching (Rutishauser and Landmesser, 1991). The first axon to emerge into a muscle region interacts only with the muscle cell surface; subsequent axons face a choice of two substrates, the muscle and the preceding axons. If L1-mediated axon-axon interaction dominates, a fascicle is formed; if NCAM-mediated axon-myotube interaction dominates, the axons branch apart.

CONCLUSION: A RATIONALE FOR THE STRUCTURAL COMPLEXITY OF NCAM

While this chapter has focussed largely on the remarkable chemical and biological properties of PSA, the extent of structural complexity and diversity of the NCAM polypeptide chain is equally unusual. These attributes include a heparin binding site, homophilic binding site, extracellular alternative splicing, a hinge region, intracellular alternative splicing, a PEST sequence, and intracellular phosphorylation. Moreover the action of NCAM has been found to influence or be influenced by intracellular calcium (Doherty et al., 1990) and protein kinase C (Cervello et al., 1992). The question is raised, therefore, as to how this complexity relates to the proposed role of NCAM and PSA as regulators of cell interaction.

Already mentioned above is the hypothesis that these attributes contribute to a *cis*-acting mechanism by which NCAM and PSA regulate cell interactions (Figure 3, bottom). While this remains an open issue, there is some experimental evidence that would restrict the nature of a more complex *cis* process. That is, effects of PSA on cell interactions are not only demonstrable with live cells, but also with purified membrane vesicles (Rutishauser et al., 1985). Moreover, the positive effects of NCAM in contact-dependent regulation of enzyme synthesis can be replaced by a plant lectin (wheat germ agglutinin), and yet this lectin-stimulated response remains susceptible to inhibition by PSA (Acheson and Rutishauser, 1988). Thus neither intact intracellular biochemistry nor NCAM adhesive function per se appear to be required to obtain regulation in at least some systems.

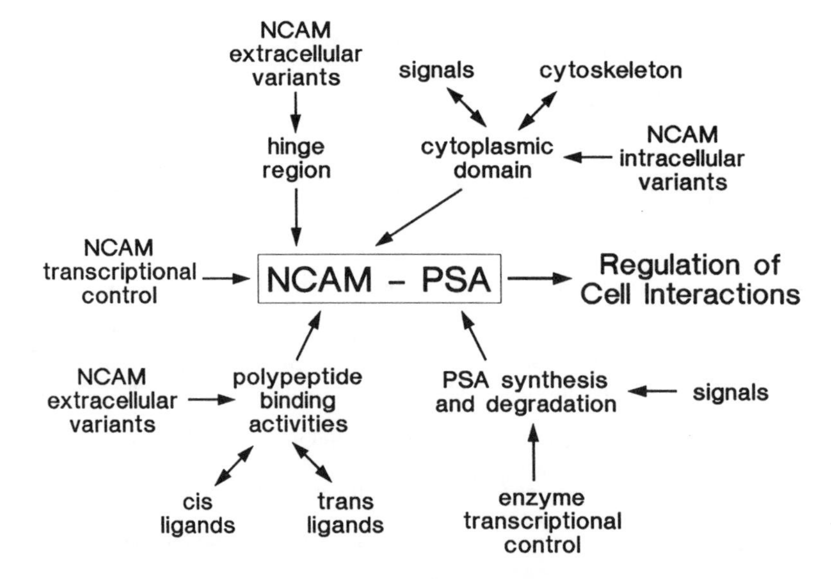

Figure 8. Potential avenues for input/output relationships between NCAM-PSA and other aspects of cell function.

If the regulation itself occurs via a *trans* mechanism (Figure 2 and Figure 3, top), then it becomes necessary to look for other uses for NCAMs multifunctional structure. An attractive possibility is available: if NCAM and PSA constitute a general system for

If the regulation itself occurs via a *trans* mechanism (Figure 2 and Figure 3, top), then it becomes necessary to look for other uses for NCAMs multifunctional structure. An attractive possibility is available: if NCAM and PSA constitute a general system for modulation of a fundamental cell function, then it would be expected that this system be able to receive from and send signals to other fundamental cell functions. In Figure 8 is illustrated a number of input/output relationships that have been suggested on the basis of known properties of NCAM. While the details are mostly speculative, the overall scene is clear: NCAM is not isolated from other cell processes and is capable of communicating with a variety of both intracellular and extracellular components. Thus the molecule appears to be well positioned to carry out the central regulatory role proposed here.

ACKNOWLEDGMENTS
This work was supported in part by NIH grants HD18369 and EY06107.

REFERENCES

Acheson, A. and Rutishauser, U. (1988) J. Cell Biol. 106, 479-486.
Acheson, A., Sunshine, J.L. and Rutishauser, U. (1991) J. Cell Biol. 48, 143-153.
Atashi, J.R., Klinz, S.G., Ingraham, C.A., Matten, W.T., Schachner, M. and Maness, P.F. (1992) Neuron 8, 831-842.
Cervello, M., Lemmon, V., Landreth, G. and Rutishauser, U. (1991) Proc. Natl. Acad. Sci, USA 88, 10548-10552.
Doherty, P., Ashton, S.V., Moore, S.E. and Walsh, F.S. (1991) Cell 67, 21-33.
Doherty, P., Rimon, G., Mann, D.A. and Walsh, F.S. (1992) J. Neurochem. 58, 2338-2341.
Drazba, J. and Lemmon, V. (1992, in press) Abstr., Soc. for Neurosci..
Landmesser, L., Dahm, L., Tang, J. and Rutishauser, U. (1990) Neuron 4, 655-667.
Rutishauser, U. (1989) In: Neurobiology of Glycoconjugates (Margolis and Margolis, eds.), Plenum Publishing, NY, pp. 367-382.
Rutishauser, U., Acheson, A., Hall, A.K., Mann, D.M. and Sunshine, J. (1988) Science 240, 53-57.
Rutishauser, U. and Jessell, T.M. (1988) Phys. Rev. 68, 819-857.
Rutishauser, U., Watanabe, M., Silver, J., Troy, F.A. and Vimr, E.R. (1985) J. Cell Biol. 101, 1842-1849.
Singer, S.J. (1992) Science 255, 1671-1677.
Rutishauser, U. and Landmesser, L. (1991) TINS 14, 528-532.
Yang, P., Yin, X., and Rutishauser, U. (1992) J. Cell Biol. 116, 1487-1496.
Zhang, H., Miller, R.H. and Rutishauser, U. (in press) Neuron.

Polysialic Acid
J. Roth, U. Rutishauser and F. A. Troy II (eds.)

THE EFFECTS OF IN-VIVO ALTERATION OF POLYSIALIC ACID ON AVIAN NERVE-MUSCLE DEVELOPMENT

Lynn Landmesser

Department of Physiology and Neurobiology, University of Connecticut,
Storrs, Connecticut, 06269, U.S.A.

SUMMARY: During development of the motor system of the chick embryo, immunostaining with an antibody specific for polysialic acid (PSA) revealed complex patterns of temporal and spatial regulation which suggested that PSA might play a role in specific axonal pathfinding, intramuscular nerve branching, and myotube cluster separation. The PSA observed appeared to be predominantly on the neural cell adhesion molecule NCAM. By removal of PSA at the relevant stages via in-ovo injection of an endoneuraminidase we obtained evidence for PSA's involvement in each of these developmental events. We also showed that PSA could be regulated by electrical activity; being up-regulated on nerve and down-regulated on muscle following blockade of neuromuscular activity with the nicotinic antagonist d-Tubocurarine.

INTRODUCTION

Adhesion molecules have been shown to play important roles during development of the nervous system, being involved in axonal pathfinding, neuronal migration, intramuscular nerve branching and synaptogenesis, and myogenesis. These molecules allow communication between specific subsets of cells at precise developmental times, either by acting as adhesive ligands or as signalling molecules. One way of controlling the cell-cell interactions mediated by these molecules is by temporal and spatial regulation of their patterns of expression. However, another way of modulating the effectiveness of at least some of these molecules is through polysialic acid (PSA). This carbohydrate modification of the NCAM molecule has been shown in a variety of in-vitro assays to alter the apparent effectiveness of NCAM in mediating adhesion and neurite outgrowth (Hoffman and Edelman, 1983; Rutishauser et al., 1988; Doherty et al., 1990b, Acheson et al., 1991). Polysialic acid on NCAM has also been shown to modulate the function of other cell surface molecules such as L1 and laminin (Acheson et al., 1991). Although PSA on NCAM has been known for some time to be developmentally regulated (Chuong and Edelman, 1984; Schlosshauer et al., 1984), the complexity of its temporal and spatial regulation could only be fully appreciated following immunostaining with PSA specific monoclonal antibodies. During our studies of development of the motor system in chick embryos, we observed that PSA was regulated in space and time 1) during initial neurite outgrowth (Tang et al., 1992), 2) during the period of intramuscular nerve branching and synaptogenesis (Landmesser et al., 1990), and 3) during the period of myogenesis when secondary myotubes separated from myotube clusters (Fredette et al., 1992). These observations strongly suggested that PSA might be involved in regulating these events. We have tested these ideas by modulating PSA directly in-vivo by the injection of an endoneuraminidase, endo-N, into developing embryos. This paper will summarize evidence for a role of PSA in each of these three events.

PATHFINDING BY AVIAN MOTONEURONS

As shown in Fig. 1, some 20,000 motoneurons destined for different target muscles exit the lumbosacral spinal cord intermingled with each other in eight separate spinal nerves. It has been known for some time that these neurons reach their target with little error (Landmesser, 1978; Tosney and Landmesser, 1985a) first by segregating into target specific bundles in the plexus region and secondly by projecting out the appropriate muscle nerve (Lance-Jones and Landmesser, 1981a; Tosney and Landmesser, 1985 a,b). The distribution of characterized cell adhesion molecules such as NCAM, L1, N-Cadherin and TAG-1 does not appear compatible with them playing the role of specific motoneuron guidance molecules. However the expression pattern of PSA, in this case carried predominantly by NCAM, suggested that it might be playing a modulatory role in axon guidance.

Specifically, PSA was observed to be low or absent on motoneuron axons (Tang et al., 1992) during their growth to the plexus region, a time when they remain tightly fasciculated and maintain relatively straight trajectories (Tosney and Landmesser, 1985b). PSA detected immunocytologically first appeared at St 23-24 (Tang et al., 1992) when axons exhibit a robust defasciculation in the plexus region and alter their neighbor relationships as they sort out into target specific bundles (Lance-Jones and Landmesser, 1981a; Tosney and Landmesser, 1985b). This sorting process has been proposed to play an essential role in the correct targeting of motor axons to their appropriate muscles. In addition, we observed that PSA levels were higher on dorsally projecting motoneurons than on those that projected to ventral targets (Tang et al., 1992; Figure 2). We also found by biochemical means that axonal PSA during this outgrowth period was on

NCAM, that all three isoforms of NCAM were present and sialylated, and that this PSA could be removed by endoneuraminidase specific for long chain length PSA (Figure 2; Tang et al., 1992).

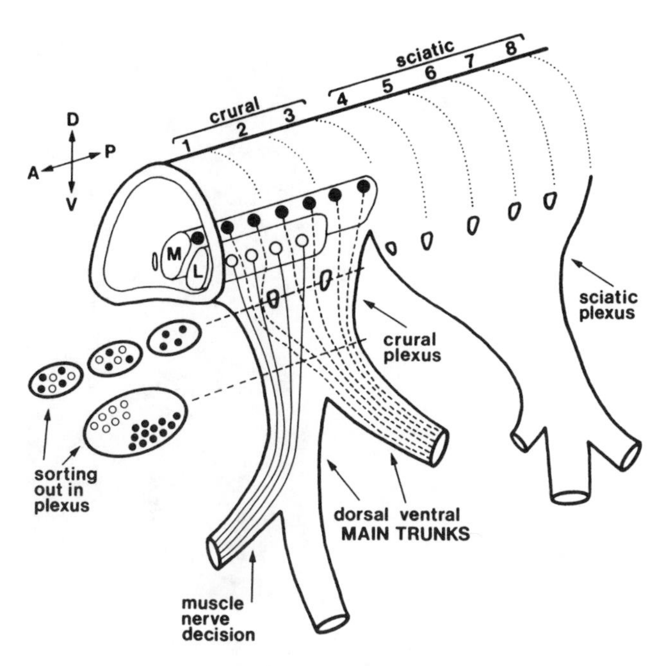

Figure 1. Diagram of pathways followed by chick lumbosacral neurons during initial outgrowth. Two motoneuron pools are indicated; dorsally projecting motoneurons (dark), ventrally projecting motoneurons (light). M, medial; L, lateral.

To test whether PSA was involved in pathfinding, we removed it with in-ovo injections of endo-N, either early so that it was absent during the period of axonal sorting out in the plexus, or later, after axons had sorted out but before they had selected the appropriate muscle nerve (Figure 2; Tang et al., 1992). We then assessed the correctness of motoneuron projections at St 30 by determining the cord positions of motoneurons projecting to different muscles, following retrograde labeling with horseradish peroxidase. Following early PSA removal it was clear that most muscles were innervated by some foreign motoneurons (10-40% of the total neurons projecting to a given muscle). 11/13 pools assayed exhibited such errors. The extent of errors was similar in pools that had high PSA levels (dorsally projecting) and those that had lower PSA levels (ventrally projecting). However, no errors were found (0/12 pools assayed) when PSA was removed later, during the period of muscle nerve selection (Tang et al., 1992).

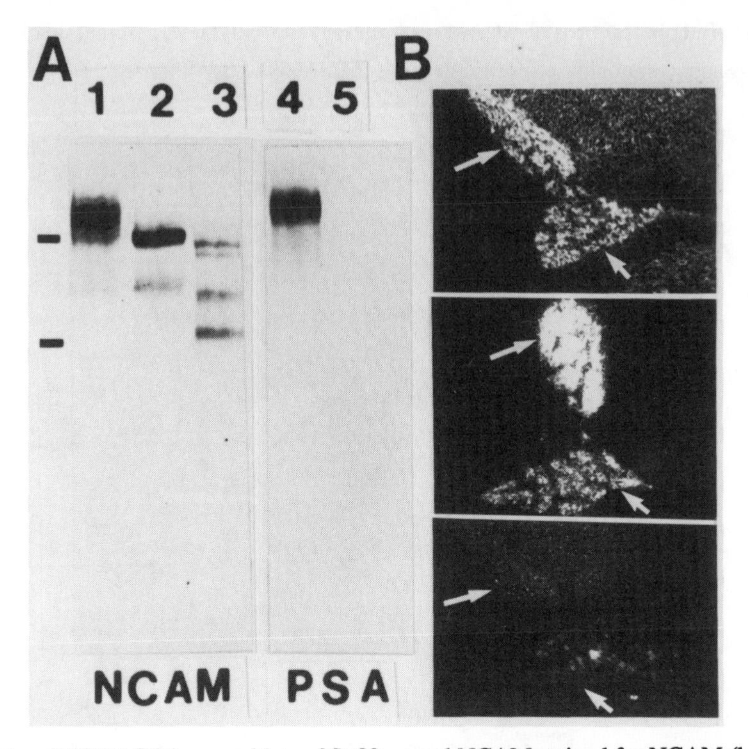

Figure 2a. SDS PAGE immunoblots of St 30 axonal NCAM stained for NCAM (lanes 1-3) or PSA (lanes 4 and 5), native sample (lane 1), treated with endo-N (lane 2 and 5), treated with neuraminidase (lane 3); Mr marks 210 and 116. 2b. Frozen cross sections of control nerve stained for NCAM (top) and PSA (middle) or from endo-N injected embryo stained for PSA (bottom); long arrow, dorsal trunk; short arrow, ventral trunk.

We conclude that the normal pattern of PSA expression is necessary for specific axon guidance, and that it appears to be acting during axonal rearrangement in the plexus, rather than during the selection of the appropriate muscle nerve. Preliminary experiments indicate that removal of PSA partially prevents the rearrangement of axons that occurs in the plexus region. Thus we feel that the results thus far obtained are most compatible with PSA playing a permissive role, possibly by reducing axon-axon adhesion sufficiently to allow axons to respond to specific guidance cues. The ligands which are modulated by PSA remain to be determined. We also can not exclude the possibility that the different levels of PSA expressed by motoneurons projecting to dorsal vs. ventral targets let PSA play a more instructive role, for example, in allowing these two populations of motor axons to sort out from one another and/or to respond differentially to other molecules acting as guidance cues. The actual effect that such differential PSA expression has on axonal behavior will have to be probed in more controlled experiments in culture, for example by determining how it affects how these axons respond to substrates of purified adhesion molecules (Lemmon et al., 1989) or of fibroblast cells transfected with different isoforms of the NCAM molecule (Doherty et al., 1990a). It is clear that in this in-vivo system with multiple adhesion systems present, PSA is not necessary for axon outgrowth (see for example Doherty et al., 1990b), since its removal did not impede either the rate or overall extent of axon outgrowth (Tang and Landmesser, unpublished observations).

THE REGULATION OF INTRAMUSCULAR NERVE BRANCHING, SYNAPTOGENESIS, AND NEURON SURVIVAL

Motoneuron ingrowth into muscle is a highly regulated process in which axons first enter muscles in larger fascicles and then establish contact with myotubes by the formation of side branches (Figure 3; Dahm and Landmesser, 1988). In chick muscle the branching patterns, which prefigure the pattern of synapses , differs between slow and fast muscle regions; in fast regions axons grow in a more defasciculated manner transverse to the plane of the myotubes, while in the slow region,

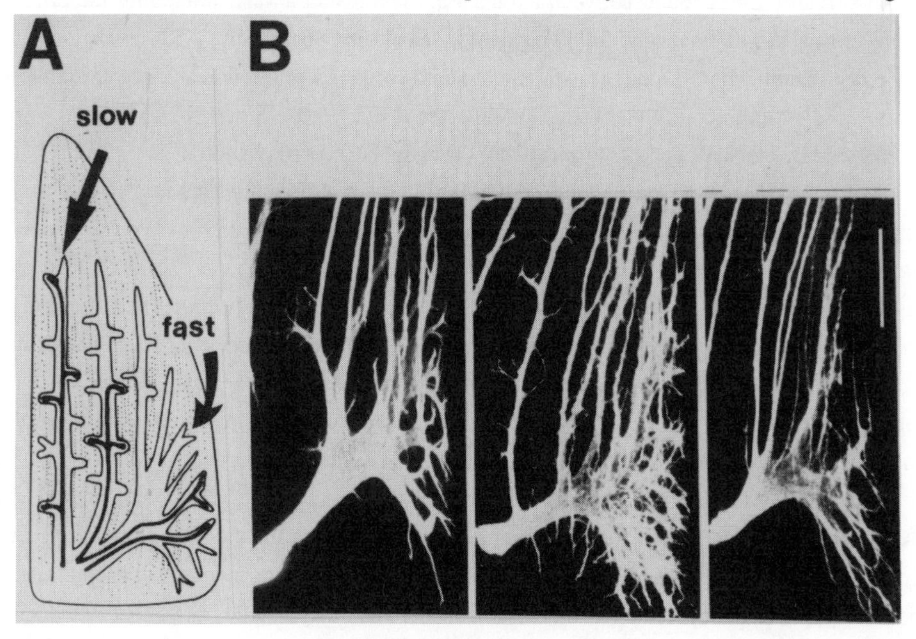

Figure 3a. Diagram of the intramuscular nerve branching pattern of the chick iliofibularis muscle at St 33. 3b. Whole mounts of fast region stained with a neuron specific antibody from a control (left) dTC treated (middle) and dTC and endo-N treated (right) embryo.

the distributed pattern of synapses on single myotubes, results from axons first growing in major nerve trunks parallel to the muscle fibers, with the subsequent formation of multiple side branches at right angles to the main nerve trunks.

Results obtained by in-ovo injection of blocking antibodies to some of the adhesion molecules suggested that the branching pattern was at least in part regulated by the balance between axon-axon fasciculation (mediated predominantly by L1) and axon-myotube interaction (mediated at least in part by NCAM) (Landmesser et al., 1988). Anti-L1 resulted in defasciculation and an increase in branching, anti-NCAM in a decrease in branching. Neuronal activity was shown to play an important role. Blocking neuromuscular activity by in-ovo injection of d-Tubocurarine, a nicotinic antagonist, produced a robust defasciculation and an increase in branching (Figure 3,4; Dahm and Landmesser, 1988). We expected that this might have been caused by a down regulation in axonal L1. However, no changes in levels of L1 or NCAM were observed following activity blockade, nor were there differences in the fast and slow muscle regions that could account for the different patterns of axonal branching (Landmesser et al., 1988).

A correlation was found between axonal PSA levels and axonal branching patterns, both during normal development and following activity blockade. Specifically, PSA levels were higher in the fast region, where axonal growth is more defasciculated, and it was sharply upregulated on all axons following activity blockade (Landmesser et al., 1990). Since NCAM levels were not altered, activity blockade appears to specifically increase NCAM sialylation.

To test whether PSA was in fact modulating the axonal branching pattern it was removed during this period by in-ovo injections of endo-N (Landmesser et al., 1990). We found that PSA removal reduced branching during normal development (Figure 4), that it caused the fast growth pattern to be more like the slow growth pattern, and that it largely prevented the increased branching produced by activity blockade (Fig 3,4). While it appeared that PSA was primarily modulating L1 function (since its effect could be antagonized by antibodies to L1), we can not rule out the participation of other molecules. For example, in neurite outgrowth assays on fibroblasts transfected with NCAM, NCAM sialylation enhances neurite outgrowth (Doherty et al., 1990b). In this context, our results could also be at least in part explained by PSA removal interfering with NCAM-mediated interactions between axons and myotubes and thus reducing branching by this means. Whatever the precise cellular mechanism and the cell ligands involved, it is clear that PSA modulates the complex process of intramuscular nerve branching.

Regulation of intramuscular nerve branching has important developmental consequences, because it can affect how developing motoneurons interact trophically with their targets during the period of naturally occurring cell death when approximately half of these motoneurons die (Oppenheim, 1991). It has been suggested that the uptake of trophic factor necessary for neuron survival might be limited by sites of uptake (for example terminal nerve branches and/or synapses).

It was in fact observed that the increased neuronal branching produced by activity blockade, was mirrored by an increase in the number of synapses (Dahm and Landmesser, 1991).

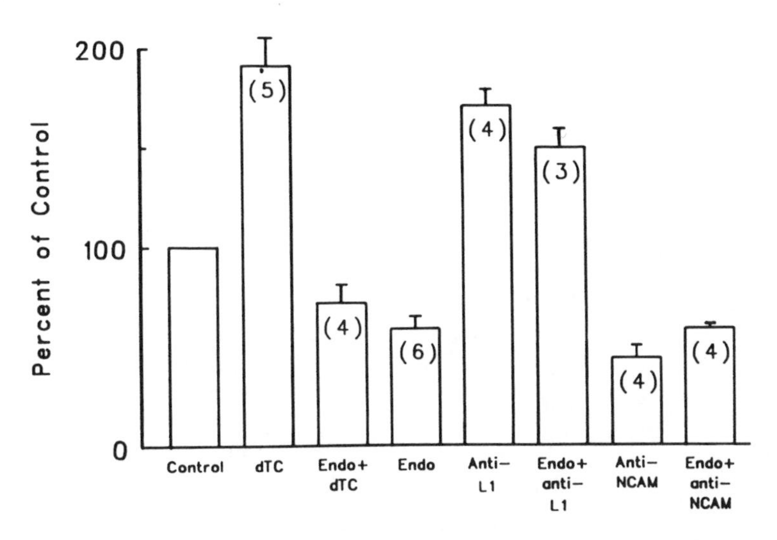

Figure 4. The number of side branches in the slow region of the iliofibularis muscle following different treatments, expressed as percent of control.

More recently we have subjected this hypothesis to another test. We found that the reduction in neuronal branching produced by endo-N during normal development, resulted in a decreased number of synapses at the onset of the motoneuron cell death period (700 vs 1,600 in the p.iliotibialis muscle). This in turn was correlated with a reduction in neuronal survival (the number of motoneurons surviving at St 36, the end of the cell death period, was only 55% of control numbers; Tang and Landmesser, 1992). Although PSA has also been observed on sodium channels (James and Agnew, 1987), its removal in-ovo via endo-N neither altered spontaneous embryonic motility nor the activation of motoneurons or muscles assessed electrophysiologically (Landmesser et al., 1990; Tang and Landmesser, in preparation). We therefore favor the idea that PSA is acting primarily by modulating the function of cell adhesion molecules.

In conclusion, these observations make the interesting point that PSA, which is capable of regulating cell-cell interactions through its influence on adhesion molecule function, can itself be regulated by functional activity. This could have important consequences in the developing nervous system where patterns of electrical activity have been implicated in the formation of neuronal circuits (Constantine-Paton et al., 1990). In addition, modulation of adhesion molecule function, could be involved in the synaptic plasticity associated with learning and memory phenomena. Interestingly, it has recently been shown in Aplysia that long term synaptic alterations are correlated with alterations in the expression of an NCAM-like molecule (Mayford et al., 1992). In our system adhesion molecule function is altered not by actual changes in expression levels but by being modulated by PSA. It will now be important to determine how PSA is regulated both during normal development and following activity blockade.

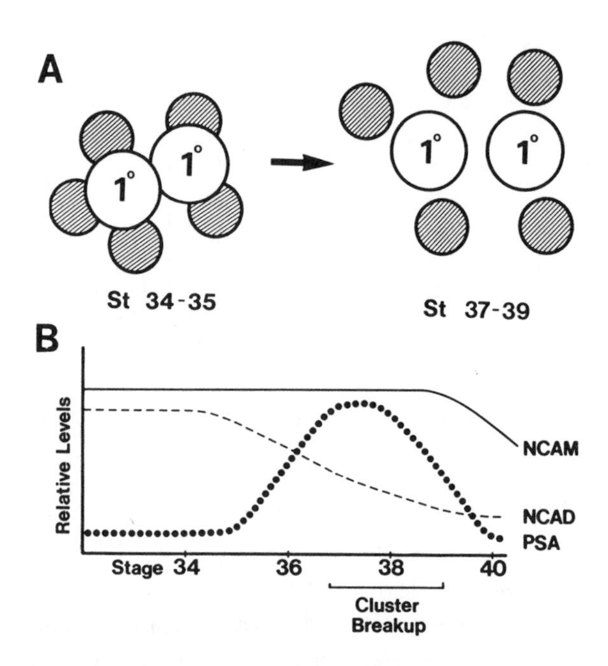

Figure 5a. Diagram of the relationship between primary (1°) and secondary (diagonal lines) myotubes during in-vivo myogenesis (cross-sectional view). 5b. Diagram representing approximate levels of adhesion molecules on myotube surface during myotube cluster breakup.

POLYSIALIC ACID EXPRESSION DURING MYOGENESIS

In-vivo myogenesis is a complex process in which cell-cell interactions between myoblasts, primary myotubes, and secondary myotubes are precisely regulated in time and space, presumably by cell surface adhesion molecules. Much work in this area has focused on the role of adhesion molecules such as NCAM and N-cadherin in early events in myogenesis such as myoblast fusion and myotube formation (Knudsen et al., 1990 a,b; Dickson et al., 1990). However, they may also be involved in later events (Covault et al., 1986). Primary myotubes when first formed (St 28-34 in the chick iliofibularis) exist as discrete clusters of electrically coupled myotubes, and in turn serve as scaffolding for a secondary generation of secondary myotubes (Figure 5a). The formation of the normal complement of muscle fibers requires that these myotubes separate from one another to form individually activated muscle fibers. In the chick iliofibularis the initial wave of secondary myotubes form between St 34-37 and shortly after separate from the clusters (Figure 5a; Fredette and Landmesser, 1991). We observed that following neuromuscular activity blockade induced by dTC, myotubes failed to separate normally from clusters, resulting in large, apparently highly adherent, clusters of myotubes.

The potential role of several adhesion molecules in myotube cluster formation and separation was investigated by determining the cellular distribution of NCAM, N-cadherin, and PSA with light and EM immunostaining during normal development and following activity blockade (Fredette et al., 1992 and in preparation). N-cadherin was preferentially expressed on myoblasts and on newly formed myotubes attached to primary myotubes, but not on the apposed surfaces of more mature myotubes. N-cadherin levels thus declined on the apposed surfaces of myotubes with time, and overall levels were quite low by St 37 when cluster separation was at its peak.

NCAM was expressed on all myoblast and myotube surfaces, but its polysialylated form was sharply and transiently up-regulated, as judged by immunostaining, between St 36-38, suggesting that the latter might play a role in myotube separation (Figure 5b). Thus during the period that myotube clusters separate, PSA is up-regulated and N-cadherin is down- regulated. If PSA is acting as a negative regulator of adhesion, as has been suggested (Rutishauser et al., 1988), its expression at this time coupled with the decline in N-cadherin could cause myotube clusters to separate. Interestingly, activity blockade, which interferes with myotube separation and results in large clusters of tightly adherant myotubes, completely prevents this transient expression of PSA and delays the down -regulation of N-cadherin (Fredette et al, 1992).The latter observations provide additional support for the role of these molecules in myogenesis, specifically in the normal separation of developing myotubes into independent myofibers.

Biochemically, the transition from primary to secondary myogenesis is accompanied by a shift from the 145 kD NCAM isoform to the lipid-linked 130 kD isoform. The 130 kD isoform appears to carry most of the PSA during the period of myotube cluster separation, while much of the 145 kD isoform is not sialylated. Following activity blockade the shift from the 145 kD to the 130 kD isoform is largely prevented (Fredette et al, 1992). Although some of the lipid-linked 130 kD isoform is still expressed , PSA expression is completely abolished. Thus in addition to selectively preventing the expression of the 130 kD isoform, activity blockade is also altering overall NCAM sialylation, possibly through an effect on the sialyltransferase. Thus developing muscle might provide a useful system in which to study the sialyltransferase responsible for this developmentally regulated expression of PSA and to determine how functional activity affects its activity.

At the EM level NCAM immunoreactivity was observed on all myotube surfaces and was especially intense on the surfaces of apposed myotubes. In contrast, PSA immunoreactivity was excluded from these apposed surfaces, and occurred primarily on free surfaces ; however it often extended for short distances into the apposed surfaces between two myotubes. Thus if PSA is contributing to myotube separation by reducing adhesive interactions between myotubes, it must be acting somewhat as a wedge, driving two adherant myotubes apart from either side. Alternatively, PSA is in a position to modulate interactions between ligands on the myotube surface and those in the extracellular matrix which forms a basal lamina around each cluster of myotubes.

Despite the correlation between PSA expresssion and myotube cluster separation, both during normal development and following activity blockade, we have not been able as yet to prevent cluster separation by removing PSA with injections of endo-N. This suggests that something in addition to PSA is required for myotube separation; one possibility is that N-cadherin must also be downregulated.

In summary, both by observing patterns of PSA expression and by altering them with in-vivo injections of endo-N , we have been able to provide evidence that PSA is involved in axonal pathfinding, in the process of intramuscular nerve branching and synaptogenesis, and in myogenesis. While many of the details of the precise cellular mechanisms of its mode of action remain to be worked out , the in-vivo systems described here, together with more simplified tissue culture systems should provide a fruitful ground for future studies.

ACKNOWLEDGEMENTS

Thanks are due to Ji-cheng Tang and Barbara Fredette who performed much of the work discussed here, to Liviu Cupceancu for his technical assistance, and to Karen Sommer for typing this manuscript ; supported by NIH grant NS 19640.

REFERENCES

Acheson, A., Sunshine, J. and Rutishauser, U. (1991) J. Cell Biol. 114: 143-153.
Chuong, C.M. and Edelman, G.M. (1984) J. Neurosci. 4: 2354-2366.
Constantine-Paton, M., Cline, H.T. and Debski, E. (1990) Ann. Rev. Neurosci. 13: 129-154.
Covault, J., Merlie, J.P., Goridis, C. and Sanes, J.R. (1986) J. Cell Biol. 102: 731-739.
Dahm, L. and Landmesser, L. (1991) J. Neurosci. 11: 238-255.
Dickson, G., Peck, D., Moore, S.E., Barton, C.H. and Walsh, F.S. (1990) Nature 344: 384-351.
Doherty, P., Fruns, M., Seaton, P., Dickson, G., Barton, C.H., Sears, T.A. and Walsh, F.S. (1990a) Nature 343: 464-466.
Doherty, P., James, C. and Walsh, F.S. (1990b) Neuron 5: 209-219.

Fredette, B.J. and Landmesser, L.T. (1991) Dev. Biol. 143: 19-35.
Fredette, B.J., Landmesser, L.T. and Rutishauser, U. (1992) Neurosci. Abst. (in press).
Hoffman, S. and Edelman, G.M. (1983) Proc. Natl. Acad. Sci. USA. 80: 5761-5766.
James, W.M. and Agnew, W.S. (1987) Biochem. Biophys. Res. Commun. 148: 817-826.
Knudsen, K.A., McElwee, S.A. and Myers, L.C. (1990a) Dev. Biol. 138: 159-168.
Knudsen, K.A., Myers, I. and McElwee, S.A. (1990b) Exp. Cell Res. 188: 175-184.
Lance-Jones, C. and Landmesser, L. (1981a) Proc. R. Soc. Lond. B. 214: 1-18.
Landmesser, L. (1978) J. Physiol. Lond. 284: 391-414.
Landmesser, L., Dahm, L., Schultz, K. and Rutishauser, U. (1988) Dev. Biol. 130: 645-670.
Landmesser, L., Dahm, L., Tang, J. and Rutishauser, U. (1990) Neuron 4: 655-677.
Lemmon, V., Farr, K.L. and Lagenaur, C. (1989) Neuron 2: 1597-1603.
Mayford, M., Barzilai, A., Keller, F., Schachner, S. and Kandel, E.R. (1992) Science 256: 638-644.
Oppenheim, R.W. (1991) Ann. Rev. Neurosci. 14: 453-501.
Reichardt, L.F. and Tomaselli, K. (1991) Ann. Rev. Neurosci. 14: 531-570.
Rutishauser, U., Acheson, A., Hall, A., Mann, D.M. and Sunshine, J. (1988) Science 240: 53-57.
Rutishauser, U. and Jessell, T.M. (1988) Physiol. Rev. 68: 819-857.
Rutishauser, U. and Landmesser, L. (1991) Trends in Neurosci. 14: 528-532.
Schlosshauer, B., Schwartz, U. and Rutishauser, U. (1984) Nature 310: 131-143.
Tang, J. and Landmesser, L. (1992) Soc. Neurosci. Abst. (in press).
Tang, J., Landmesser, L. and Rutishauser, U. (1992) Neuron 8: 1031-1044.
Tosney, K.W. and Landmesser, L. (1985a) J. Neurosci. 5: 2366-2344.
Tosney, K.W. and Landmesser, L. (1985b) J. Neurosci. 5: 2345-2358.
Vimr, E.R., McCoy, R.D., Vollger, H.F., Wilkinson, N.C. and Troy, F.A. (1984) Proc. Natl. Acad. Sci. USA. 81; 1971-1975.

Polysialic Acid
J. Roth, U. Rutishauser and F. A. Troy II (eds.)

POLYSIALIC ACID AS A SPECIFIC AND POSITIVE MODULATOR OF NCAM DEPENDENT AXONAL GROWTH

Patrick Doherty and Frank. S. Walsh

Department of Experimental Pathology, UMDS, Guy's Hospital, London SE1 9RT, England.

SUMMARY : Antibody perturbation studies suggest that cell-contact dependent axonal growth can be mediated in part by the binding of NCAM on neurons to NCAM expressed on a variety of non-neuronal cells. In this review we consider differences in the ability of various isoforms of NCAM to stimulate cell contact dependent neurite outgrowth and recent evidence that suggests this function cannot be accounted for by NCAM dependent adhesion *per se*. In contrast NCAM dependent neurite outgrowth can solely be accounted for by the activation of a cell adhesion molecule (CAM) specific second messenger pathway in neurons that culminates in a calcium influx into neurons. We also discuss recent data that shows that polysialic acid (PSA) on neuronal NCAM can act as a highly specific positive modulator of the cell-contact dependent axonal growth stimulated by NCAM. This result is not readily reconcilable with PSA inhibiting the trans-binding of NCAM on neurons to NCAM on non-neuronal cells and an alternative model is discussed.

INTRODUCTION

One of the intriguing aspects of long polymer chains of α 2-8 linked sialic acid (PSA) is the fact that they are carried on very few glycoproteins in higher vertebrates. To date convincing evidence has been obtained for the presence of PSA on the neural cell adhesion molecule NCAM (Finne 1982, Finne et al., 1985) and on the sodium channel (see chapter by Zuber, Lackie and Roth). NCAM is one of a number of cell surface glycoproteins that has been shown to be involved in contact-dependent recognition between cells (reviewed in Edelman 1986, Walsh and Doherty 1991). Many of these molecules operate by a self or homophilic binding mechanism; that is NCAM on one cell binds directly to NCAM on a second cell (e.g. see Hoffman and Edelman, 1983; Rathjen and Jessel, 1991; Doherty & Walsh 1992). Using *in vitro* assays that measure cell aggregation rates antibodies to NCAM have been shown to inhibit neuronal aggregation (e.g. see Edelman, 1986). Also, the introduction of NCAM by transfection into non-neuronal cells can be correlated with increased adhesion of brain membranes to the transfected cells (Edelman et al, 1987) or by increased adhesion between transfected cells (Edelman et al., 1987; Pizzey et al., 1989). Thus by both old (antibody perturbation) and new (transfection) criteria, NCAM can be operationally defined as being a cell adhesion molecule (CAM).

However, the question as to whether NCAM actually contributes to adhesion *in vivo* remains unclear. In *in vitro* assays an NCAM adhesive component is generally measured following inactivation of calcium-dependent adhesive mechanisms (i.e. under non-physiological conditions). Even under these conditions the relative strength of the adhesive interactions mediated by NCAM expressed by transfection in non-neuronal cells is very weak with cell aggregates of NCAM transfected cells being disrupted by the smallest of shear forces (Pizzey et al., 1989). These results contrast with those obtained with calcium-dependent molecules such as the cadherins which in similar experimental paradigms clearly mediate robust adhesion between cells (reviewed in Takeichi, 1988, 1991). Thus it appears possible that the calcium-dependent adhesion molecules (i.e. cadherins and integrins) rather than the calcium-independent molecules (i.e. NCAM and other immunoglobulin superfamily members such as L1) mediate true adhesion between cells *in vivo*.

Liposomes containing purified NCAM bind better to neuroblastoma cells following removal of PSA (Sadoul et al., 1983). In addition removal of PSA substantially increases the aggregation rate between brain membranes (Hoffman and Edelman, 1983) and also aggregation between intact neuroblastoma cells (Acheson et al., 1991). In the latter study removal of PSA had an effect even when NCAM function was blocked by antibodies supporting the hypothesis that PSA on NCAM may act as a global inhibitor of cell adhesion (Rutishauser et al., 1988).

Adhesion *per se* is a measurement that cannot be readily translated into physiological function. The interest of our laboratory is in the molecular basis of axonal growth and guidance during development and following injury. Considerable evidence suggests that contact between the filopodia of the neuronal growth cone with matrix and cellular elements present in its local microenvironment underlie much of this growth and guidance (e.g. see O'Connor et al., 1990) Direct evidence for NCAM playing a role in contact dependent axonal growth comes from a large number of antibody perturbation studies (reviewed in Doherty & Walsh, 1989). Briefly, work from a number of laboratories has shown that the ability of neurons to extend neurites over monolayers of astrocytes, myotubes and Schwann cells can in some instances be inhibited (in part) by the addition of antibodies that block the function of NCAM. The relative importance of NCAM depended on both the type of neuron and monolayer cell and full inhibition often required the additional perturbation of the function of integrins, N-cadherin and the L1 glycoprotein. For example, antibodies to NCAM contribute to an inhibition of neurite outgrowth from chick retinal neurons grown on astrocyte monolayers but had no effect on neurite outgrowth from chick ciliary ganglion neurons cultured on the same astrocyte monolayers. Yet the same antibodies could inhibit growth from chick ciliary ganglion neurons cultured on muscle cells. Thus NCAM's ability to promote neurite outgrowth appears to be dynamically regulated. Many questions relating to the ability of NCAM to stimulate neurite outgrowth require direct functional assays and the one used in our studies is described below.

MEASUREMENT OF CELL-CONTACT DEPENDENT AXONAL GROWTH STIMULATED BY TRANSFECTED CAMs

A single gene encodes for at least 20-30 (and theoretically up to 300) unique isoforms of NCAM (reviewed in Walsh and Doherty 1991). Differential alternative splicing determines whether NCAM is directly secreted from cells, linked to the membrane by a GPI-anchor or spans the membrane with a small or large cytoplasmic domain. Membrane anchored isoforms of NCAM share a common extracellular structure that consists of five immunoglobulin like domains with two further domains that share homology to fibronectin type III (FNIII) repeats proximal to the membrane. This extracellular structure can be modified at only two sites. The use of up to four small exons (15, 48, 42 and 3 base-pairs in humans) can alter the structure by insertion of up to 37 amino acids between the FNIII repeats. Also, a 30 base pair exon named VASE can be spliced between the two exons that encode the fourth immunoglobulin domain.

The complexity of NCAM isoforms precludes a biochemical approach to the study of NCAM function. We have therefore adopted a transfection based strategy to study the ability of NCAM and other CAMs (N-cadherin and L1) to stimulate cell-contact dependent neurite growth. In transfection models the control substratum (i.e. untransfected cells) can be selected for its ability to support neuronal adhesion *per se*, and this allows for the study of the neurite outgrowth promoting activity of the transfected CAM to be studied in a physiologically relevant model. More importantly, facets of function relating to the lateral diffusion of CAMs in membranes and/or their ability to interact with cytoskeletal elements in a cellular substratum are obviously lost when the CAM is studied as a purified molecule. Our selection of NIH-3T3 fibroblasts for transfection based studies was based on three criterion; firstly they do not express functional levels of any of the above CAMs, secondly they form excellent monolayers and finally they support neuronal adhesion and to a reasonable extent basal neurite outgrowth.

We have transfected 3T3 fibroblasts with plasmid vectors containing cDNAs encoding several isoforms of human NCAM or a single isoform of human L1 or chicken N-cadherin. Stable clones have been isolated and characterised for expression of the transfected CAM. In a large number of studies we have compared the ability of neurons isolated from the chick retina or the rat hippocampus or cerebellum to extend neurites over confluent monolayers of parental 3T3 cells and 3T3 cells expressing characterised levels of the various isoforms of

NCAM, or the other CAMs (reviewed in Walsh and Doherty, 1992; Doherty and Walsh 1992)

ALTERNATIVE SPLICING MODULATES NCAM FUNCTION

Neurite outgrowth from E6 chick retinal ganglion cells, PND1-6 rat cerebellar neurons and E17 rat hippocampal neurons is significantly increased on cell monolayers that express the 120kDa GPI anchored isoform of NCAM or the 140kDa transmembrane isoform with the small cytoplasmic domain (e.g. see Doherty et al., 1990a,b. 1992a-c). In each instance the longest neurite was increased by a factor of 2 - 3 fold over a 16 - 24 hour period of culture. Neurite outgrowth stimulated by transfected NCAM is abolished by antibodies that inhibit only the function of the transfected human NCAM, or by enzymatic removal of GPI-anchored human NCAM from the monolayer. In the case of chick neurons the response was also inhibited by species-specific antibodies that bind exclusively to NCAM in the neurons providing substantive evidence for a homophilic binding mechanism. In general the same neurons responded by a similar extent to chick N-cadherin (Doherty et al., 1991b) and human L1 (Williams et al., 1992) stably expressed in clones of transfected 3T3 cells. There was a highly co-operative relationship between the level of NCAM expressed by the 3T3 cells and neurite outgrowth. Up to a given threshold value NCAM did not stimulate neurite outgrowth, but relatively small increases above this value induced substantial increases in neurite outgrowth. (Doherty et al., 1990b)

Not all NCAM isoforms stimulate cell-contact-dependent neurite growth. The 180 kDa isoform that uses exon 18 and differs only in the size of its cytoplasmic domain is relatively poor at promoting neurite outgrowth (Doherty et al., 1992c). As the extracellular structure of this isoform is identical to others that function extremely well in this bioassay we would suggest that mobility in the membrane of the monolayer cells may be important for NCAM dependent axonal growth. It has previously been reported that the 180 kDa isoform shows a reduced lateral mobility in neuronal membranes due perhaps to a specific interaction with cytoskeletal components (Pollerberg et al., 1987). The above results caution against using CAMs coated to a substratum in order to study their function.

As discussed above, there are two sites where alternative splicing changes the structure of membrane-bound NCAM. Insertion of 37 amino acids between the FN III repeats does not affect NCAM's ability to stimulate neurite outgrowth (Doherty et al., 1990a). In contrast, use of the VASE exon which results in a ten amino acid sequence being inserted into the fourth Immunoglobulin-like domain substantially reduces NCAM's ability to stimulate neurite outgrowth (Doherty et al., 1992a). The pattern of expression of NCAM transcripts containing VASE in the developing brain also suggests that usage of this exon may change the function of NCAM from a molecule that promotes axonal growth to one that functions in strictly adhesive interactions between cells in the mature brain (Small and Akeson, 1990; Walsh et al., 1992).

ACTIVATION OF A SECOND MESSENGER PATHWAY RATHER THAN ADHESION *PER SE* UNDERLIES NCAM'S ABILITY TO STIMULATE NEURITE OUTGROWTH

The isoforms of NCAM that do not stimulate neurite outgrowth retain their ability to mediate adhesion and it follows that adhesion *per se* cannot be the stimulus for neurite outgrowth. CAM-dependent (NCAM, N-cadherin and L1) neurite outgrowth can be completely inhibited by (1) reducing extracellular calcium to 0.25 mM (2) blocking N- and L- type calcium channels on neurons and (3) by pre-loading neurons with a chelator of calcium (see Doherty et al., 1991b; Williams et al., 1992). None of these affects integrin dependent neurite outgrowth or that stimulated by NGF and cAMP. These data suggest the existence of a CAM specific second messenger pathway underlying cell-contact dependent neurite outgrowth. Activation of this pathway is sufficient to stimulate neurite growth even in the absence of any presumptive CAM-mediated adhesion step (Saffell et al., 1992; Williams et al., 1992). For example potassium depolarisation activates calcium influx into neurons via N- and L- type calcium channels and this can fully mimic CAM stimulated neurite outgrowth. The above data clearly suggest that adhesion *per se* does not directly contribute in any way to cell-contact dependent neurite outgrowth stimulated by NCAM, N-cadherin or L1. Rather, the ability of CAMs to induce calcium influx into neurons can solely account for their ability to promote neurite outgrowth.

PSA CAN OPERATE AS A POSITIVE AND HIGHLY SPECIFIC MODULATOR OF NCAM DEPENDENT NEURITE OUTGROWTH

For historical reasons the function of PSA has largely been considered in terms of modulation of NCAM's adhesive properties. However as discussed above adhesion *per se* cannot readily be translated into physiological function and we have established that the ability of NCAM and other CAMs to stimulate neurite outgrowth is not directly related to their ability to mediate adhesion. There is also no correlation between the ability of components of the extracellular matrix to support adhesion and their ability to stimulate neurite outgrowth (Gundersen, 1987). Thus the role of PSA in more complex biological responses such as cell-contact dependent axonal growth cannot be assumed from studies that have elucidated its role in adhesion.

Our work has established that the interaction of NCAM on neurons with some, but not all, NCAM isoforms expressed by monolayer cells can result in cell-contact dependent neurite outgrowth. If, as has been suggested, PSA antagonises such trans-binding then its removal might be expected to promote NCAM dependent neurite outgrowth. All of the neurons that respond to transfected NCAM by extending longer neurites express PSA and this can be specifically removed using the bacterial enzyme endoneuraminidase N (Endo-N) (e.g. see Rutishauser et al., 1988, Doherty et al., 1990b). We have found that removal of PSA from E6 chick retinal ganglion cells (Doherty et al., 1990b), E17 rat hippocampal neurons (Doherty et al., 1992b) and PND 1-6 rat cerebellar granule cells (Doherty et al., 1992a) substantially (by up to 80%) inhibits their ability to respond to transfected NCAM. This is a very specific effect of removal of PSA as neurite outgrowth over control 3T3 cells and 3T3 cells that express NCAM isoforms that support only adhesion is not affected. Also removal of PSA did not impair the same neurons ability to respond to transfected N-cadherin or L1 expressed in the same 3T3 cells. (e.g. see Williams et al., 1992). Thus we can conclude that in instances where PSA is expressed on neuronal NCAM, but not on NCAM in the substratum (as is the case for axonal contact with many cell types such as astrocytes, Schwann cells and muscle cells), then it acts specifically to modulate NCAM dependent and not general neurite outgrowth. In the case of axonal growth along other axons the function of PSA may not be so simple as PSA is present on both the advancing growth cone and also on the axonal substratum. Also this type of axonal growth cannot be inhibited by antibodies to NCAM (Chang et al., 1987) and this most likely reflects the fact that axonal substratum expresses the 180 kDa isoform of NCAM which does not stimulate neurite outgrowth (see above).

The data on PSA removal promoting general adhesion but specifically inhibiting NCAM stimulated neurite outgrowth cannot readily be explained even by an inverse relationship between neurite outgrowth and adhesion as we have only ever observed positive co-operative (Doherty et al., 1990a) and saturable (Doherty et al., 1992a) dose-response curves for NCAM level and neurite outgrowth. This is despite the fact that small increases in NCAM level have much more dramatic effects on adhesion than removal of PSA (Hoffman and Edelman, 1983). Also we know of no example where increasing the level of a physiologically relevant adhesion molecule (e.g. an extracellular matrix molecule or a CAM) actually reduces neurite outgrowth.

The above data are not consistent with PSA inhibiting by steric or other means the trans-binding of NCAM on neurons to NCAM on the cellular substratum. We have proposed that cis-interactions between NCAM-NCAM and/or NCAM with other molecules may affect the ability of NCAM to cluster in the neuronal membrane and that <u>transient</u> clustering may be a key step for activation of the second messenger pathway that stimulates neurite outgrowth (Doherty & Walsh, 1992). Adhesion *per se* may involve the formation of stable rather than transient clusters of NCAM. One possibility is that PSA promotes neurite outgrowth by promoting cis-interactions that favour the formation of transient rather than stable clusters of NCAM.

PHYSIOLOGICAL SIGNIFICANCE OF PSA ON NCAM

The expression of PSA on NCAM is both temporally and spatially regulated during development. In general expression of PSA is at its highest during periods of axonal growth and synaptic plasticity with loss of expression following shortly after formation of stable synapses (e.g. see Chuong and Edelman, 1984; Schosshauer et al, 1984; Doherty et al., 1990b and Tang et al, 1992). Although PSA is lost from most regions of the developed brain, it remains expressed in areas where neurogenesis and associated axonal growth continue into adulthood. Expression of NCAM isoforms containing the product of the VASE exon show a reciprocal pattern of expression to NCAM isoforms that contain the highest level of PSA (Small and Akeson, 1990). These patterns of expression suggest that two independent but complementary mechanisms have evolved to control NCAM function. The presence of PSA and absence of VASE is associated with areas of plasticity, whereas loss of PSA and expression of VASE is associated with stability. Results from studies on neurite outgrowth *in vitro* provide direct evidence for the primary purpose of PSA and isoform switching being to change the function of NCAM from a molecule that promotes plasticity (via activation of second messenger pathways) to one that promotes stability (through direct adhesive interactions).

The ability of chick retinal ganglion cells (Doherty et al., 1990b), rat cerebellar neurons (Doherty et al., 1992a) and rat hippocampal neurons (Doherty et al., 1992b) to respond to a fixed level of transfected NCAM (the 140 kDa isoform) by increased neurite outgrowth is dramatically reduced over relatively short developmental periods (reviewed in Walsh and Doherty 1992). Broadly speaking, these neurons are at their most responsive if isolated at that time when they would normally by extending axons to their targets *in vivo*. In contrast the older neurons show maintained or increased responsiveness to N-cadherin and/or L1 demonstrating that responsiveness to CAMs is not co-ordinately regulated. Loss of responsiveness is not associated with loss of neuronal NCAM expression, but can be substantially mimicked by the experimental removal of PSA from the younger responsive neurons (see above). Although this data indicates an important role for PSA in modulating neurons ability to extend axons in response to NCAM it was also clear that complete removal of PSA did not totally inhibit this function. Also whereas NCAM dependent neurite outgrowth from chick retinal ganglion cells is down regulated between E6 and E9, PSA expression on axons in the fibre tract is down regulated between E10 and E14. (Schlosshauer et al., 1984). Similarly the differences in PSA in the rat hippocampus and cerebellum are not substantial over the period of loss of neuronal responsiveness to transfected NCAM. Using the polymerase chain reaction we have recently obtained evidence that use of the VASE exon by cerebellar and hippocampus neurons may be the initial trigger that down-regulates neuronal responsiveness to NCAM during normal development of the central nervous system. We find rapid increases in the ratio of NCAM transcripts containing the product of the VASE exon over the PND6-PND8 period in the cerebellum and the E18-PND4 period in the hippocampus and this correlates with loss of the ability of these neurons to respond to NCAM *in vitro* (Walsh et al., 1992). One possibility is that whereas the use of VASE might be the initial trigger, longer term changes in NCAM's PSA content could reinforce a change in NCAM's function from a molecule that promotes plasticity (e.g. all migration and axonal growth) to one that promotes and/or maintains stable connections between cells.

VASE is generally not expressed in the peripheral nervous system and we find no differences in the ability of the NCAM isoforms expressed in the non-neuronal cells of the periphery to promote neurite outgrowth. Thus changes in NCAM's level of expression and PSA content rather than isoform switching may primarily regulate NCAM dependant axonal growth during innervation of peripheral targets such as skeletal muscle. Evidence for this has been obtained by Landmesser's laboratory and is discussed in detail elsewhere in this book. However points of particular relevance to this chapter include the observation that an upregulation in the expression of NCAM on myotubes (rather than changes in NCAM on motor neurons) appears to trigger the side-branching from axon fascicles into developing muscle. What is of particular interest is that side-branching could be substantially inhibited not only by antibodies that block NCAM function, but also by the removal of PSA from neuronal NCAM. The effects of the antibodies and removal of PSA were not additive suggesting (to us) a similar mechanism (Landmesser et al., 1990). These data are consistant with the above *in vitro* studies that have shown that PSA on neuronal NCAM may be required for NCAM-dependent axonal growth. Removal of PSA from motor neurons at earlier stages of development, just as they are about to enter and sort-out in the plexus resulted in projection errors on emergence from the plexus (Tang et al., 1992). These observations underscore the importance of NCAM (via its PSA content) for axonal growth and pathway selection during development of the peripheral nervous system. Projection errors in the retino-tectal pathway can also be observed following the injection of antibodies that block the function of NCAM in *xenopus* (Fraser et al., 1988), suggesting a similar role for NCAM in the developing central nervous system.

CONCLUSIONS

Using a transfection-based bioassay we have provided the first direct evidence that NCAM can stimulate cell-contact dependent axonal growth. These studies also revealed that modification of the cytoplasmic domain by use of a single exon, or the extracellular structure by use of the VASE exon, impairs NCAM's ability to stimulate neurite outgrowth without obviously altering its ability to support adhesion. NCAM dependent neurite outgrowth can be solely accounted for by the activation of a CAM specific second messenger pathway that culminates in calcium influx into neurons. Older neurons lose their ability to extend longer neurites in response to transfected NCAM and removal of PSA from younger neurons can substantially mimic this naturally occuring change. Biochemical studies on responsive and non-responsive neurons suggest that in the central nervous system loss of neuronal responsiveness might initially be triggered by an increase in the proportion of NCAM transcripts containing the VASE exon. Loss of PSA might then serve to reinforce a change in NCAM's function from a molecule that promotes plasticity through the activation of a second messenger pathway, to one that promotes or maintains stable contacts between cells through direct adhesive interactions. The fact that NCAM's ability to support adhesion and neurite outgrowth cannot be correlated, and that PSA acts as a positive modulator of the latter function suggests to us that PSA does not inhibit the trans-binding of NCAM on the neuron to NCAM in the substratum. An alternative possibility is that PSA affects the ability of NCAM to transiently cluster (and thereby activate a second messenger pathway) in the neuronal membrane via a cis-mechanism.

ACKNOWLEDGEMENTS

We would like to thank Jurgen Roth for the gift of ENDO-N and Hendrika Rickard for typing the manuscript. The work was supported by the Wellcome Trust and the Medical Research Council.

REFERENCES

Acheson, A., Sunshine, J.L. and Rutishauser, U. (1991). NCAM polysialic acid can regulate both cell-cell and cell-substrate interactions. J. Cell Biol., 114, 143-153.

Chang, S., Rathjen F.G. and Raper J.A. (1987). Extension of Neurites on Axons is Impaired by antibodies against Specific Cell Surface Glycoproteins. J. Cell Biol. 104, 355-362.

Chuong, C-M & Edelman, G.M. (1984). Alterations in neural cell adhesion molecules during development of different regions of the nervous system. J. Neurosci 4, 2354-68.

Doherty, P and Walsh, F.S. (1989). Neurite guidance molecules. Current Opinion in Cell Biology. 1: 1102-1106.

Doherty, P., Fruns, M., Seaton, P., Dickson, G., Barton, C.H., Sears, T.A. and Walsh, F.S. (1990a). A threshold effect of the major isoforms of NCAM on neurite outgrowth. Nature 343, 464-466.

Doherty, P., Cohen, J and Walsh, F.S. (1990b). Neurite outgrowth in response to transfected NCAM changes during development and is modulated by polysialic acid. Neuron, 5: 209-219.

Doherty, P., Rowett, L.H., Moore, S.E., Mann, D.A. and Walsh, F.S. (1991b) Neurite outgrowth in response to transfected NCAM and N-cadherin reveals fundamental differences in neuronal responsiveness to CAMs. Neuron, 6: 247-258.

Doherty, P., Ashton, S.V., Moore S.E., and Walsh, F.S. (1991a). Morphoregulatory activities of NCAM and N-cadherin can be accounted for by G-protein dependent activation of L- and N- type neuronal calcium channels. Cell 67: 21-33.

Doherty, P., Moolenaar C.E.C.K., Ashton, S.V., Michalides R.J.A.M., and Walsh, F.S. (1992a). Use of the VASE exon down regulates the neurite growth promoting activity of NCAM 140. Nature 356; 791-793.

Doherty, P., Skaper, S.D., Moore S.E., Leon. A., and Walsh, F.S. (1992b). A developmentally regulated switch in neuronal responsiveness to NCAM and N-cadherin in the rat Hippocampus. Development (in press).

Doherty, P., Rimon, G., Mann, D.A., and Walsh, F.S. (1992c). Alternative Splicing of the Cytoplasmic Domain of Neural Cell Adhesion Molecule Alters Its Ability to Act as a substrate for Neurite Outgrowth. J. Neurochem. 58, 2338-2341.

Doherty, P. and Walsh, F.S. (1992). CAMs, Second Messengers and Axonal Growth. Current Opinion in Neurobiology 2 (in press).

Edelman, G. (1986). Cell adhesion molecules and the regulation of animal form and tissue pattern. A. Rev. Cell. Biol. 2, 81-116.

Edelman, G.M., Murray, B.A., Mege, R-M., Cunningham, B.A. and Gallin, W.J. (1987). Cellular expression of liver and neural cell adhesion molecules after transfection with their cDNAs results in specific cell-cell binding. Procl. Natl. Acad. Sci. 84: 8502-8506.

Finne, J. (1982). Occurrence of unique polysialosyl carbohydrate units in glycoproteins of developing brain. J. Biol. Chem. 257: 11966-11970.

Finne, J. and Makela, P.H. (1985). Cleavage of the polysialosyl units of brain glycoproteins by bacteriophage endosialidase. J. Biol. Chem. 260: 1265-1270.

Fraser, S.E., Carhart, M.S., Murray, B.A., Cheng-Ming Chuong and Edelman G.M. (1988). Alterations in the Xenopus Retinotectal Projection by Antibodies to Xenopus N-CAM. Developmental Biology. 129, 217-230.

Gundersen, R.W. (1987). Response of sensory neurites and growth cones to patterned substrata of laminin and fibronectin *in vitro*. Dev. Biol. 121, 423-431.

Hoffman, S. and Edelman, G.M. (1983). Kinetics of neuronal binding by E and A forms of neural cell adhesion molecule. Procl. Natl. Acad. Sci. 80: 5762-5766.

Landmesser, L., Dahm, L., Tang, J.C., and Rutishauser U. (1990). Polysialic acid as a regulator of intramuscular nerve branching during embryonic development. Neuron 4: 655-667.

O'Connor T.P. Duerr, J.S, Bentley, D. (1990) Pioneer growth cone steering decisions mediated by single filopodial contacts *in situ*. J. Neurosci 10: 3955-3946.

Pizzey, J.A., Rowett, L.H., Barton, C.H., Dickson, G. and Walsh, F.S. (1989). Intercellular Adhesion Mediated by Human Muscle Neural Cell Adhesion Molecule: Effects of Alternative Exon Use. J. Cell. Biol. 109, 3465-3476.

Pollerberg, G.E., Burridge, K., Krebs, K.E., Goodman, S.R., Schachner, M. (1987). The 180-KD component of the neural cell adhesion molecule N-CAM is involved in a cell-cell contacts and cytoskeleton-membrane interactions. Cell. Tissue. Res. 250: 227-236.

Rutishauser, U., Acheson, A., Hall, A.K., Mann, D.M., Sunshine, J. (1988). The neural cell adhesion molecule (NCAM) as a regulator of cell-cell interactions. Science 240: 53-57.

Rathjen, F.G., Jessell, T.M. (1991). Glycoproteins that regulate the growth and guidance of vertebrate axons: domains and dynamics of the immunoglobulin/fibronectin type III subfamily. Seminars in The Neurosciences 3, 297-308.

Sadoul, R., Hirn, M., Deagostini-Bazin, H., Rougon, G. & Goridis, C. (1983). Adult and embryonic mouse neural cell adhesion molecules have different binding properties. Nature 304, 347-349.

Saffell, J.L., Walsh, F.S. and Doherty, P. (1992). Direct activation of second messenger pathways mimics cell adhesion molecule dependent neurite outgrowth. J. Cell. Biol. 118 (in press).

Schlosshauer, B., Schwartz, U., and Rutishauser, U. (1984). Topological distribution of different forms of neural cell adhesion molecule in the devloping chick visual system. Nature 310, 141-143.

Small, S.J. and Akeson, R. (1990). Expression of the unique NCAM VASE exon is independently regulated in distinct tissues during development. J. Cell. Biol. 111: 2089-2096.

Takeichi, M. (1988). The cadherins: cell-cell adhesion molecules controlling animal morphogenesis. Development 10, 639-655.

Takeichi, M. (1991). Cadherin cell adhesion receptors as a morphogenetic regulator. Science 251: 1451-1455.

Tang, J. Landmesser, L. and Rutishauser,U. (1992). Polysialic Acid Influences specific pathfinding by Avian Motoneurons. Neuron 8, 1031-1044.

Walsh, F.S. and Doherty, P. (1991). Structure and Function of the Gene for Neural Cell Adhesion Molecule. Seminars in the Neurosciences 3, 271-284.

Walsh, F.S. and Doherty, P. (1992). Alternative splicing and polysialic acid modulates the ability of NCAM to activate a second messenger pathway that stimulates neurite growth. Cold. Spring. Harbor. Symp. Quant. Biol. Vol. 57. (in press).

Walsh, F.S., Furness, J., Moore, S.E., Ashton, S. and Doherty, P. (1992). Use of the NCAM VASE exon by Neurons is associated with a specific down regulation of NCAM dependent neurite outgrowth in the developing cerebellum and hippocampus. J. Neurochem. (in press).

Williams, E.J., Doherty, P., Turner, G., Reid, R.A., Hemperley, J.J., Walsh, F.S. (1992). Calcium Influx into Neurons can Solely Account for cell-contact dependent Neurite outgrowth stimulated by Transfected L1. J. Cell. Biol. (in press).

Polysialic Acid
J. Roth, U. Rutishauser and F. A. Troy II (eds.)

INTERNALIZATION AND NEUROTROPHIC ACTIVITY OF AN HOMEOBOX PEPTIDE

A. Joliot, I.Le Roux, M. Volovitch, E. Bloch-Gallego and A. Prochiantz

CNRS USA1414, Ecole Normale Supérieure, 46 Rue d'Ulm, 75230 Paris Cedex 05, France.

SUMMARY: The sixty amino acid long peptide corresponding to the homeobox of Antennapedia (pAntp) translocates through the membrane of neurons in culture and reaches their nuclei. This energy-independent process is followed by an enhanced morphological differentiation of the neurons. Internalization by neurons is 4-fold that observed with fibroblasts, a difference abolished upon treatment with Endo-N or by preincubating the cells with an antibody directed against NCAM-specific α 2,8-linked polysialic acid (PSA). A structural model suggests that PSA mimics a DNA motif normally recognized by the third helix of homeodomains. To understand the mode of action of the peptide, we constructed three mutants modified in their capacity to specifically bind promoters and/or to translocate through the cell membrane. The biological properties of the mutants suggest that the amphipatic structure of helix 3 is necessary for pAntp translocation. They also demonstrate that the neurotrophic action of pAntp requires its internalization and the integrity of its specific DNA-binding capacity.

INTRODUCTION

Homeobox proteins are trans-activating factors highly expressed during development. Mutations affecting this class of factors produce morphological abnormalities in all species (review in McGinnis and Krumlauf, 1992). Homeoproteins bind specific sequences present in the promoters of several genes, including other homeogenes and genes coding for molecules with morphoregulatory activity such as adhesion molecules (Hirsch et al., 1990; Jones et al., 1992a, b). The DNA motifs recognized by the members of a given family of homeoproteins are very similar. For example, class I homeoproteins bind repeated sequences encompassing the ATTA motif. The part of the homeoprotein that binds the promoters is a 60 amino acid long region of the homeoproteins, called the homeobox. All homeoboxes share the same structure with 3 α helices, the third helix being responsible for binding specificity. Within a given class, sequence identity is very high even across highly divergent species.

Classically, the role of homeoproteins is to define the shape of organs during early development. However, the late expression of some homeogenes, in particular in the nervous system, suggests that they have other functions as well. This possibility is further indicated by mutations affecting the late expression of some homeogenes and resulting in axonal misguiding and aberrant synaptogenesis (Doe et al., 1988a, b; Miller et al., 1992; White et al., 1992).
In this laboratory we have proposed that homeoproteins might influence the shape of post mitotic neurons (Prochiantz, 1990; Prochiantz et al., 1992). This proposal is based on the fact that many proteins important in organ morphogenesis, such as adhesion molecules, are also involved in the morphological differentiation of neurons.

RESULTS

To test the hypothesis that homeoproteins influence neuronal morphology, we introduced into embryonic neurons in culture, high amounts of the 60 amino acid long peptide corresponding to the homeobox of Antennapedia. The rational was that, due to the conservation of sequences among class I homeboxes and among the DNA motifs recognized by these homeoboxes, the peptide called pAntp, should bind many promoters and antagonize, at once, the activity of all class I homeoproteins present in the nerve cells.

The mechanical introduction of pAntp into E15 to E17 rat cells derived from different regions of the embryonic brain resulted in its nuclear accumulation and produced a clearcut effect on neuronal morphogenesis (Joliot et al., 1991a). This effect, however, could find several explanations other than the inhibition of homeoprotein binding to specific promoters. However, before going into the details of the actual mechanism of action of pAntp we have to report on a very peculiar property of the homeobox peptide.

A very surprising observation, indeed, was that the homeobox peptide added in the culture medium penetrated into the nerve cells, was conveyed to their nuclei and promoted neurite growth (Joliot et al., 1991a, b). This translocation is very efficient, energy-independent and the peptide can be recovered in an undegraded form from the neuronal nuclei. Interestingly, although all cell types can internalize pAntp, translocation through neuronal membranes is particularly efficient. For example, after one hour, the amount of pAntp found in neuronal nuclei is 4 times that found in fibroblasts.

This relative cell specificity led us to look for receptors that might bind aAntp at the neuronal surface. By analogy with other molecules (e.g. toxins) capable of trans-locating through the membranes of eukaryotic cells, we looked at the effect of various neuraminidases on pAntp internalization. We found that only endoneuraminidase N, an enzyme which cleaves α 2,8-linked polysialic acid (PSA) reduces the level of pAntp neuronal internalization to that observed in fibroblasts. This role of PSA was confirmed by experiments in which adding antibodies against the complex sugar inhibited its translocation. Interestingly, all treatments reducing pAntp internalization also abolished its morphogenetic effects on neurons in culture (Joliot et al., 1991b). The capability of PSA to accelerate the internalization of pAntp suggested that the complex sugar might mimic some DNA traits normally recognized by homeodomains. The latter hypothesis was reasonable since affinity for complex sugars (e.g. heparin) is a well-know property of some transactivating factors. We thus computerized a threedimensional model of PSA, based on RMN studies, and found a regular and repeated motif of 4 oxygen residues which is also present in dsDNA and known to participate in the recognition of the third helix of the homeodomain (Joliot et al., 1991b).

To pursue the work on the mode of action of pAntp and on its mode of penetration, we generated pAntp mutants that were modified in their DNA-binding and/or trans-location properties. Three mutants were produced. In a first mutant, pAntp50A, glutamine in position 50 was replaced by an alanine, thus modifying a very important site for promoter specific recognition. In a second mutant, pAntp40P2, two amino acids present in the turn between helices two and three were replaced by two proline residues. This mutation modifies and rigidifies the angle between the two helices. In the third mutant, pAntp48S, two hydrophobic residues of the third helix were removed, resulting in a modification of the hydrophobic properties of the helix. All mutants were produced in E.coli purified to homogeneity and eventually metabolically labelled or fluoresceinated.
Compared to wild type pAntp, pAntp50A had a reduced affinity for the cognate homeobox binding site present in the Hox-1.3 promoter and pAnt48S or pAntp40P2 were totally unable to bind this site. When tested for their capacity to translocate through the membrane, we found no difference between pAntp, pAntp40P2 and pAntp50A, all three being internalized and conveyed to the nuclei at 37°C and 10°C. In contrast, pAntp48S was not internalized, suggesting that the structure of the third helix is crucial for pAntp internalization. Finally none of the mutants enhanced neurite growth demonstrating that pAntp neurotrophicity requires its internalization and the integrity of its specific DNA-binding properties (Bloch-Gallego et al., unpublished; Le Roux et al.,unpublished).

DISCUSSION

That pAntp, in order to promote neurite differentiation, should penetrate into the cells and should conserve its specific DNA-binding properties strongly suggests that it acts by interfering with endogenous homeoproteins present in the cell nucleus. In view of the complexity of the network of interactions between homeoproteins, it is not possible to speculate further on its precise mode of action. However, the competitive inhibition model on which the first experiments were based seems rather reasonable at the present stage of our study. In any case it can be precluded that pAntp acts unspecifically, for example by coating the culture substratum or interacting non-specifically with cytoplasmic or nuclear elements. The hypothesis that some homeoproteins play a role in the morphological differentiation of postmitotic neurons is therefore reinforced by our experiments.

The computerized design of the third helix indicates that it contains separate hydrophobic and highly charged domains, suggesting that the helix is endowed with amphipatic properties that might be responsible for pAntp internalization. This latter point is in good agreement with the results obtained with pAnt48S which lacks two hydrophobic amino acids in the third helix and is not capable of translocating through the cell membrane.

The role of PSA is not to permit the internalization but to enhance it. It seems that PSA acts by trapping the homeobox at the surface of the neurons. In addition it might provide the acidic environment allowing for a hydrophilic-hydrophobic transition necessary for pAntp insertion in the membrane. If this scheme is correct, passage into the cytoplasm would necessitate a reversion to an hydrophilic conformation permitted by the neutral pH of the cytoplasm. Such a mechanism is reminiscent of that described for bacterial toxins, except that in the case of toxins, acidification occurs in the endosomes and not at the cell surface (Jiang et al., 1991; Olnes et al., 1988). It will be interesting to study whether bacterial toxins can trans-locate through neuronal membranes independently of the formation of endosomes. Conversely it should be of interest to analyse whether pAntp internalization by cells without PSA requires an acidification step in the endosomal compartment.

The pAntp internalization suggests that molecules with an higher molecular weight and encompassing the homeobox sequence might translocate through the cell membrane and reach the nuclei. This was demonstrated in a study where we verified that such chimeric proteins are, indeed, internalized (Perez et al., 1992). Since, homeoproteins can be viewed as chimeric proteins encompassing a homeobox, it is not impossible that homeoprotein isoforms might act as paracrine trans-activating regulators. This hypothesis is presently studied in our laboratory.

REFERENCES

Doe C.Q., Hiromi Y., Gehring W.J. and Goodman C.S. (1988) Expression and function of the segmentation gene *fushi tarazu* during Drosophila neurogenesis. Science 239, 170-175.

Doe C.Q., Smouse D. and Goodman C.S. (1988) Control of neuronal fate by the Drosophila segmentation gene *evenskipped.* Nature 333, 376-378.

Hirsch M.R., Gaugler L., Deagostini-Bazin H., Bally-Cuif L. and Goridis C. (1990) Identification of positive and negative regulatory elements governing cell-type-specific expression of the neural cell adhesion molecule gene. Mol. Cell. Biol. 10, 1959-1968.

Jiang J.X., Chung L.A. and London E (1991) Self translocation of diphtheria toxin across model membranes. J. Biol. Chem. 266, 24003-24010.

Joliot A., Pernelle C., Deagostini-Bazin H. and Prochiantz A. (1991) Antennapedia homeobox peptide regulates neural morphogenesis. Proc. Natl. Acad. Sci. USA 88, 1864-1868.

Joliot A.H., Triller A., Volovitch M., Pernelle C. and Prochiantz A. (1991) α 2,8-polysialic acid is the neuronal surface receptor of Antennapedia homeobox peptide. New Biol. 3, 1121-1134.

Jones F.S., Prediger E.A., Bittner D.A., De Robertis E.M. and Edelman G.M. (1992) Cell adhesion molecules as targets for Hox genes: neural cell adhesion molecule promoter activity is modulated by cotransfection with Hox-2.5 and Hox-2.4. Proc. Natl. Acad. Sci. USA 89, 2086-2090.

Jones F.S., Chalepakis G., Gruss P. and Edelman G.M. (1992) Activation of the cytotactin promoter by the homeobox-containing gene Evx-1. Proc. Natl. Acad. Sci. USA 89, 2091-2095.

McGinnis W. and Krumlauf R. (1992) Homeobox genes and axial patterning. Cell 68, 283-302.
Miller D.M., Shen M.M., Shamu C.E., Bürglin T.R., Ruvkun G., Dubois M.L., Ghee M. and Wilson L. (1992) C. elegans *unc-4* gene encodes a homeodomain protein that determines the pattern of synaptic motor neurons. Nature 355, 841-845.
Olsnes S., Moskaug J.O., Stenmark H. and Sandvig K. (1988) Diphtheria toxin entry: protein translocation in the reverse direction. Trends Biochem. Sci. 13, 349-351
Perez F., Joliot A., Bloch-Gallego E., Zahraoui A., Triller A. and Prochiantz A. (1992) Antennapedia homeobox as a signal for the cellular internalization and nuclear addressing of a small exogenous peptide. J. Cell Sci. in press.
Prochiantz A. (1990) Morphogenesis of the nerve cell. Comments Dev. Neurobiol. 1, 143-155.
Prochiantz A., Joliot A., Volovitch M. and Triller A. (1992) Have homeoproteins autocrine and paracrine activities? Implications for our understanding of cellular recognition during development. Comm. Dev. Neurobiol. in press.
White J.G., Southgate E. and Thomson J.N. (1992) Mutations in the Caenorhabditis elegans *unc-4* gene alter the synaptic input to ventral cord motor neurons. Nature 355, 838-841.

EXPRESSION PATTERNS OF POLYSIALIC ACID DURING VERTEBRATE ORGANOGENESIS

Peter Lackie*, Christian Zuber and Jürgen Roth

Division of Cell and Molecular Pathology, Department of Pathology, University of Zürich, Schmelzbergstr 12, CH-8091 Zürich, Switzerland. *Present address: Southampton University, Medicine 1, Level D, Centre Block, Southampton General Hospital, Southampton SO9 4XY, England

SUMMARY: Polysialic acid (poly Sia) is widely expressed during vertebrate organogenesis, most frequently concomitantly with the neural cell adhesion molecule N-CAM but also with sodium channels. Cells of the three primary germ layers retain their ability to express poly Sia during organogenesis. During this period, poly Sia is found predominantly in neuroectodermally and mesodermally derived cells whereas it is expressed more transiently in ectodermally and endodermally derived cells. In epithelia, poly Sia is generally baso-laterally sorted. Expression of poly Sia is often associated with inductive changes, and transformations between cell phenotypes. The highly regulated pattern of expression is consistent with an important role during organogenesis while the widespread expression suggests that poly Sia is a general modulator of cell surface mediated processes.

INTRODUCTION

Early in development, the process of gastrulation leads to the establishment of three primitive germ cell layers in the embryo. Organogenesis may be defined as the processes by which these layers interact and rearrange themselves into the complex structures of the adult organs (for detailed review see Sadler, 1990 or Hooper and Hart, 1985). The formation of the neural tube (neurulation) by folding of the external (ectoderm) layer, marks the start of this process in vertebrates. At this stage, five cell compartments can be distinguished. These consist of the primary germ layers; the ectoderm, mesoderm and endoderm, together with the newly formed neuroectoderm and neural crest. The ectoderm forms the main elements of the outer lining of the organism, while the endoderm is the progenitor of the epithelia of the tubular and glandular digestive tracts and the respiratory system. The mesodermal cells are the origin of much of the circulatory system, musculature, urogenital tract and large portions of the digestive and respiratory tracts, contributing mesenchymal elements to many organ systems. The neuroectoderm is the progenitor of the central and peripheral nervous systems, whereas the transient neural crest undergoes an epithelial to mesenchymal transformation, ultimately contributing to portions of the endocrine system, peripheral nervous system, skeleton and skin.

Central to the processes of organogenesis is the concept of induction. This is the interaction of two groups of cells ("anlagen") to initiate, de novo, the development of a new structure. The development of the metanephros, or permanent kidney, is a paradigm of reciprocal induction (Saxen, 1987 for review), in this case by two anlagen of common origin. Similar processes occur in many other organ systems during development. An analogous process is the establishment of specific cell-cell contact between cells from different origins, for example during innervation. During such processes, which necessarily must involve the "invasive" growth of one cell type through tissue of a different origin, lack of adhesion or cell interaction may be important to prevent one process interfering with another.

The developmental fate of cells is often "determined" relatively early in embryogenesis and elements of specific organs can be attributed to particular regions of the early embryo. Thus development involves the establishment of programmed differences between cells and their ability to respond to certain stimuli. By the onset of organogenesis the fate of cells in the developing embryo is largely determined, or at least restricted, but local cell interactions and induction events are still required to produce the functional organ structures. Cell migration, adhesion and differentiation are the basic processes whereby the specific structures of the component organs of each organism are constructed. Each of these processes involves cell surface molecules, which act either as receptors for diffusible factors or mediate cell-cell or cell-substrate interaction. The exquisite control of organogenesis via cell surface molecules and the tremendous potential to modulate the overall structure of the organism, is amply demonstrated by the evolutionary range of vertebrate structure. Despite such variation, there is an underlying homology in the overall organ plan and its development, suggesting that similar control mechanisms are active. Thus, any cell surface molecule which is developmentally expressed is of potential interest as a candidate modulator of cell behaviour during organogenesis.

The neural cell adhesion molecule (N-CAM), originally reported in neural structures (Jorgensen and Bock, 1974, Thiery et al., 1977), was the first molecule to be isolated and fully characterised as a cell adhesion molecule (Cunningham et al., 1987). N-CAM is one of the best characterised developmentally regulated adhesion molecules (see Edelman and Crossin, 1991 and Rutishauser and Jessell, 1988 for review). The widespread expression of N-CAM in non-neural tissues is now well established (Crossin et al., 1985, Chuong and Edelman, 1985, Roth et al.,

1987). The genetic control of pattern formation during development involves the expression of homeobox genes (Cooke, 1991) and it has recently been reported that the N-CAM gene is likely to be a target for regulation by homeobox gene products (Jones et al., 1992). Although the primary information regarding molecular structure is genetically determined, secondary modification, for example by carbohydrates, can be controlled independently. Thus the presence of variable amounts of homopolymers of α 2,8-linked sialic acid (poly Sia) on N-CAM (Rothbard et al., 1982), provides an additional possible level for the surface modulation of cell properties and makes poly Sia an especially interesting molecule in the context of organogenesis. The effect of poly Sia on the properties of N-CAM has been studied in some detail (Hoffman and Edelman, 1983, Sadoul et al., 1985, Rutishauser et al., 1988, Hall et al., 1990, Yang et al., 1992) and physiologically probably results in reduced N-CAM mediated adhesion. In mammals, poly Sia expression is restricted and to date has only been reported to be on N-CAM and the α-subunit of sodium channels (Zuber et al., 1992).

Regeneration in response to injury and oncogenesis often reflect patterns of cell behaviour occurring during organogenesis. N-CAM and poly Sia are expressed during muscle and limb bud regeneration (Booth and Brown, 1988, Maier et al., 1986), in Wilm's tumour, a embryonic type of kidney tumour (Roth et al., 1988a, Roth et al., 1988b, Roth and Zuber, 1990) and in small cell carcinoma of the lung (Komminoth et al., 1991, Moolenaar et al., 1990). This aspect of poly Sia expression will be reviewed in more detail elsewhere in this volume.

The more general aspects of poly Sia expression and biochemistry have been recently been reviewed (Troy, 1992), and further details will be found elsewhere in this volume. This article will therefore be restricted to discussing and reviewing some of the most notable developmentally regulated expression patterns of poly Sia and its possible functions during organogenesis and the analogous processes of tissue repair.

ORGANOGENESIS AND POLYSIALIC ACID EXPRESSION

We have mapped the distribution of poly Sia during organogenesis (Lackie et al., 1990a, Lackie et al., 1991, Lackie et al., 1990b, Lackie et al., 1992) using the monoclonal antibody mAb 735 in conjunction with immunocytochemistry and immuno-gold silver staining (Roth, 1989). This antibody is immunoreactive with only homopolymers of 8 or more α2,8-linked sialic acid residues (Frosch et al., 1985, Roth et al., 1987, Häyrinen et al., 1989, Husmann et al., 1990). Therefore, in this article, poly Sia is used to refer specifically to polymers of 8 or more α2,8-linked sialic acid residues. Most of the information reviewed here relates to the expression of poly Sia as detected by mAb 735 using immunocytochemistry and immunoblotting. Another important tool for the investigation of poly Sia expression and function is the bacteriophage endoneuraminidase N which reduces the chain length of poly Sia to less than 8 (Tomlison and Taylor, 1985; Finne and Makela, 1985, Pelkonen et al., 1989) and abolishes immunoreactivity with mAb 735 (Roth et al., 1987,

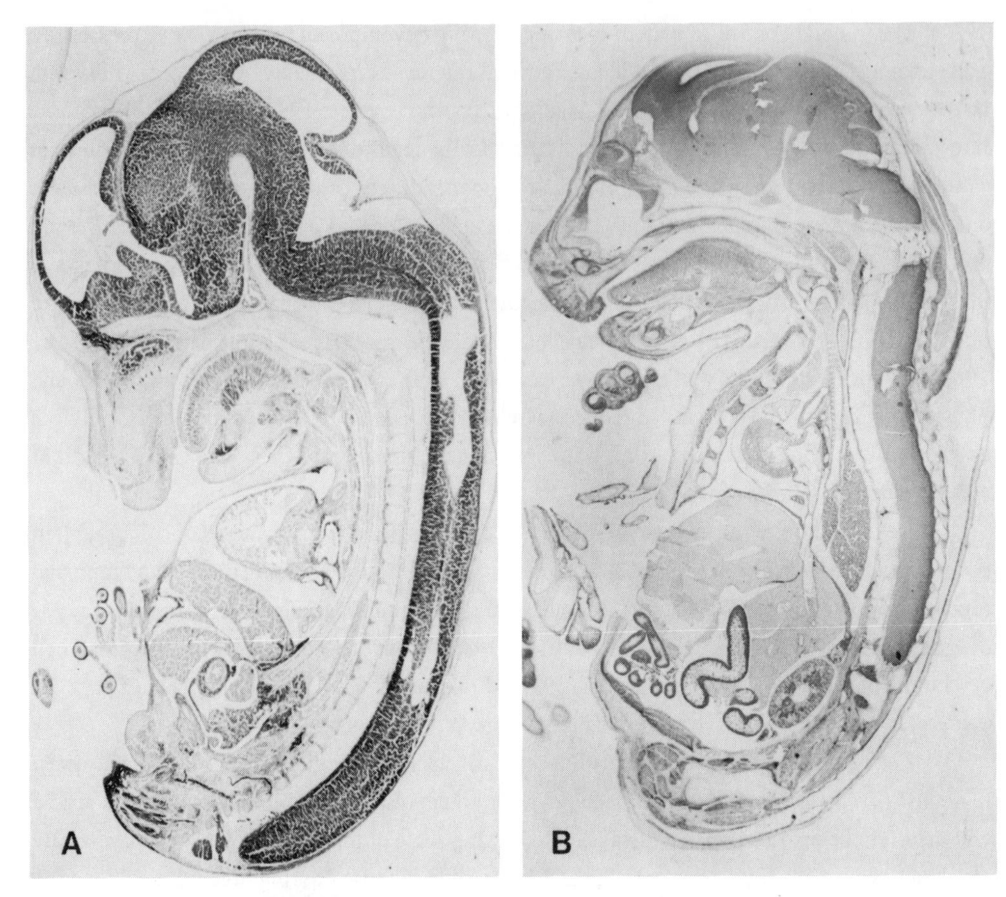

Fig. 1. Immunolocalization of polysialic acid with directly gold-labeled mAb 735 in a parasaggital section through entire rat embryos of embryonic day 16 (A) and 18 (B). Due to silver intensification, the immunolabeling appears black and can be detected in the entire central nervous system and a number of organs. In the photographs the two embryos appear of similar size but it should be noted that the 16 day old embryo measures half the size of the 18 day old embryo.

Roth et al., 1988, Lackie et al., 1990a). The expression of shorter chain length α2,8-linked sialic acid is not well characterised, although it is possible that most N-CAM molecules carry α2,8-linked sialic acid in one or other form. Results from immunoblotting using mAb 735 and N-CAM antibodies, indicate that, even within one organ system, there is great variation in the amount of poly Sia expressed on N-CAM molecules (Lackie et al., 1991).

NEUROECTODERM INTERACTIONS

During the period of organogenesis, poly Sia is most abundant in the ectodermally derived central and peripheral nervous systems. As the signalling mechanism used to respond rapidly and specifically to environmental stimuli, the neurons are the "wires" of the nervous system. It is important that the correct neural connections are made. Consequently the development of the nervous system requires an unrivalled degree of control and specificity.

Poly Sia expression is not uniform in either the central or peripheral nervous system (Theodosis et al., 1991, Zuber et al., 1992, Lackie et al., 1991). During rat brain development, the ratio of poly Sia to N-CAM has been shown to decrease and in adult brain could not be detected on NCAM by chemical analysis (Hoffman et al., 1982, Finne, 1982). However is not clear whether this is due to a change in the proportion of N-CAM without poly Sia or to a reduction in the average chain length of poly Sia on N-CAM. In fact both probably occur. Immunocytochemical studies have shown that poly Sia is still present in adult rat brain (Theodosis et al., 1991, Zuber et al., 1992) and is also expressed on sodium channels in this tissue (Zuber et al., 1992). This raises the question as to whether individual cell surface expression of poly Sia is on sodium channels, on N-CAM of one or more isoforms or a combination of these possibilities. Further, is the degree of polysialylation consistent on each cell? Electron microscopic studies indicate that in adult rat brain, the cell surface distribution of poly Sia is very heterogeneous and does not necessarily coincide with the presence, or even the local surface distribution, of N-CAM or sodium channels (Zuber et al., 1992). This suggests that there is molecular heterogeneity of poly Sia expression on cells.

Functionally, poly Sia is important in the development of the nervous system (Fraser et al., 1984, Landmesser et al., 1990, Rutishauser and Landmesser, 1991). Its importance in neural organogenesis has been demonstrated by the overexpression of α2,3 sialyltransferase in Xenopus which causes competitive down-regulation of the α2,3-linked sialic acid substrate required for poly Sia addition and leads to abnormal neural development (Livingston et al., 1990). The detailed expression of poly Sia during the development of the nervous system is outside the scope of this article (see article by Zuber et al, this volume).

MESODERM-MESODERM INTERACTIONS

HEART

The heart is one of the first organs to be formed in vertebrate development and is unique in being functionally active from an early stage. This ability is obligatory for further development of the embryo, since it overcomes the limits of diffusion for the supply of nutritional and respiratory elements. The heart, in common with the rest of the circulatory system, is mesodermally derived. Its development is conceptually complex, largely due to the intricacy of the final organ and the requirement to remain fully functional throughout the life of the embryo and adult (for review of

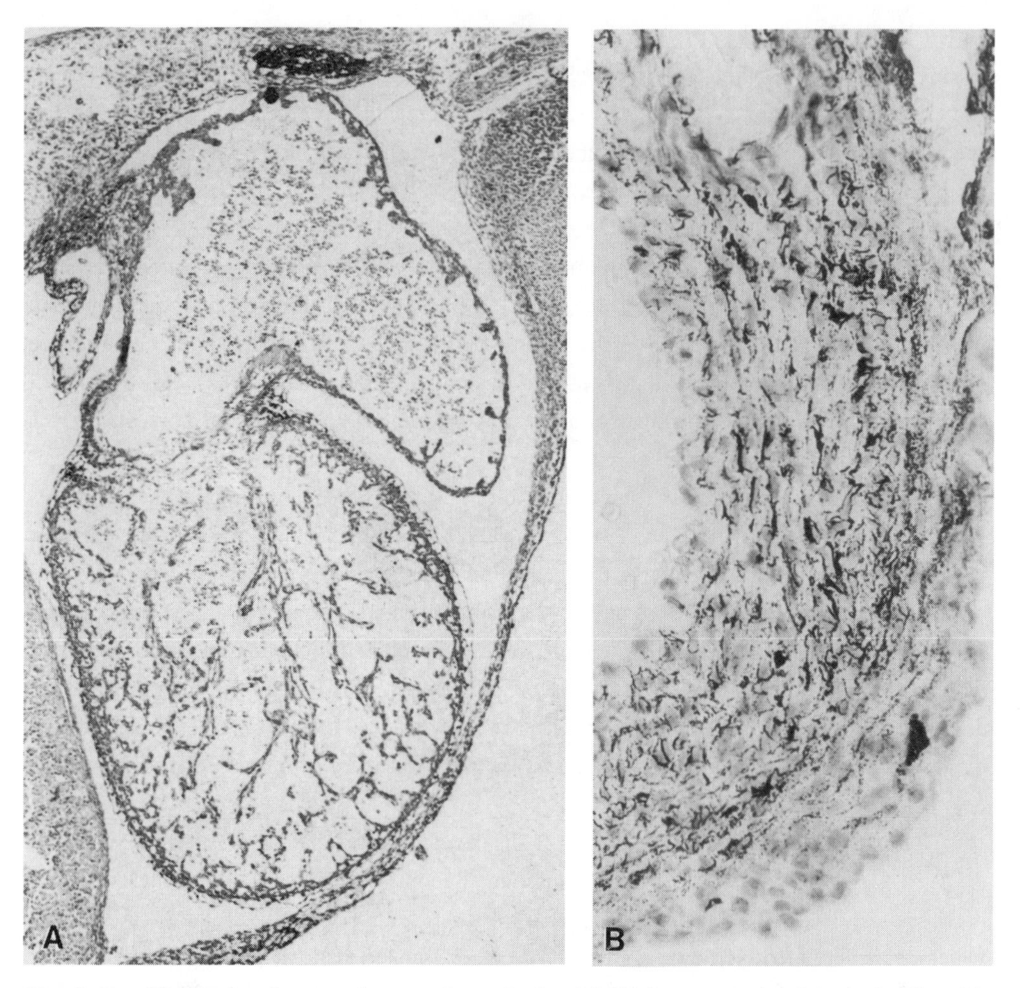

Fig. 2. Paraffin section from rat heart embryonic day 16 (A) immunostained for poly Sia with directly gold-labeled mAb 735 shows the overall structure and immunostaining. At higher magnification (B), the immunostaining of the ventricular wall can be clearly seen.

development see Hooper and Hart, 1985). I will deal here with only two interesting aspects of the formation of the heart in relation to poly Sia expression. First, the expression pattern of poly Sia in the epicardium and second the transformation of endothelial cells to mesenchymal cells during the formation of the cardiac cushions and ultimately of the cardiac valves.

The epicardium is thought to be formed by the migration of a sheet of cells from the mesothelium, adjacent to the sinus venosus, over the outer surface of the primitive myocardium.

During this process the epicardial cells show significantly stronger poly Sia reactivity than does the underlying myocardium (Lackie et al., 1991). poly Sia reactivity is reduced as the epicardial layer is completed and although this layer continues to expand to account for the subsequent growth of the heart relatively little poly Sia is expressed.

The mesenchymal cells which are the progenitors of the heart valves are derived from endothelial cells which line the primitive chambers of the heart (Markwald et al., 1977). Cells in the endothelium migrate into the endocardial cushions, concomitantly changing their characteristics to mesenchymal cells, contributing to the transient structure of the cushion and further differentiating to form the heart valves (Markwald, 1987). In this process, poly Sia and NCAM immunoreactivity is seen on endocardial cells (Mjaatvedt and Markwald, 1989, Lackie et al., 1991), and on the transformed cells in the endocardial cushions but not on peripheral cells in the cushions nor in the area which should include migrating cells. The predominant NCAM isoforms in heart development seem to be the 140/160 KDa peptides, but both 120 and 180 KDa peptides have been seen (Wharton et al., 1989, Lackie et al., 1991). At all stages of heart development, N-CAM with and without poly Sia are found but poly Sia forms predominated until a week after birth in rat.

In skeletal muscle, myotubes are formed by the fusion of myoblasts resulting in syncitial multinucleate cells. N-CAM (Grumet et al., 1982, Knudsen et al., 1990, Dickson et al., 1990) and poly Sia (Rutishauser and Landmesser, 1991) are known to be important in this process. Since cell fusion is not important in cardiac muscle development the analogous expression patterns of N-CAM and poly Sia suggest that these molecules may have additional functions in the developing heart, which may also be important in skeletal muscle.

KIDNEY

In developmental terms, the progenitor of the urogenital system is the pronephros which arises from intermediate mesoderm in the cranial region (see Saxen, 1987 for a review of kidney development). In mammals, these mesoderm cells cluster to form segmentally arranged nephrotomes which develop into poly Sia positive primitive pronephric tubules. These tubules grow in a caudal direction, concomitant with the development of further new nephrotomes and tubules which fuse with the extant tubules to form the pronephric duct. Although the early pronephric units degenerate even before the last are formed, the pronephric duct continues its caudal growth and interacts with a caudal region of unsegmented intermediate mesoderm to form the mesonephros. In the mesonephros unsegmented intermediate mesoderm forms cell clusters which are strongly poly Sia immunoreactive and condensations around the branches of the less poly Sia reactive pronephric duct. The ureteric bud arises from bifurcation of the mesodermally derived mesonephric duct and grows into the metanephric mesenchyme. The uretic bud then branches and around the later branches loose, strongly poly Sia immunoreactive mesenchyme condenses and subsequently differentiates to form the epithelium of the nephron. Classical studies

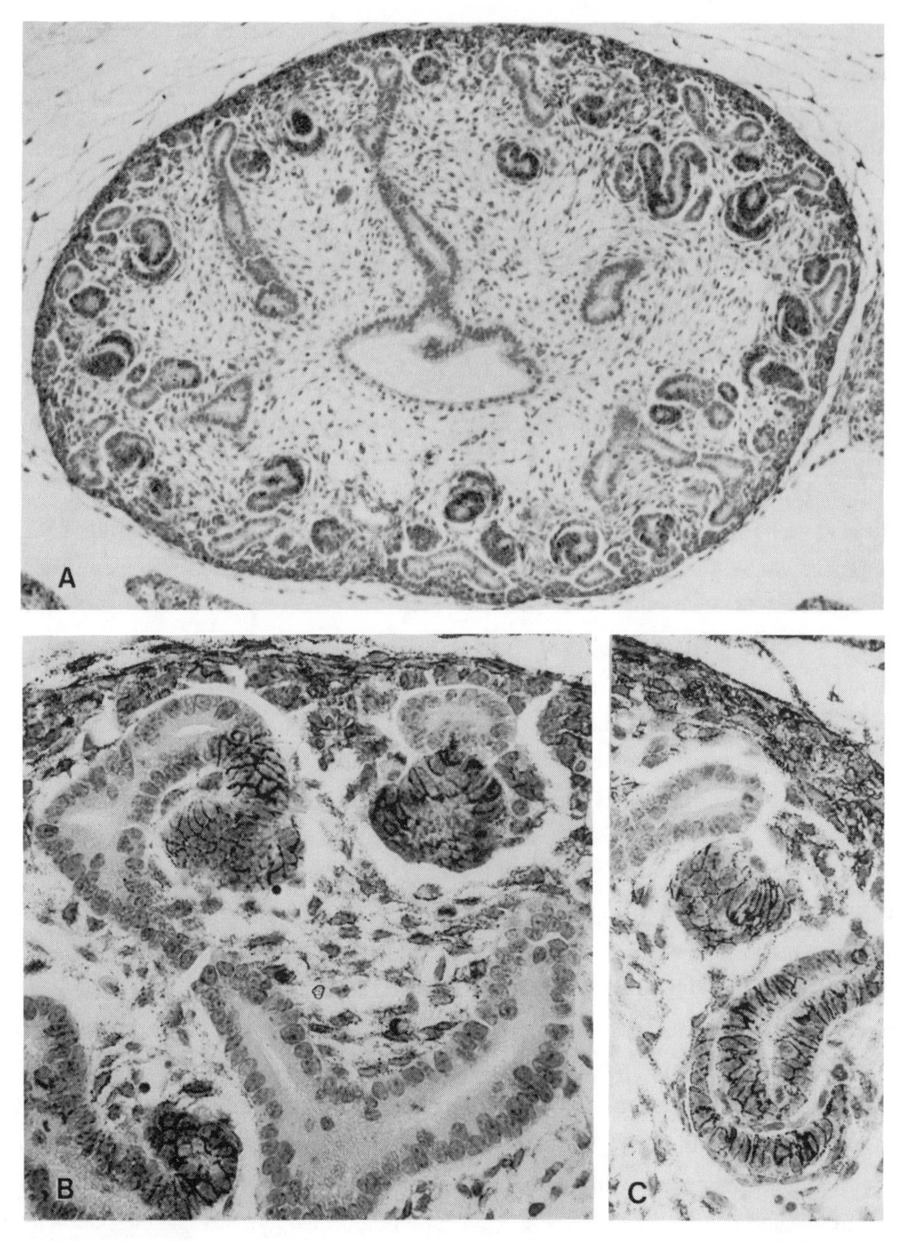

Fig. 3. Paraffin section of metanephros at embryonic day 18, immunostaining for poly Sia with directly gold-labeled mAb 735. (A) Sagittal section through entire metanephros shows poly Sia staining in a variety of stages of nephron development with the pelvis, major and peripheral collecting ducts being not immunoreactive. (B, C) Details of poly Sia immunostaining with positive renal vesicles in A and S-shaped body in B. The condensed mtanephric mesenchyme shows intense immunostaining and the loose mesenchyme exhibits reduced intensity of immunostaining. The collecting duct epithelium is not stained.

of kidney development (see Saxen, 1987 for review), have shown that the branching of the uretic bud and the mesenchyme to epithelium transformation will not occur if the two anlagen are cultured separately but will continue if they are cultured together. Further, the mesenchyme will be induced to start producing kidney epithelial structures if other inducers (eg: spinal chord) are used. Thus, the mesenchymal cells are primed to become kidney epithelium but to realise this potential need a signal usually from cell-cell contact with the ureteric bud. Although both anlagen initially express poly Sia, as individual differentiated nephrons form, poly Sia reactivity is lost (Lackie et al., 1990a). Since nephron formation is spread over a considerable period of development, often continuing into postnatal life, there may be fully functional nephrons in the metanephros which are poly Sia unreactive, whilst other tubules are still being induced (Roth et al., 1987). This induction process also allows the development of nephrons to continue as the overall size of the kidney (and organism) increases such that some mesenchymal cells may remain as "latent nephron epithelium" until early in postnatal life when kidney nephron induction is finally complete.

The occurrence of N-CAM poly Sia on both of the reciprocally inducing anlagen (Lackie et al., 1990a) suggests that neither poly Sia nor N-CAM are inductive signals. The tight developmental control of their expression does however suggest that they play some role in induction. Further, this role occurs relatively early while cells are condensing, aggregating to form an epithelium, forming tight junctions, invading the surrounding mesenchyme and making contact with the branch of the ureteric bud. The latter will go on to form the collecting ducts. Once the basic structures of the nephron are formed, poly Sia expression declines, before the growth and maturation of the nephron is complete (Roth et al., 1987; Lackie et al., 1990a). This in turn suggests that loss of poly Sia does not itself signal terminal cell differentiation.

The formation of kidney nephron epithelium from mesenchymal condensates, involves the establishment of apical and baso-lateral membrane domains in the epithelial cells. poly Sia immunoreactivity was not found on the apical surface of these newly formed epithelia (Lackie et al., 1990a). Since phoshatidylinositol (Pi) linked membrane proteins, including the Pi linked 120KDa isoform of N-CAM, are known to be sorted to the apical domain of epithelia, this suggest that either the Pi linked N-CAM isoform is not expressed in these cells or it is not polysialylated.

CARTILAGE

As well as being an important structural material in its own right, cartilage is also the developmental progenitor of the ossified skeleton, in the form of hyaline cartilage models laid down at the putative site of the calcified endoskeleton (see Hooper and Hart, 1985 for review). If we consider the formation of cartilage in the development of the rib cage or the production of the cartilage rings of the trachea, in both cases, cells from the somatic mesoderm aggregate to form a precartilagenous mesenchymal condensation. While mesenchymal cell recruitment into the cartilage condensation continues, this condensation is strongly poly Sia positive, but as centrally located cells start to differentiate and form proteoglycans, poly Sia immunoreactivity is lost (Lackie et al., 1992). An outer layer of poly Sia positive cells remains. In this example, loss of poly Sia immunoreactivity closely corresponds to the onset of cell differentiation. Further studies to look at the effect of removing poly Sia by endoneuraminidase treatment on cartilage formation in vitro show that cartilage formation is not prevented by removal of poly Sia but the pattern of laying down cartilage may be affected (Lackie et al, 1992 in prep.).

ENDODERM-MESODERM INTERACTIONS

INTESTINE

The endoderm of the early embryo gives rise to the epithelial lining of the tubular digestive tract, the digestive glands including the liver and pancreas and the epithelium of the respiratory tract. Early in its development, the tubular portions of the digestive tract have been found to be poly Sia immunoreactive (Lackie et al., 1992). However, this reactivity is lost comparatively soon after the basic structural differentiation of the tract. Pancreatic epithelial cells for example, are poly Sia unreactive a short period (ca. 1 day in rat) after the pancreas forms. This is also true of liver which shows no poly Sia immunoreactivity even when the adjoining tubular digestive tract epithelium (which has the same origin) is reactive. poly Sia immunoreactivity is lost at different times along the tubular digestive tract, persisting longest in the caudal regions of the hindgut derived epithelium (Lackie et al., 1992). The mesenchymal, connective and muscular elements of the digestive tract are contributed by mesodermally derived cells which strongly express poly Sia into early postnatal development - long after poly Sia is lost in the epithelium.

RESPIRATORY TRACT

The respiratory diverticulum buds from the putative oesophageal portion of the foregut as the vitelline duct forms. This is the progenitor of the entire epithelial lining of the respiratory tract which is therefore of endodermal origin. The epithelium of the lung bud requires the presence of a mesenchymal inducer to differentiate into the branched structures of the respiratory tract (Wessells, 1970; Deuchar, 1975 for review). In vitro studies have shown that the endodermally derived epithelial bud is pluripotent at this stage in development. If no mesenchyme is present then no branching or development occurs. If mesenchyme from the putative tracheal region is present the epithelium elongates but does not branch, whereas if putative bronchial mesenchyme is used the normal branching network is produced. Furthermore, if intestinal or liver mesenchyme is used then the epithelium differentiates according to the structure expected in the organ of the inducing mesenchyme rather than lung. In the developing rat poly Sia immunoreactivity has been found in the developing trachea and major bronchi. In man, cells positive for poly Sia are also found in intermediate bronchi (Lackie et al., 1992). In neither species has poly Sia immunoreactivity been found in later generations of the branched respiratory tract. Indeed poly Sia is lost early in lung development while the lung is still a tubular structure. In contrast, the inducing mesenchyme is strongly immunoreactive for poly Sia and remains so until early in postnatal life. A further species difference is seen in avian lung development, where the trachea and the larger branches of the bronchi and bronchioles are poly Sia immunoreactive until late in embryonic development (Lackie et al., 1992). In the respiratory epithelium poly Sia is also basolaterally sorted during its period of expression.

In summary, in respiratory tract development poly Sia is expressed in the epithelial lining of rat and human while it is growing, presumably under the primary inductive influence of the tracheal mesenchyme. Expression in the mesenchyme is more prolonged as is generally true of the organs with endodermal components. Loss of poly Sia coincides with terminal determination, if not differentiation, of the cell populations. Chick lung shows a qualitatively different pattern of expression.

ECTODERM-MESODERM INTERACTIONS

HAIR FOLLICLES

Hair follicles are formed by the interaction of the endodermally derived epidermis and the mesodermally derived dermis. The first stage in this process is the formation of N-CAM and poly Sia positive condensates, analogous to those reported in chicken feather formation which is thought to result from N-CAM induced cell aggregation (Chuong and Edelman, 1985 a, b). At the site of the condensed cell group, poly Sia negative epidermal bud invaginates into the poly Sia positive dermis which forms a condensation of mesenchymal cells around the epidermal bud and by a process of reciprocal induction contributes to the formation of the hair papilla. During induction, the epidermal cells express poly Sia which is lost as the hair papilla starts to form, with the exception of cells in the "neck" of the epithelial root sheath which remain reactive. Recent studies have shown that the main source of proliferating cells is in the neck region of the epithelial root sheath (Cotsarelis et al., 1990) which remain poly Sia positive later in development (Lackie et al., 1992).

FUNCTIONAL CORRELATES OF POLYSIALIC ACID EXPRESSION- CLUES FOR A GENERAL ROLE?

Although there is now clear evidence that expression poly Sia is closely controlled during development and may also be expressed in embryonic type tumours such as Wilm's tumour and in tumours with neural or endocrine characteristics (Roth et al., 1988a, Roth et al., 1988b, Komminoth et al., 1991) the developmental and cellular functions of poly Sia remain elusive. In the studies detailed above certain common features of poly Sia expression can be seen:

(1) Poly Sia immunoreactivity is frequently present on most mesodermally and

neuroectodermally derived structures. In mesenchyme, loss of poly Sia occurs as, or soon after, cells are recruited into more specific cell compartments and differentiate, for example to form nephrons, cartilage, or the connective elements of organs. poly Sia is expressed early in the development of endodermally derived structures, characteristically during induction but is lost as fully determined cell populations are formed. However, expression is often down-regulated before growth finishes and during induction both anlagen may be poly Sia positive. Ectodermally derived structures can also express poly Sia, as exemplified by hair follicles where the germinative cell population remains poly Sia immunoreactive. A common factor between the cells expressing poly Sia in these situations is that they all retain a certain developmental plasticity and potential to change cell phenotype. Many cells undergoing these processes also form into condensed groups before or during induction, which is assumed to reflect increased intracellular adhesion. Thus in these cases poly Sia expression would coincide with induction and cell transformation and may correlate with a subsequent increase in cell adhesion - often as poly Sia is down-regulated.

(2) Poly Sia is often baso-laterally sorted in epithelia. If required developmentally, apical poly Sia expression could presumably be achieved by expression of Pi-linked polysialylated N-CAM isoforms. This being the case, exclusive baso-lateral expression suggests that the functional role of poly Sia, if any, resides in the baso-lateral domain of the epithelium. This portion of the epithelium is involved in anchoring the epithelium, establishing and maintaining contact with basal and sub-epithelial cells and further, in ensuring that the epithelial lining covers the full extent of the appropriate tubular structure. These processes are all characterised by more or less specific cellular interactions.

(3) On neuroectodermally derived neuronal structures such as nerves, poly Sia is found for extended periods, in certain cases even persisting into adulthood. A characteristic of these cells is the ability to grow through regions consisting of other cell types, but to subsequently make, and sustain, contact with target cells. Invasive growth requires reduced or more transient cell adhesion while innervation conversely requires more stable adhesion and specific recognition.

Thus poly Sia expression seems to correlate with situations in which cell-cell interactions are critical, as exemplified by induction, epithelial establishment or innervation. The diversity of patterns of developmentally regulated poly Sia expression suggests that poly Sia functions more as a modulator of other signalling processes rather than as a primary signalling pathway in its own right.

SURFACE MODULATION; MECHANISMS OF POLYSIALIC ACID ACTION

As detailed above, there does not seem to be a single, clearly defined function or process in organogenesis which is mediated by poly Sia. Rather, there are several apparently contradictory processes which may be affected.

The cell surface acts as a selective clearing house between the cell, its immediate neighbours and the local environment. Cell surface molecules share the plasma membrane and often interrelated signalling pathways. These molecules thus interact and their distribution is often heterogeneous, being affected by other molecules present in the plasma membrane. The presence of a large, highly charged molecule such as poly Sia, may significantly interfere with (or enhance) the binding of other molecules either specifically or my modifying the general cell surface environment. Such effects have been shown for N-CAM and Ll interaction (Kadmon et al., 1990a, Kadmon et al., 1990b). It has been suggested that cell surface properties may be generally modified by poly Sia (Rutishauser et al., 1988, Yang et al., 1992). In the context of poly Sia expression during organogenesis, the concept that poly Sia on the cell surface enhances intracellular spacing and thus reduces specific interaction via other molecules (Rutishauser et al., 1988, Yang et al., 1992, see also article of Rutishauser in this volume) could provide an explanation for the expression and function of poly Sia. However, to confirm that poly Sia is physiologically important in, for example, induction and to explain how this is brought about functionally, a more detailed knowledge of the other factors controlling these processes is required.

In conclusion, poly Sia has many of the properties required for a developmentally controlled modulator of cell surface mediated cellular interactions. Such properties are central to the process of organogenesis and may thus hold the key to our understanding of poly Sia function and its putative role in the modulation of other cell-surface mediated effects.

ACKNOWLEDGEMENTS

We would like to thank Angela Grau and Esther Ackerman for their expert technical assistance in our studies detailed in this article and Dr A Semper for critical reading of the manuscript. We are also indebted to Drs. E. Bock, C. Goridis, D. Bitter-Suermann and C. Weisgerber for the kind gifts of N-CAM and poly Sia antibodies and endoneuraminidase used in our studies. This work was supported by grants from The Wellcome Trust, the Swiss National Science Foundation grant nr. 37-26273.89, the Cancer League of the Kanton Zürich, and the Kanton Zürich.

REFERENCES

Booth, C. M., and M. C. Brown. (1988). Localization of neural cell adhesion molecule in denervated muscle to both the plasma membrane and extracellular compartments by immuno-electron microscopy. Neuroscience 27: 699-709.

Chuong, C.-M., and G. M. Edelman. (1985). Expression of cell-adhesion molecules in embryonic induction I: Morphogenesis of nestling feathers. J. Cell Biol. 101: 1009-1026.

Chuong, C. -M., and G. M. Edelman. (1985). Expression of cell adhesion molecules in embryonic induction II: Morphogenesis of adult feathers. J. Cell Biol. 101: 1027-1043.

Cooke, J. (1991). Inducing Factors and the Mechanism of Body Pattern Formation in Vertebrate Embryos. Current Topics Develop. Biol. 25: 45-75.

Cotsarelis, G., T. T. Sun, and R. M. Lavker. 1990. Label retaining cells reside in the bulge area of pilosebaceous unit: implications for follicular stem cells, hair cycle, and skin carcinogenesis. Cell 61: 1329-1337.

Crossin, K. L., C-M. Chuong, and G. M. Edelman. 1985. Expression sequences of cell adhesion

molecules. Proc. Natl. Acad. Sci. U S A 82: 6942-6946.
Cunningham, B. A., J. J. Hemperly, B. A. Murray, E. A. Prediger, R. Brackenbury, and G. M. Edelman. 1987. Neural cell adhesion molecule: Structure, immunoglobin-like domains, cell surface modulation and alternative RNA splicing. Science 236: 799-806.
Deuchar, E. M. 1975. Cellular Interactions in Animal Development. Chapman and Hall, London.
Dickson, G., D. Peck, S. E. Moore, C. H. Barton, and F. S. Walsh. 1990. Enhanced myogenesis in N-CAM-transfected mouse myoblasts. Nature 344: 348-351.
Edelman, G. M., and K. L. Crossin. 1991. Cell adhesion molecules - implications for a molecular histology. Ann. Rev. Biochem. 60: 155-190.
Finne, J., and P. H. Makela. 1985. Cleavage of the polysialosyl units of brain glycoproteins by a bacteriophage endosialidase: Involvement of a long oligosaccharide segment in molecular interactions of polysialic acid. J. Biol. Chem. 260: 1265-1270.
Finne, J. 1982. Occurrence of unique polysialosyl carbohydrate units in glycoproteins of developing brain. J. Biol. Chem. 257: 11966-11970.
Fraser, S. E., B. A. Murray, C. -M. Chuong, and G. M. Edelman. 1984. Alterations of the retinotectal map in Xenopus by antibodies to neural cell adhesion molecules. Proc. Natl. Acad. Sci. U S A 81: 4222-4226.
Frosch, M., J. Görgen, G. J. Boulnois, K. M. Timmis, and D. Bitter-Suermann. 1985. NZB mouse system for production of monoclonal antibodies to weak bacterial antigens- Isolation of an IgG antibody to the polysaccharide capsules of Escherichia coli Kl and group B meningococci. Proc. Natl. Acad. Sci. U S A 82:1194-1 198.
Grumet, M., U. Rutishauser, and G. M. Edelman. 1982. Neural cell adhesion molecule is on embryonic muscle cells and mediates adhesion to nerve cells in vitro. Nature 295: 693-695.
Hall, A. K., R. Nelson, and U. Rutishauser. 1990. Binding properties of detergent-Solubilized NCAM. J. Cell Biol. 110: 817-824.
Hayrinen, J., D. Bitter-Suermann, and J. Finne. 1989. Interaction of meningococcal group-B monoclonal antibody and its FAB fragment with α-2-8-linked sialic acid polymers requirement of a long oligosaccharide segment for binding. Molec. Immunol. 26: 523529.
Hoffman, S., and G. M. Edelman. 1983. Kinetics of homophilic binding by E and A forms of the neural cell adhesion molecule. Proc. Natl. Acad. Sci. U S A 80: 5762-5766.
Hoffman, S., B. C. Sorkin, P. C. White, R. Brackenbury, R. Mailhammer, U. Rutishauser, B. A. Cunningham, and G. M. Edelman. 1982. Chemical characterization of a neural cell adhesion molecule (N-CAM) purified from embryonic brain membranes. J. Biol. Chem. 257: 7720-7729.
Hooper, A. F., and N. H. Hart. 1985. Foundations of animal development. Oxford Universtiy Press.
Husmann, M., J. Roth, E. A. Kabat, C. Weisgerber, M. Frosch, and D. Bitter-Suermann. 1990. Immunohistochemical localization of polysialic acid in tissue sections - differential binding to polynucleotides and DNA of a murine IgG and a human IgM monoclonal antibody. J. Histochem. Cytochem. 38: 209-215.
Jones, F. S., E. A. Prediger, D. A. Bittner, E. M. De Robertis, and G. M. Edelman. 1992. Cell adhesion molecules as targets for Hox genes - Neural cell adhesion molecule promoter activity is modulated by cotransfection with Hox-2.5 and Hox-2.4. Proc. Natl. Acad. Sci. U S A 89: 2086-2090.
Jorgensen, 0. S., and E. Bock. 1974. Brain specific synaptosomal membrane proteins demonstrated by crossed immunoelectrophoresis. J. Neurochem. 23 :879-880.
Kadmon, G., A. Kowitz, P. Altevogt, and M. Schachner. l990a. The neural cell adhesion molecule N-CAM enhances Ll-dependent cell cell interactions. J. Cell Biol. 110: 193-208.
Kadmon, G., A. Kowitz, P. Altevogt, and M. Schachner. l990b. Functional cooperation between the neural adhesion molecules Ll and N-CAM is carbohydrate dependent. J. Cell Biol. 110: 209-218.
Knudsen, K. A., S. A. McElwee, and L. Myres. 1990. A role for the neural cell adhesion molecule, N-CAM, in myoblast interaction during myogenesis. Devel. Biol. 138: 159-168.
Komminoth, P., J. Roth, P. M. Lackie, D. Bitter-Suermann, and P. U. Heitz. 1991. Polysialic acid of the neural cell adhesion molecule distinguishes small cell lung carcinoma from carcinoids. Am. J. Pathol. 139: 297-304.
Lackie, P. M., C. Zuber, and J. Roth. l990a. Polysialic acid and N-CAM in embryonic rat

kidney: mesenchymal and epithelial elements show different patterns of expression. Development 110: 933-947.
Lackie, P. M., C. Zuber, and J. Roth. 1991. Expression of polysialylated N-CAM during rat heart development. Differentiation 47: 85-98.
Lackie, P. M., C. Zuber, and J. Roth. 1990b. Highly sialylated neural cell adhesion molecules localised in derivatives of the three primary germ layers throughout embryonic development. Trans. Royal Microsc. Soc. 1: 657-660.
Lackie, P. M., C. Zuber, and J. Roth. 1992. Polysialic acid of the neural cell adhesion molecule (N-CAM) is widely expressed during organogenesis in mesodermal and ectodermal derivatives. Submitted for publication.
Landmesser, L., L. Dahm, J. Tang, and U. Rutishauser. 1990. Polysialic acid as a regulator of intramuscular nerve branching during embryonic development. Neuron 4: 655-667.
Livingston, B. D., E. M. De Robertis, and J. C. Paulson. 1990. Expression of betagalactoside alpha2,6 sialyltransferase blocks synthesis of polysialic acid in Xenopus embryos. Glycobiology 1: 39-44.
Maier, C. E., M. Watanabe, I. Singer, I. G. McQuarrie, J. Sunshine, and U. Rutishauser. 1986. Expression and function of neural cell adhesion molecule during limb regeneration. Proc. Natl. Acad. Sci. U S A 83: 8395-8399.
Markwald, R. R., T. P. Fitzharris, and F. J. Manasek. 1977. Structural development of endocardial cushions. Am. J. Anat. 148: 85-120.
Markwald, R. R. 1987. Role of the extracellular matrix in morphogenesis. Mead Johson Symposium No. 29: Perinatology Press, New York pp7-13.
Mjaatvedt, C. H., and R. R. Markwald. 1989. Induction of epithelial-mesenchymal transition by an in vivo adheron-like complex. Devel. Biol. 136: 118-128.
Moolenaar, C. E. C. K., E. J. Muller, D. J. Schol, C. G. Figdor, E. Bock, D. Bitter-Suermann, and R. J. A. M. Michalides. 1990. Expression of neural cell adhesion molecule related sialoglycoprotein in small cell lung cancer and neuroblastoma cell lines H69 and chp-212. Cancer Res. 50: 1102-1106.
Pelkonen, S., J. Pelkonen, and J. Finne. 1989. Common cleavage pattern of polysialic acid by bacteriophage endosialidases of different properties and origins. J. Virol. 63: 4409-4416.
Roth, J., D. J. Taatjes, D. Bitter-Suermann, and J. Finne. 1987. Polysialic acid units are spatially and temporally expressed in developing postnatal rat kidney. Proc. Natl. Acad. Sci. U S A 84: 1969-1973.
Roth, J., C. Zuber, P. Wagner, D. J. Taatjes, C. Weisgerber, P. U. Heitz, C. Goridis, and D. Bitter-Suermann. 1988a. Reexpression of poly(sialic acid) units of the neural cell adhesion molecule in Wilms tumor. Proc. Natl. Acad. Sci. U S A 85: 2999-3003.
Roth, J., C. Zuber, P. Wagner, I. Blaha, D. Bitter-Suermann, and P. U. Heitz. 1988b. Presence of the long chain form of polysialic acid of the neural cell adhesion molecule in Wilms tumor: Identification of a cell adhesion molecule as an onco-developmental antigen and its implications for tumor histogenesis. Am. J. Pathol. 133: 227-240.
Roth, J., and C. Zuber. 1990. Immunoelectron microscopic investigation of the surface coat of wilms tumor cells the dense lamina is composed of highly sialylated neural cell adhesion molecule. Lab. Invest. 62: 55-60.
Roth, J. 1989. Postembedding labeling on lowicryl K4M tissue sections: Detection and modification of cellular components. Meth. Cell Biol. 31: 513-551.
Rothbard, J. B., R. Brackenbury, B. A. Cunningham, and G. M. Edelman. 1982. Differences in the carbohydrate structures of neural cell adhesion molecules from adult and embryonic chicken brains. J. Biol. Chem. 157: 11064-11068.
Rutishauser, U., and T. M. Jessell. 1988. Cell adhesion molecules in vertebrate neural development. Physiol. Rev. 68:819-850.
Rutishauser, U., A. Acheson, A. K. Hall, and J. Sunshine. 1988. N-CAM as a regulator of cell-cell interactions. Science 240: 53-57.
Rutishauser, U., and L. Landmesser. 1991. Polysialic acid on the surface of axons regulates patterns of normal and activity-dependent innervation. Trends in Neurosciences 14:528-532.
Sadler, T. W. 1990. Langman's Medical Embryology (6th edition). Williams and Wilkins, Baltimore.

Sadoul, R. V., M. Hirn, H. Deagostini-Bazin, G. Rougon, and C. Goridis. 1985. Adult and embryonic mouse neural cell adhesion molecules have different binding properties. Nature 304: 347-349.
Saxen, L. 1987. Organogenesis of the kidney. Cambridge University Press Cambridge UK.
Theodosis, D. T., G. Rougon, and D. A. Poulain. 1991. Retention of embryonic features by an adult neuronal system capable of plasticity: polysialylated neural cell adhesion molecule in the hypothalamo-neurohypophysial system. Proc. Natl. Acad. Sci. U S A 88: 5494-5499.
Thiery, J. P., R. Brackenbury, U. Rutishauser, and G. M. Edelman. 1977. Adhesion among neural cells of the chick embryo. II Purification of a cell adhesion molecule from neural retina. J. Biol. Chem. 252: 6841-6845.
Tomlison, S., and P. W. Taylor. 1985. Neuraminidase associated with Coliphage E that specifically depolymerises the Escherichia coli Kl capsular polysaccharide. J. Virol. 55: 374-378.
Troy, F. A. 1992. Polysialylation - From Bacteria to Brains. Glycobiology 2: 5-23.
Wessells, N. K. 1970. Mammalian lung development: Interactions in formulation and morphogenesis of tracheal buds. J. Exp. Zool. 175: 455-466.
Wharton, J., L. Gordon, F. S. Walsh, T. P. Flanigan, S. E. Moore, and J. M. Polak. 1989. Neural cell adhesion molecule (N-CAM) expression during cardiac development in the rat. Brain Res. 483: 170-176.
Yang, P. F., X. H. Yin, and U. Rutishauser. 1992. Intercellular space is affected by the polysialic acid content of NCAM. J. Cell Biol. 116: 1487-1496.
Zuber, C., P. M. Lackie, W. A. Catterall, and J. Roth. 1992. Polysialic acid is associated with sodium channels and the neural cell adhesion molecule (N-CAM) in adult rat brain. J. Biol. Chem. 267: 9965-9971..

Polysialic Acid
J. Roth, U. Rutishauser and F. A. Troy II (eds.)

DEVELOPMENTAL EXPRESSION OF GANGLIOSIDES IN VIVO AND IN VITRO

Harald Rösner

Institute of Zoology of the University of Hohenheim-Stuttgart, Garbenstr. 30,
7000 Stuttgart 70, Germany

SUMMARY: The expression of gangliosides is developmentally regulated in nervous as well as in extraneural tissues. Data obtained by means of biochemical, immunocytochemical and enzymatical approaches in vivo and in vitro revealed for the developing avian and mammalian (including human) brain: (1) no or very low ganglioside expression by the neural tube, (2) increase of GD3 in glial- and neural progenitor cells in parallel with the spatiotemporal increase of proliferation activity, (3) decrease of GD3 (and GT3) and expression of higher sialylated b- and c-gangliosides in "new-born" neurons migrating or extending neurites, (4) several-fold intensification of total neuronal ganglioside synthesis and a shift from b- (and c-) in favour of a-gangliosides during growth spurt and synaptogenesis, (5) accretion of GM1 and GM4 during central myelination and of LM1 during peripheral myelination, (6) establishment of the final area- and function-specific pattern with GM1, GD1a, GD1b, and GT1b as major components and only traces of c-gangliosides, (7) slow decrease of mainly GD1a during aging and prevalence of b-gangliosides in the old brain.

Gangliosides are sialic acid containing glycosphingolipids structurally defined as hematosides, lacto-, neolacto-, globo-,and ganglio-series gangliosides (for rev. Ledeen and Yu, 1982; Ledeen, 1978; Wiegandt, 1982). They are principle membrane constituents of vertebrate cells, synthesized by step-wise addition of carbohydrate moieties to ceramide during its transfer from the ER through the cis- and trans-Golgi compartments (for rev. Tettamanti et al., 1987) and degraded via lysosomal pathways (Sandhoff et al., 1987). Their functional significance still remains obscure. However, numerous experimental data suggest that apart from their basic function as structural membrane components per se these glycolipids play an important role in receptor-mediated signal transduction (for rev. Hakomori, 1981) as well as direct cell/cell and cell/extracellular matrix interactions (Cheresh et al., 1986). The mature brain, for example, contains about 20-fold higher amounts of gangliosides than extraneural tissues (for rev. Ledeen and Yu, 1982; Svennerholm, 1984). Neuronal membranes are especially enriched in ganglio-series gangliosides synthesized along different pathways (Fig. 1). While a- and b- gangliosides predominate in the adult avian and mammalian brain, in fish nervous tissue b- and mainly c-gangliosides are abundant, showing a high degree of diversity (Hilbig and Rahmann, 1987). In addition, striking regional differences of the ganglioside composition have been established in the mammalian (including human) brain (Suzuki, 1965; Rösner, 1977; Kracun et al., 1984). Unlike neurons, ganglioside pattern of astrocytes resemble that of most non-neural cells containing much less gangliosides and a prevalence of GM3 and GD3 (Robert et al., 1977; Asou et al., 1989; Sbashing-Agler et al., 1988). Likewise, oligodendroglia and central myelin seem to express less highly sialylated compounds, and instead contain GM1 and GM4 in addition to GM3 and GD3 (Cochran et al.,1982; Kim et al., 1986). Schwann cells and peripheral myelin were shown to be enriched in ganglioside LM1 (Chou et al. 1982). Taken into consideration different sialic acid structures (n-/o-acetyl, n-glycolyl), lactonization, fucosylation, and different ceramide moieties one can exspect a high degree of cell- and species-specific variability in ganglioside pattern.

In addition to this, numerous studies have shown changes in ganglioside expression during embryonic differentiation and aging (Table I), as well as after oncogenic transformation (Hakomori 1985). The aim of the present paper is to review developmental and aging related changes in gangliosides. Since only few data exist with respect to extraneural organs, I will focuss mainly on nervous tissue gangliosides. Special aim of this presentation is to point out that, in spite of striking cell- and species- specific differences, there are some general principles of developmental changes of brain ganglioside expression common to all higher vertebrates including humans.

DEVELOPMENTAL CHANGES OF GANGLIOSIDES OF EXTRANEURAL ORGANS

Studies concerning developmental changes of gangliosides in extraneural tissues are up to now very scattered. These studies suffer from the principal fact that extraneural gangliosides are more heterogeneous, are present in small amounts, and that their structur is poorly known. For example, human hemapoietic cell lines were found to have cell line-specific ganglioside pattern, which, however, are in many cases very complex, containing more than 100 different components (Rosenfelder et al., 1992). Nevertheless, studies performed with rat stomach (Bouhours et al., 1987), human liver (Riboni et al., 1992), and mouse liver (Nakamura et al., 1988, Lorke et al., 1990, 1992 in prep.) revealed distinct devopmental changes in the content and composition of gangliosides. Like in the brain, a developmental change of a complex pattern at birth to a "simplified" pattern after 3 weeks was observed in the liver of C57BL/10 and NMRI mice (Lorke et al., 1992 in prep.)

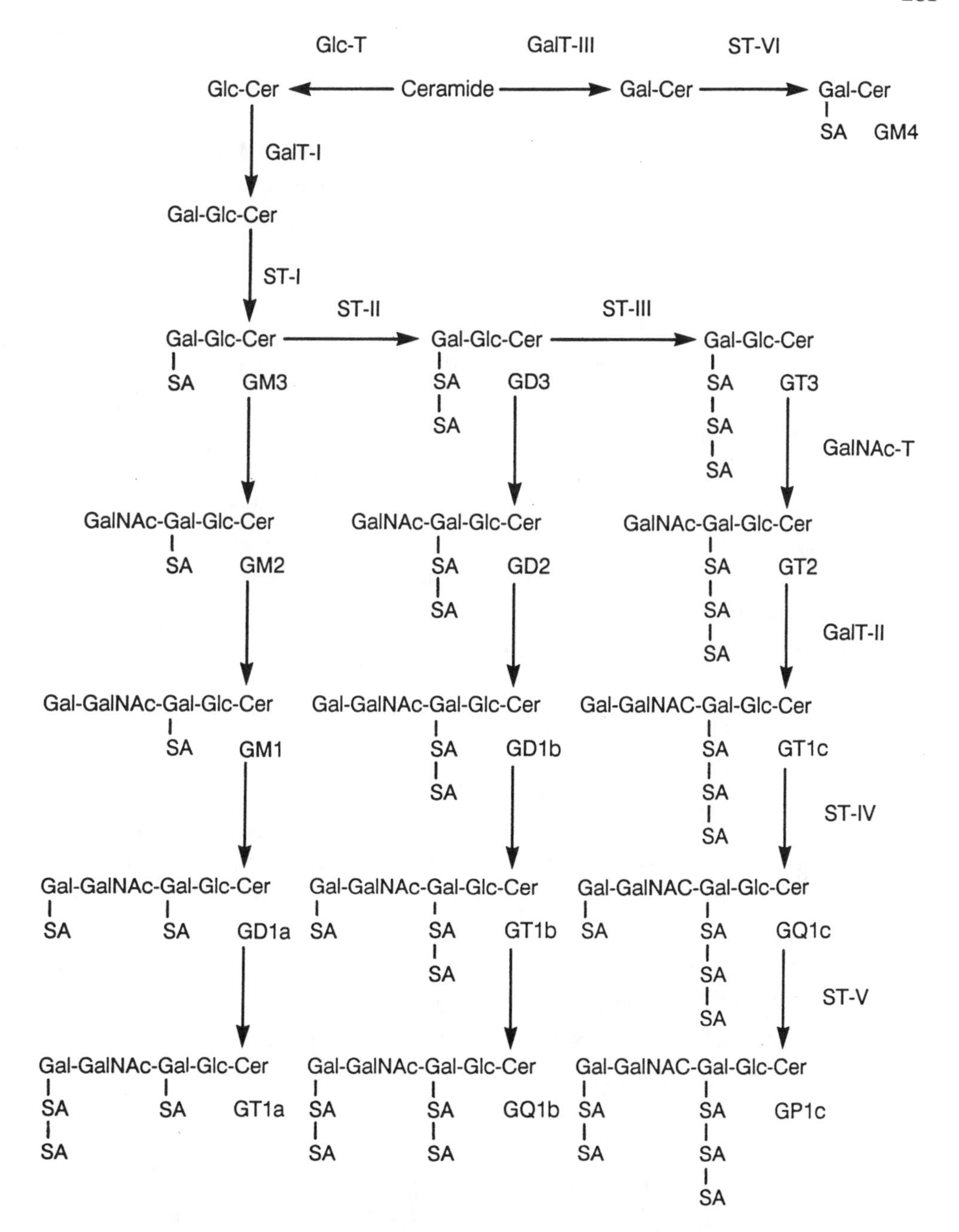

Fig. 1. Biosynthesis pathways of gangliosides; nomenclature of gangliosides according to Svennerholm (1977), IUPAC-IUPCommission; ST, sialyltransferase; GalT, galactosyltransferase; Glc T, glycosyltransferase; GalNAc T, N-acetylgalactosaminyltransferase.

GANGLIOSIDES AND BRAIN DEVELOPMENT

Stimulance for the assumption that gangliosides may play a crucial role in nervous tissue development goes back to their discovery in the human brain (Klenk 1942) and the detection of inherited disorders of ganglioside metabolism leading to abnormal brain development and neuronal disfunction (for rev. Fishman and Brady, 1976, Sandhoff and Christomanou, 1979). Since then, numerous in vivo- and in vitro-studies (Table I) using biochemical, enzymatical, and immunological approaches have shown that neuronal (and glial) ganglioside expression is strictly developmentally regulated in the sense that changes in gangliosides concentration and composition coincide with neuronal (and glial) differentiation and maturation.

Morphologically, brain development and aging can be considered as a multiple-stage process defined by more and less irreversible spatial and temporal cellular differentiation and growth (Fig. 5): (I) neural tube formation, (II) proliferation of neuronal and glial progenitor cells, (III) neurogenesis, neuron migration, and neuritogenesis, (IV) fiber tract mapping, arborization of dendrites and axons, synaptogenesis, embryonic neuron death (V) oligodendroglia proliferation, myelination, functional determination of final connections, (VI) period of the mature brain conserving function-dependend structural plasticity, (VII) aging and neuronal degeneration.

NEURAL TUBE FORMATION (PERIOD I)

Only few studies provide data concerning gangliosides during this very early stage of brain development. Felding-Habermann et al. (1987) found by radiolabeling that the "prebrain"-chicken embryo synthesizes mainly GD3 and several other gangliosides. Thierfelder et al. (1992) reported an immuno- staining of GD3 by mAb R24 of the whole blastoderm of stage 4-5 according to Hamburger and Hamilton (1951). At stage 12-13 the staining was restricted to the endophyllic crescent and cranial part of the notochord. These observation corresponds to results by Rösner et al. (1985,1992) who found up to around stage 8 an immuno-expression of GD3 which was restricted to heavily multiplying cells of mainly mesodermal origin (Fig. 2). Neural stem cells of the forming neural groove and tube, however, did not react with mAb R24 (GD3) or other antibodies directed at GD2,GM1, GD1b or c-gangliosides at this early developmental stage.

Table I. Developmental expression of gangliosides in nervous tissue (list of references)

Reference	species	in vivo/in vitro
Svennerholm, 1964	human	x/-
Suzuki, 1965	human, rat	x/-
Vanier et al., 1971	human, rat	x/-
Merat and Dickerson, 1973	rat, pig	x/-
Rösner, 1975	chicken	x/-
Dreyfus et al., 1975	chicken	x/-
Rösner, 1977	mouse	x/-
Yusuf et al., 1977	human	x/-
Mansson et al., 1978	human	x/-
Yavin and Yavin, 1979	rat	-/x
Engel et al., 1979	chicken	x/x
Irwin et al., 1980	rat, mouse	x/-
Dreyfus et al., 1980	chicken	-/x
Rösner, 1980	chicken	x/-
Panzetta et al., 1980	chicken	x/-
Hilbig et al., 1982	mouse, rat	x/-
Rösner, 1982	chicken	x/-
Segler-Stahl et al., 1983	human	x/-
Hilbig et al., 1984	rat	x/-
Goldman et al., 1984	rat	x/-
Landa et al., 1984	chicken	-/x
Maccioni et al., 1984	rat	x/-
Seybold and Rahmann, 1985	fish	x/-
Landa and Moskona, 1985	chicken	x/x
Rösner et al., 1985	chicken	x/-
Schaal and Wille, 1985	rat	x/-
Kracun et al., 1986	human	x/-
Constantine-Paton et al., 1986	rat	x/-
Panzetta et al., 1987	chicken	-/x
Seyfried , 1987	mouse	x/-
Felding-Habermann et al., 1987	chicken	x/-
Rösner et al., 1988 a,b	chicken, rat	x/-
Hirabayashi et al., 1988	chicken	x/-
Rohrer et al., 1988	chicken	-/x
Schlosshauer et al., 1988	rat	x/-
Yu et al., 1988	rat	x/-
Svennerholm et al., 1989	human	x/-
Bouvier and Seyfried, 1989	mouse	x/-
Maccioni et al., 1989	chicken	x/-
Sonnino et al., 1990	chicken	x/-
Drazba et al., 1991	chicken	x/-
Daniotti et al., 1991	chicken	x/-
Thierfelder et al., 1992	chicken	x/-
Kracun et al., 1992	human	x/-
Thangnipon and Balasz, 1992	rat	-/x
Rösner et al., 1992	chicken	x/-

These data suggest that during neurulation of the chicken embryo (up to around stage 8, 26-29 hrs) the progenitor cells of glia and neurons express either no or undetectable (by mAbs) amounts of GD3 and ganglio-series gangliosides.

PROLIFERATION OF NEURAL AND GLIAL PROGENITOR CELLS (PERIOD II)

Numerous biochemical studies (Table I) indicate that GD3 is the predominant ganglioside of the early, immature nervous system of birds and mammals (including humans), but a minor glycolipid of the mature brain. More detailed information concerning the spatiotemporal cellular expression of GD3 was recently obtained by means of immunohistochemical staining using specific mAbs. (Rösner et al., 1985, 1992). We observed that first expression of GD3 in chicken neuroepithelium occurs in the anterior prosencephalic vesicle of stage 9-10 (1 1/2 days). During the next 3 days (up to stage 25-27) immuno-expression of GD3 increases dramatically from rostral to caudal in all regions of the forming brain and spinal cord in parallel with the spatiotemporal increase in proliferation activity. Likewise, in the retina and otic anlage GD3, labeling follows the temporal gradient of proliferation from the center to the periphery. Also, neuronal crest and multiplying non neural cells (e.g. lens, endothelia of forming blood vessels) heavily express GD3. Similar results have been revealed for the embryonic rat brain demonstrating an intense immuno-expression of GD3 by growing endothelia, by immature neuro- and glioblasts of the ventricular and subventricular layers of the prenatal forebrain (Fig 2; Rösner et al., 1986,1988), by germinal layers of the rat cerebellum (Schaal and Wille, 1983; Goldman et al., 1984), dentate gyrus, and hippocampus (Goldman et al., 1984), as well as by the primordium of the human brain of 5 weeks and the hippocampus and pallium of 17 weeks of gestation (Kracun et al., 1992). It is important to note that at this early stage of development mAbs to ganglio-series gangliosides , GD1b, or c-polysialogangliosides never stained GD3+ mitotic germinal zones neither in embryonic chicken nor rat brain (Rösner et al., 1985, 1986, 1988). These gangliosides seem to appear for the first time on "new born" neurons (see below). Biochemical data by Rösner (1980,1982) showed that in the chicken neuroepithelium of E5, consisting of more than 80% of immature cells, apart from GD3, GT3 is also abundant. Although enzymatic data of "pre-neurogenesis" stages are lacking, these last results suggest a high activity of ST-II and ST-III (Fig. 1) of neuronal and glial precursors cells. Enzymatic data of "pre-neurogenesis" stages are lacking.

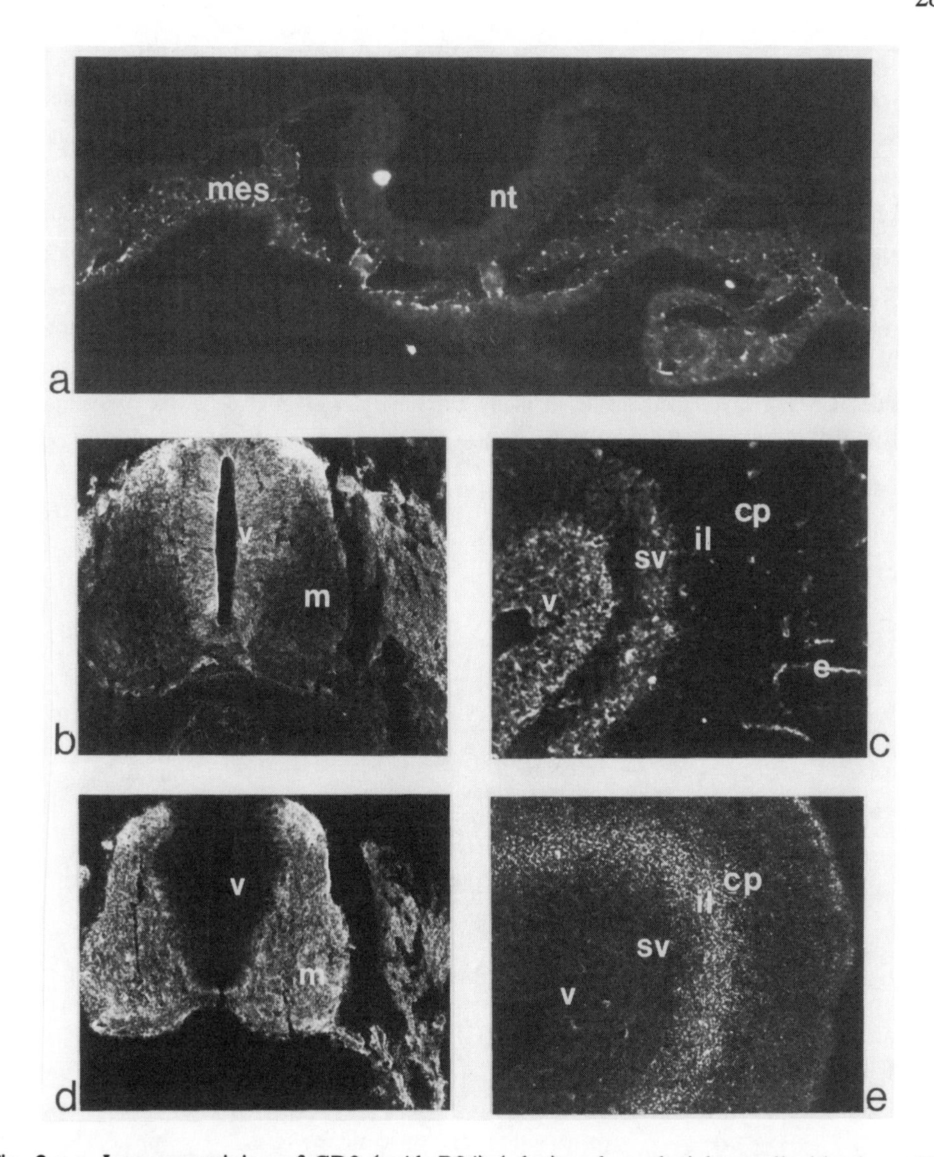

Fig. 2 a-e. Immuno-staining of GD3 (mAb R24) (a,b,c) and c-polysialogangliosides by mAb Q211 (d,e): stage 8 chicken neural tube unstained, GD3-stained (GD3+) mesodermal cells (a); E5 chicken spinal cord with GD3+ ventricular layer (b) and c-gangliosides+ mantle layer (d); E19 rat cerebrum with GD3+ endothelia, ventricular and subventricular layer (c) and c-gangliosides+ neurons in the intermediate layer (thalamo-cortical fibre tract) and forming cortical plate (e); nt, neural tube; mes, mesoderm; v, ventricular layer; sv, subventricular layer; m, mantle layer; e,endothelia; il, intermediate layer; cp, cortical plate.

From the above data the general conclusion seems to be justified that heavily multiplying embryonic neuroectodermal as well as most non neural stem cells predominantly and intensely express ganglioside GD3 (and GT3 ?). In this respect they behave similar to reactive glia (Seyfried et al., 1982; Levine et al., 1986; Yu et al., 1974) and many neuroectodermal transformed cell lines as melanomas (Pukel et al., 1982), gliomas and astrocytomas (Eto and Shinoda, 1982) and meningiomas (Fredmann et al., 1990). Interestingly, induction of ST-II activity and accretion of GD3 in rat fibroblasts by transfection with c-myc DNA, correlated with the ability of anchorage-independent growth known to be a reliable feature of tumorigenicity in vitro (Nakaishi et al., 1988). At this point it should be remembered that GD3 is besides GM3 a main ganglioside in many extraneural and glial cells. Its concentration, however, differs strikingly but is much lower than in embryonic, reactive or transformed cells.

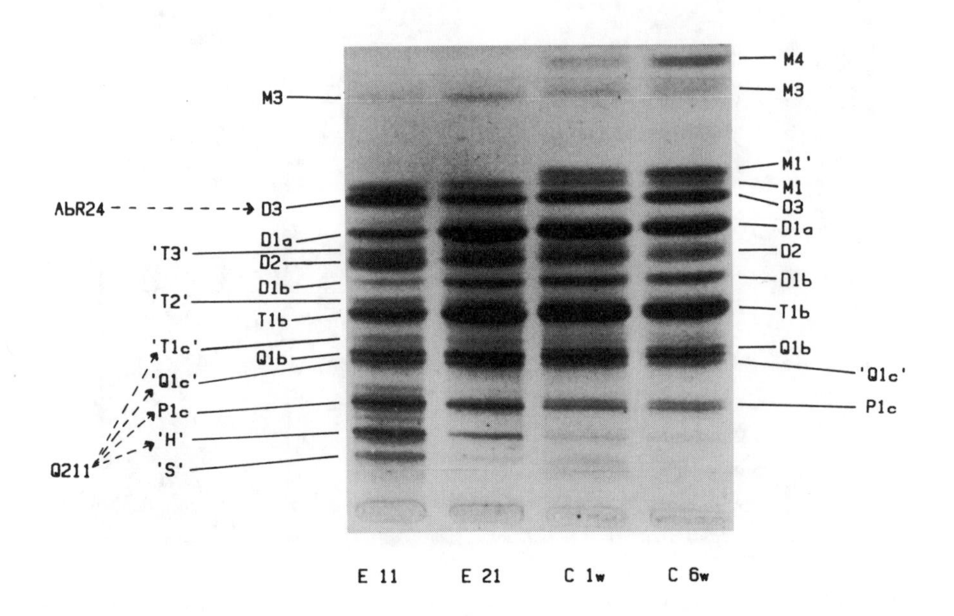

Fig. 3. Ganglioside pattern of chicken optic lobes from E11 to 6 weeks (6w) after hatch; c-pathway gangliosides T3, T2, T1c, Q1c, P1c, H, S decrease; D1b, T1b, D1a (neurons) and M1', M4 (myelin) increase; D3 is recognized by mAb R24; T1c, Q1c, P1c, H, are recognized by mAb Q211.

NEUROGENESIS, NEURON MIGRATION AND NEURITOGENESIS (PERIOD III)

Proceeding in brain development, neuroblasts differentiate to postmitotic neurons, which migrate from the ventricular into the forming mantle layers and extend their neuritic processes. Numerous biochemical studies working with different areas of the developing avian and mammalian (including human) brain (Table I) have revealed the general phenomenon that during this developmental stage the content of GD3 decreases and that of ganglio-series gangliosides with up to 5 sialic acid residues increases. Similar changes were found to occur during differentiation of cell and tissue cultures from embryonic chick retina (Panzetta et al., 1987; Landa and Moscona, 1985) and optic tectum (Engel et al., 1979). As a further general phenomenon it was observed that developmental accumulation of b-pathway gangliosides preceedes the increase of a-pathway gangliosides (see next chapter). In addition to this, Rösner (1980,1982) found that several "novel" highly sialylated gangliosides increase very early in parallel with b- gangliosides in the embryonic chicken brain (Fig. 3). Sialic acid to sphingosine ratios, tlc-identification of enzyme-digested derivatives and FAB-spectrometric identification of the main component as GP1c (Rösner et al., 1981, 1985) established that these gangliosides represent glycosphingolipids of the c-pathway characterized by 3 sialic acid residues linked to the inner galactose (Fig. 1). C-gangliosides had been first identified in goldfish brain (Ishizuka and Wiegandt, 1972). As major brain gangliosides of bony fishes and rajiform elasmobranchs (Rahmann and Hilbig, 1983) they are synthesized via a separate pathway (Yu and Ando, 1980). The presence of the c-gangliosides GT3, GT2, GT1c, GQ1c, GP1c, and GH in embryonic chicken brain has been confirmed by immuno-tlc by means of the mAbs Q211 (Rösner et al., 1985,) and M6704 and M7103 (Hirabayashi et al., 1988). Subsequently, some of these gangliosides were detected in the perinatal rat cerebrum (Rösner et al., 1988) and retina (Daniotti et al., 1992), the fetal human brain (Greis and Rösner, 1990; Kracun et al., 1992) and in the adult brains of turtles (Greis and Rösner, 1990) and bovine (Hirabayashi et al., 1988). Interestingly, small amounts of c-gangliosides are still present in the adult human brain and seem to accumulate in Alzheimer's disease fibrillary plaques and neurofibrillary tangles (Fig. 4). Immuno-histochemical staining by means of mAb Q211 revealed that in embryonic chicken brain c-polysialogangliosides are neuron specific. Their appearance coincides with differentiation of mitotic neuroblasts to postmitotic young neurons in both the central and peripheral nervous system in vivo (Rösner et al. 1985) and in vitro (Rohrer et al., 1988,Rösner et al., 1992).

Immunostaining with mAbs to GD1b showed that also gangliosides of the b-pathway are for the first time expressed by newborn chicken neurons, leaving the germinal layer and migrate

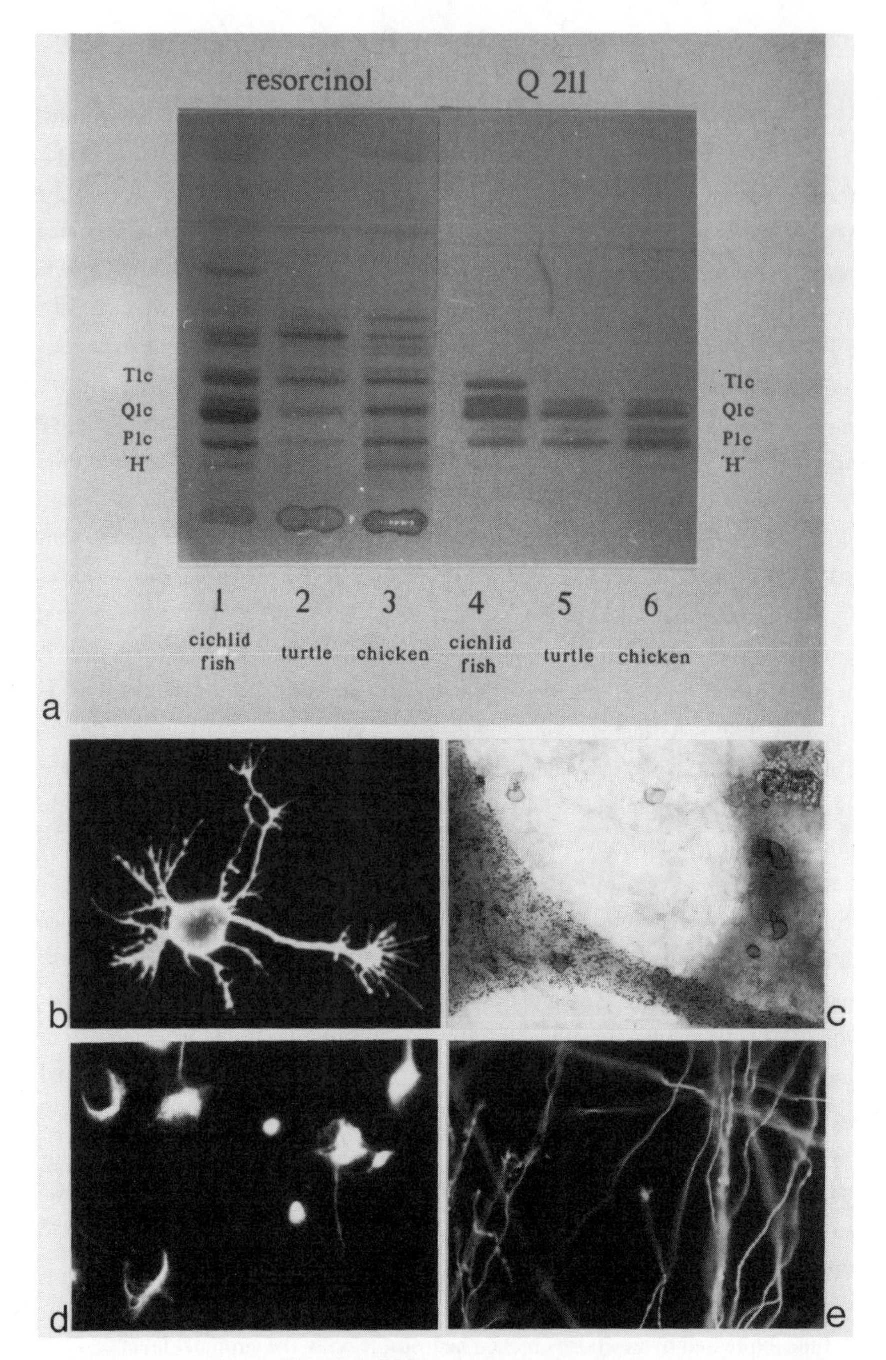

Fig. 4 a-e. Tlc-immunostaining by mAb Q211 of c-polysialogangliosides from the brains of adult cichlid fish (1,4), turtle (2,5), and embryonic chicken (3,6); Q211-staining of an embryonic chicken neuron by fluoresceine-labeled (b) or gold-labeled second antibody (c), of Alzheimer neurofibrillary tangles in retinoic acid differentiated SH-SY5Y neuroblastoma cells (d), and of goldfish retinal axons in explant culture (e).

into the mantle layer (Rösner et al., 1992). At the same time these young neurons decrease the expression of GD3 (Fig. 2).

These data show that a decrease of GD3 expression and an increase of higher sialylated ganglio-series gangliosides, biochemically established as a general developmental phenomenon of avian brain (Table I), is initially related to neurogenesis. It is yet uncertain wether this concept of neurogenesis associated changes of gangliosides can be extended from the avian to the mammalian brain, but it seem very likely. Thus, an early developmental decrease of GD3 and a simultaneous increase of b-gangliosides has been shown in numerous reports to occur in the mammalian brain, too (Table I). It was also found in primary cultures (Thangnipon and Balazs, 1992). Evidence for the significance of b-gangliosides in early mammalian neuronal differentiation mainly stems from experiments with the recessive twt/twt mouse. This mutant was shown to have a marked b-ganglioside deficiency of mainly Q1b (Seyfried, 1987) and will die before reaching E18 from failed neural differentiation (Bouvier and Seyfried, 1989). Furthermore, Rösner et al. (1988); and Greis and Rösner (1990) found in rat embryonic brain between E17 and E20 increasing immuno- expression of c-polysialogangliosides on new-born cortical neurons and ingrowing fibres of the thalamo-cortical/hippocampal tract (Fig. 2) At the same developmental stage, immuno- expression of GD3 was lacking or very low in these rat neurons, but high in dividing progenitor cells (Fig. 2) (Rösner et al., 1988; Goldman et al., 1984). A decrease of GD3 and an increase of b- and c-gangliosides was also found to occur in very early development of the fish brain (Seybold and Rahmann, 1985).

At this point, it is important to note that in the chicken brain not only GD3 but also GD2, GT3, and GT2 decrease during neurogenesis (Rösner, 1980, 1982).

Taken together, these data suggest that neurogenesis in the vertebrate brain is associated with rapid maturation of the golgi complex including transfer mechanisms for GD3 and GT3 from the cis- through the trans-golgi compartments where they are glycosylated to b- and c-gangliosides (Fig. 1). Concerning enzymes, according to the current concept of ganglioside biosynthesis (Pohlentz et al., 1988; Iber et al., 1990; van Echten et al., 1990) a maintainance of high activities of sialyltransferases II and III (ST-II, ST-III, fig. 1) as compared to GalNAc T and a rise of Gal T-II as well as ST-IV and ST-V could explain a drop of GD3 and GT3 due to an increased synthesis of tetraosyl b- and c-gangliosides, while formation of a-gangliosides (via GM2) would remain low . In this model, in addition to a golgi-transport mediated regulation, Gal T-II would be a key enzyme in "opening" the pathway to complete b- and c-gangliosides.

Enzymatic studies have been performed with chicken retina of E7 in vivo (Maccioni et al., 1989) and in vitro (Landa et al., 1984), the youngest stage investigated so far. These data demonstrate a higher activity of ST-II compared to Gal T-II and GalNAc T, which fits well with the above model. However, because the E7-retina contains about 30% neurons, a discrimination between immature cells and neurons with respect to enzyme activities was not possible. In later developmental stages, ST-II activity of chicken retina decreases and that of GalNAc T and Gal T-II increase (Panzetta et al., 1980; Maccioni et al., 1989; Landa et al., 1984). In embryonic rat brain, ST-II has been reported to remain on a high level up to E14 (Maccioni et al., 1984), respectively up to E18 (Yu et al., 1988).

The above mentioned phenomenon that intense expression of GD3 is reduced to immunologically undetectable amounts in mature neurons seems not to be shared by all neuron types of rat brain. Thus, cerebellar Purkinje cells are reported to maintain intense expression of GD3 on their dendritic trees up to the adult stage (Reynolds and Wilkin, 1988). The same was shown for differentiated rat retinal neurons (Daniotti et al., 1990, 1992). Furthermore, alkali-labile forms of GD3, which are immuno-recognized by the Jones antibody (Reinhardt-Maelike et al., 1990) have been found to be abundant in the perinatal rat retina (Constantine-Paton et al., 1986). The Jones antigen(s) seems to be expressed by glial cells and neurons (Mendez-Otero and Constantine-Paton, 1990) and are found to be developmentally regulated elsewhere in the rat brain in correlation to periods of maximal cell migration (Schlosshauer et al., 1988).

GROWTH SPURT AND SYNAPTOGENESIS (PERIOD IV)

After reaching appropriate targets, neurons enter a period of accelerated growth (axonal and dendritic arborization) leading to an increase in volume and surface area and differentiation of synaptic contacts. To what extent these metabolic and structural changes, which include more dependency on trophic (survival, probably target-derived) factors, are genetically time-controlled or epigenetically triggered by neuron-target contacts is yet unknown. Concerning gangliosides, again remarkable changes have been observed to occur during this period (Fig. 5). The expression of c-polysialogangliosides was found to decrease in embryonic chicken (Rösner et al., 1980, 1982, 1992) and postnatal rat brain (Rösner et al., 1988; Greis and Rösner 1990). Synthesis of gangliosides via the b-pathway and even more strikingly via the a-pathway is enhanced leading to a further increase of GD1b and GT1b and to a several-fold accumulation of GM1, GD1a (and GT1a). This increase in total gangliosides and simultaneous shift in favour of a-gangliosides was shown to be a general feature of the developing avian and mammalian brain during the period of so-called "growth spurt" and synaptogenesis (Fig. 5). On the enzymatic level (Fig. 1) these changes could be caused by a decrease of the activities of ST-II and ST-III and an increase of the GalNAc T activity leading to a shift in favour of GM2 and subsequently of GM1 and GD1a. In fact, a decrease of ST-II activity concomittantly with a rise of GalNac T activity have been shown to occur in chicken retina (Pancetta et al., 1980; Landa et al., 1984; Maccioni et al., 1989) and embryonic rat brain (Maccioni et al., 1984; Yu et al., 1988). The significance of the relative activities of ST-II versus GalNAc T in regulation of both the lactosyl-/gangliotetraosyl-ganglioside and the a-/b-ganglioside ratio was confirmed by Daniotti et al. (1991). These authors demonstrated that unlike chicken the rat retina maintains a high ST-II/GalNac T activity ratio up to adulthood, resulting in a much lower ratio of a- to b-gangliosides and a much higher content of GD3. Moreover, a shift in synthesis from b- to a-gangliosides and vis-versa was shown in primary culture of murine cerebellar cells to depend directly on the activities of ST-II and GalNAc T, which could be experimentally modulated by change of pH (Iber et al., 1990). The modulation of membrane flow by drugs, lowering of ambient temperature, or application of metabolic inhibitors (van Echten and Sandhoff, 1989; van Echten et al., 1990) further supports the concept that sialylation of GM3 to GD3 (ST-II) (and of GD3 to GT3, ST-III) and glycosylation of GM3 to GM2 (GalNAc T) are the most important regulatory steps in promoting ganglioside synthesis through either the b- (and c-) or the a-pathway. A transport-mediated regulation of the later steps of glyosylation within the trans golgi complex seems to be unlikely but cannot be excluded. Another open question is whether the enzyme-acitivities are regulated on the transcriptional or post-transcriptional level or on both. Data obtained from oncogene-transfected cell lines suggested that both regulatory mechanisms are effective (Nagai et al., 1987).

MYELINATION, STRUCTURAL AND FUNCTIONAL MATURATION (PERIOD V)

Turning back to the development of the avian and mammalian brain the period of growth spurt and synaptogenesis is immediately followed by myelination. On the ganglioside level this is indicated by (1) a second rise of GD3, probably due to an increased oligodendroglia proliferation (Rösner, 1982), (2) an accretion of GM4 (Fig. 5; Rösner, 1982), shown to be enriched in oligodendroglia and central myelin (Cochran et al., 1982). In the chicken brain the increase in GM4 is paralleled by a rise of a neuraminidase resistant form of GM1 (Rösner, 1982), probably specific for avian central myelin. During peripheral myelination increasing accretion of LM1 has been shown (Chou et al., 1982). Proceeding in development, in addition to species - also region-specific changes of the brain ganglioside composition occur depending on the time-course and degree of morphological and functional differentiation. Thus, it is generally known (Table I) that cerebellar gangliosides change temporally different as compared to the cerebrum due to a later and different temporal and morphological development. In addition, the high degree of local differences in the ganglioside composition of the adult human brain (Kracun et al., 1984) suggests local changes during final morphological and functional maturation. In fact, similar developmental profiles of gangliosides in the human frontal and occipital cortex and hippocampus up to 4 month after birth were observed (Kracun et al., 1986, 1992). Thereafter, there was a drop of GD1a in the occipital cortex leading to a preponderance of b-gangliosides, especially in the visual cortex. In the frontal cortex , however, there was only a small decrease and in the hippocampus no decrease of a-gangliosides, which resulted in a prevalence of GD1a and GM1 in the adult pattern.

At this point it is important to note that apart from the main ganglio-series gangliosides many other components have been described to occur more or less transiently throughout development. Thus, the lacto-series gangliosides 3'-LM1, 3'-iso LM1, and LD1 were demonstrated to constitute a significant portion with different profiles in fetal human brain and fuc-GM1 was shown to increase in the postnatal human brain (Svennerholm et al., 1989). In the chicken brain, the gangliosides GM1b and GD1 (Fig. 1; Hirabayashi et al., 1990) and in the chicken retina certain o-acetylated compounds, which react with mAb 8A2 have been reported to be developmentally regulated (Drazba et al., 1991).

ADULT NEURONAL PLASTICITY AND AGING (PERIODS VI AND VII)

Only few groups have investigated profiles of brain gangliosides throughout adulthood and aging, revealing partly different results. Whereas Segler-Stahl et al. (1983) reported a 50% decrease of total gangliosides of the human whole brain (based on fresh weight) between age 25 and 50, Svennerholm et al. (1989) found in the human frontal cortex no remarkable change during this period. Likewise, Kracun et al. (1992) detected no dramatic changes of total ganglioside-content of the frontal, occipital, and cerebellar cortex, and the hippocampus between age 20 and 50. Thereafter, up to age 90, these authors found a more or less pronounced decrease in the ganglioside content. With respect to individual gangliosides, all three groups found a substantial decrease of GD1a. GD1b and GT1b, however, remained rather constant. As a consequence of these slow changes, a shift of prevalence of a-gangliosides (GM1, GD1a) to a prevalence of b-gangliosides (GD1b, GT1b) becomes manifest in the old human brain (Fig. 5).

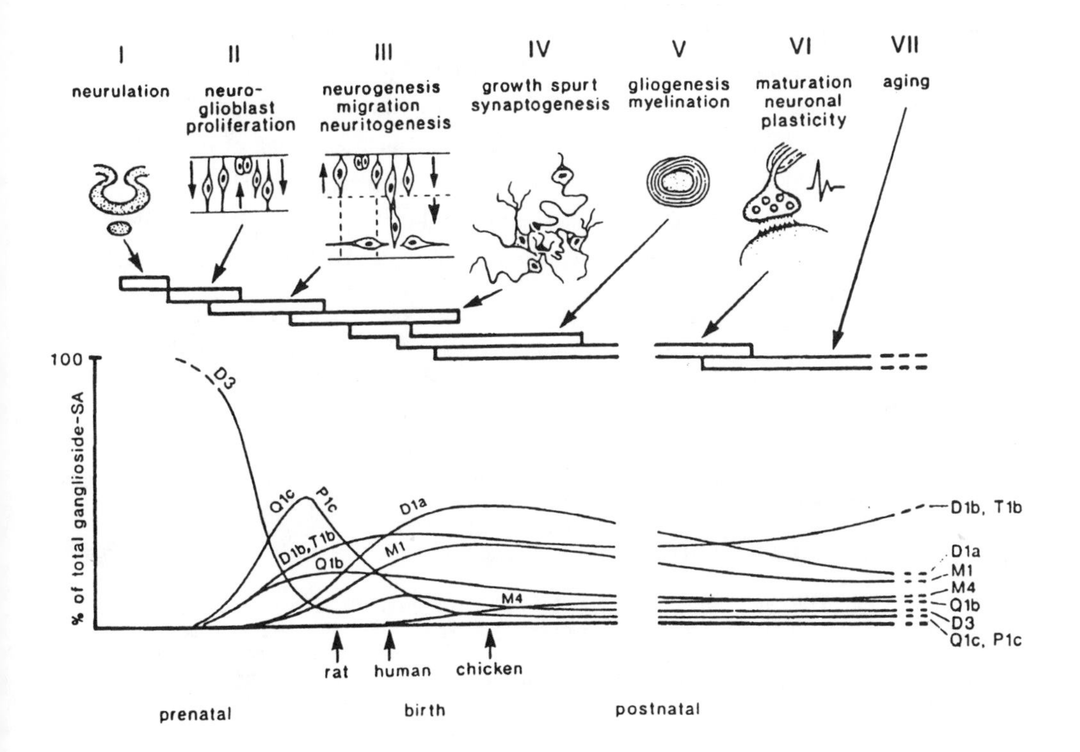

Fig. 5. Schematic presentation of profiles of brain gangliosides (change of the ganglioside pattern) during development and aging.

A simlar shift from a- to b-gangliosides was observed in the aged avian (Rösner, 1982) and senescent rat brain (Hilbig et al., 1984), and seems to be a general feature of the aging mammalian brain (Fig. 5). Interestingly, increasing evidence favours the view that c-polysialogangliosides, detectable in small amounts in human brain throughout life-span, accumulate in fibrillary senile plaques and neurofibrillary tangles of brains with Alzheimer's disease. The c-ganglioside epitope was immuno-histochemically detected by means of the m Abs A2B5 (Emoy et al., 1987), M 7103 and M 6704 (Takahashi et al., 1991) and Q211 (Rösner and Bayreuther, 1992). In addition, we found an intense immuno-staining of neurofibrillary tangles of retinoic acid differentiated SH-SY5Y neuroblastoma cells by Q211 (Rösner and Bayreuther, 1992) and identified two c-gangliosides in lipid-extracts of isolated neurofibrillary tangles of Alzheimer brains by means of immuno-tlc (Rösner and Bayreuther, 1992; Heffer-Lauc jet al., 1992 inj prep.).

Acknowledgement: I thank the European Communities (CI1-0579) and the Fidia Comp., Abano for supporting part of this work and W. Jansen for reading the manuscript and improving the english.

Asou, H., Hirano, S. and Kyemura, K. (1989) Cell Struc. Funct. 14: 561-568.
Bouhours, J.F., Bouhours, D. and Hansson, G.C. (1987) J. Biol. Chem. 262: 16370-16375.
Bouvier, J.D. and Seyfried, T.N. (1989) J. Neurochem. 52: 460-466.
Cheresh, D.A., Pierschbacher, M.D., Herzig, M.A. and Mujoo, K. (1986) J. Cell Biol. 102: 688-696.
Chou, K.H., Nolan, C.E. and Jungalwala, F.B. (1982) J. Neurochem. 39: 1547-1558.
Chou, K.H., Nolan, C.E. and Jungalwala, F.B. (1985) J. Neurochem. 44: 1905-1912.
Cochran, F.B., Yu, R.K. and Ledeen, R.W. (1982) J. Neurochem. 39: 773-779.
Constantine-Paton, M., Blum, A.S., Mendez-Otero, R. and Barnstable, c.J. (1986) Nature 324: 459-462.
Daniotti, J.L., Landa, C.A., Gravotta, D. and Maccioni, H.J.F. (1990) J. Neurosci. Res. 26: 436-446.
Daniotti, J.L., Landa, C.A., Rösner, H. and Maccioni, H.J.F. (1991) J. Neurochem. 57: 2054-2058.
Daniotti, J.L., Landa, C.A., Rösner, H. and Maccioni, H.J.F. (1992) J. Neurochm. 59: 107-117.
Drazba, J., Pierce, M. and Lemmon, V. (1991) Dev. Biol. 145: 154-163.
Dreyfus, H., Louis, J.C., Harth, S. and Mandel, P. (1980) Neurosci. 6: 1647-1655.
Dreyfus, H., Urban, P.F., Edelharth, S. and Mandel, P. (1975) J. Neurochem. 25: 245-250.
Emory, C.R., Ala, T.A. and Frey, W.H. (1987) Neurology 37: 768-772.
Engel, E.L., Wood, J.G. and Byrd, F.I. (1979) J. Neurobiol. 10: 429-440.
Eto, Y. and Shinoda, S. (1982) In: New Vistas in Glycolipid Research (A. Makita Ed.) Plenum, New York pp 279-290.
Felding-Habermann, B. and Wiegandt, H. (1987) In: Gangliosides and Modulation of Neuronal Function (H. Rahmann Ed.) Springer NATO ASI Series H7 pp 359-372.
Fishman, P.H. and Brady, R.O. (1976) Science 194: 906-915.
Fredman, P., Dumanski, J., Davidsson, P., Svennerholm, L. and Collins, V.P. (1990) J. Neurochem. 55: 1838-1841.
Goldman, J.E., Hirano, M., Yu, R.K. and Seyfried, T.N. (1984) J. Neurochem. 7: 179-192.
Greis, Ch. and Rösner, H. (1990a) Brain Res. 517: 105-110.

Greis, Ch. and Rösner, H. (1990b) Dev. Brain Res. 57: 223-234.
Hakomori, S. (1981) Ann. Rev. Biochenm. 50: 733-764.
Hakomori, S. (1985) In: Molecular Biology of Tumor Cells (B. Wahren et al. Eds.) Raven Press, New York pp. 139-156.
Hamburger, V. and Hamilton, H.L. (1951) J. Morpho. 50: 49-92.
Hilbig, R., Lauke, G. and Rahmann, H. (1984) Dev. Neurosci. 6: 260-270.
Hilbig, R., Rahmann, H. (1987) In: Gangliosides and Modulation of Neuronal Functions (H. Rahmann Ed) Springer NATO ASI SERIES H7 PP. 333-350.
Hilbig, R., Rösner, H., Merz, G., Segler-Stahl, C. and Rahmann, H. (1982) Roux's Arch. Dev. Biol. 191: 281-284.
Hirabayashi, Y., Hirota, M., Matsumoto, M., Tanaka, H., Obata, K. and Ando, S. (1988) J. Biochem. 4: 973-979.
Hirabayashi, Y.., Hyogo, A., Nakao, T., Tsuchiya, S., Suzuki, Y., Matsumoto, M., Kon. K. and Ando, S. (1990) J. Biol. Chem. 265: 8144-8151.
Hirabayashi, Y., Nakao, T. and Matsumoto, M. (1988) Chrom. 20: 485-490.
Iber, H., van Echten, G., Klein, D. and Sandhoff, K. (1990) Eur. J. Cell. Biol. 52: 236-240.
Irwin, L.N., Michael, D.B. and Irwin, C.C. (1980) J. Neurochem. 34: 1527-1530.
Ishizuka, J. and Wiegandt, H. (1972) Biochem. Biophys. Acta 260: 279-289.
Kim, S.U., Moretto, G. and Yu, R.K. (1986) J. Neurochem. 15: 303-321.
Klenk, E. (1942) Hoppe-Seyler's Z. physiol. Chem. 273: 76-86.
Kracun, I., Rösner, H. and Cosovic, C. (1986) In: Neuronal Plasticity and Gangliosides (G. Tettamanti, R.W. Ledeen, K. Sandhoff, J. Nagai, G. Toffano, Eds) Plenum Press pp 339-348.
Kracun, I., Rösner, H., Cosovic, C. and Stavljenic, A. (1984) J. Neurochem. 43: 979-989.
Kracun, I., Rösner, H., Drnovsek, V., Heffer-Lauc, M., Cosovic, C. and Lauc, G. (1991) Int. J. Dev. Biol. 35: 289-295. Kracun, I., Rösner, H., Drnovsek, V., Vukelic, Z., Cosovic, C.,Trbojevic-Cepe, M. and Kubat, M. (1992) Neurochem. Int. 20: 421-431.
Kracun, I., Rösner, H., Kostovic, J. and Rahmann, H. (1983) Roux's Arch. Dev. Biol. 192: 108-112.
Landa, C.H. and Moskona, A.A. (1985) Int. J. Dev. Neurosci. 3: 77-78.
Landa, C.H., Panzetta, P. and Maccioni, H.J.F. (1984) Dev. Brain Res. 14: 83-92.
Ledeen, R.W. (1978) J. Supramol. Structure 8: 1-17.
Ledeen, R.W. and Yu, R.K. (1982) In: Methods in Enzymology (V. Ginsburg Ed) Academic Press, New York pp. 139-191.
Levine, S.M., Seyfried, T.N., Yu, R.K. and Goldman, J.E. (1986) Brain Res. 374: 260-269.
Lorke, D.E., Sonnentag, U. and Rösner, H. (1990) Dev. Biol. 142: 194-202.
Maccioni, H.J.F., Landa, C.A. and Panzetta, P. (1989) Neurol. Neurobiol. 49: 117-127.
Maccioni, H.J.F., Panzetta, P., Arrieta, D. and Caputto, R. (1984) Int. J. Dev. Neurosci. 2: 13-19.
Mansson, J.E., Vanier, M.T. and Svennerholm, L. (1978) J. Neurochem. 30: 273-275.
Mendez-Otero, R. and Constantine-Paton, M. (1990) Dev. Biol. 138: 400-409.
Merat, A. and Dickerson, J.W.T. (1973) J. Neurochem. 20: 873-880.
Nagai, Y., Sanai, Y. and Nakaishi, H. (1987) In: Gangliosides and Modulation of Neuronal Functions (H. Rahmann Ed) Springer NATO ASI Series H7 pp. 275-292.
Nakaishi, H., Sanai, Y., Shiroki, K. and Nagai, Y. (1988) Bioche. Biophys. Res. Comm. 150: 760-765.
Nakamura, Y., Hashimoto, Y., Yamakawa, T. and Suzuki, A. (1988) J. Biochem. 103: 396-398.
Panzetta, P.D., Gravotta, D. and Maccioni, H.J.F. (1987) J. Neurochem. 49: 1763-1771.
Panzetta, P., Maccioni, H.J.F. and Caputto, R. (1980) J. Neurochem. 35, 1001-08.
Pohlentz, G., Klein, D., Schwarzmann, G., Schmitz, D. and Sandhoff, K. (1988) Proc. Natl. Acad. Sci. USA 85: 7044-7048.
Pukel, C.S., Lloyd, K.O., Travassos, L.R., Dippold, W.G., Oettgen, H.F. and Lloyd, J.O. (1982) J. Exp. Med. 155: 1133-1147.

Rahmann, H. and Hilbig, R. (1983) Com. Biochem. Physiol. 77B: 151-160.
Reinhardt-Maelike, S., Cleeves, V., Kindler-Rörborn, A. and Rajewsky, M.F. (1990) Brain Res. 51: 279-282.
Reynolds, R. and Wilkin, G.P. (1988) J. Neurosci. 20: 311-319.
Riboni,L., Acquotti, D., Casellato, R., Ghidoni, R., Montagnolo, G., Benevento, A., Zecca, L. Rubino, F. and Sonnino, S. (1992) Eur. J. Biochem. 203: 197-113.
Robert, J., Rebel, G. and Mandel, P. (1977) J. Lipid Res. 18: 517-522.
Rohrer, H., Henke-Fahle, S., El-Sharkawey, T., Lux, H.D. and Thoenen, H.C. (1985) EMBO J. 4: 1709-1714.
Rosenfelder, G., Ziegler, A., Wernet, P. and Braun, D.G. (1982) JNCI 68: 203-209.
Rösner, H. (1975) J. Neurochem. 24: 815-816.
Rösner, H. (1977) Roux's Arch. Dev. Biol. 183: 325-335.
Rösner, H. (1980) Roux's Arch. Dev. Biol. 188: 205-213.
Rösner, H. (1981) J. Neurochem. 37: 993-997.
Rösner, H. (1982) Dev. Brain Res. 236: 49-61.
Rösner, H., Al-Aqtum, M. and Henke-Fahle, S. (1985) Dev. Brain Res. 18: 85-95.
Rösner, H., Al-Aqtum,. M. and Rahmann, H. (1992) Neurochem. Int. 20: 339-351.
Rösner, H. and Bayreuther, K. (1992) Int. Symp. Polysialic Acid Rigi-Kaltbad
Rösner, H., Greis, C. and Henke-Fahle, S (1988a) Dev. Brain Res. 42: 161-171.
Rösner, H., Greis, C., Willibald, C.J. and Henke-Fahle, S. (1988b) In: New Trends in Ganglioside Research (R. Ledeen, G.Tettamanti, R.K. Yu, E. Hogan, A. Yates Eds) Springer pp 435-448.
Rösner, H., Rahmann, H., Reuter, G., Schauer, R., Katalinic, J.P. and Egge, H. (1985) Biol. Chem. Hoppe-Seyler 366: 1177- 1181.
Rösner, H., Willibald, C.J. and Henke-Fahle, S. (1986) In: Neuronal Plasticity and Gangliosides (G. Tettamanti, R.W. Ledeen, K. Sandhoff, J. Nagai, G. Toffano Eds) Plenum Press pp. 330-338.
Sandhoff, K. and Christomanou, H. (1979) Humangenetic 50: 107-143.
Sandhoff, K., Schwarzmann, G., Sarmientos, F. and Conzelmann, E. (1987) In: Gangliosides and Modulation of Neuronal Function (H. Rahmann Ed) Springer NATO ASI Series H7 pp. 231-250.
Sbashing-Agler, M., Dreyfus, H., Norton, W.T., Sensenbrenner, M., Farooq, M., Burne, M.C. and Ledeen, R.W. Brain Res. 461: 98-106.
Schaal, H., Wille, C. and Wille, W. (1985) J. Neurochem. 45: 544-551.
Schlosshauer, B., Blum, A.S., Mendez-Otero, R., Barnstable, C.J. and Constantine-Paton, M. (1988) J. Neurosci. 8: 580-592.
Segler-Stahl, K., Webster, J.C. and Brunngraber, E.G. (1983) Gerontology 29: 161-168.
Seybold, V. and Rahmann, H. (1985) Roux's Arch. Dev. Biol. 194: 166-172.
Seybold, U. and Rahmann, H. (1987) Zool. Jb Anat. 116: 409-420.
Seyfried, T.N. (1987) Dev. Biol. 123: 286-291.
Seyfried, T.N. and Yu, R.K. (1985) Mol. Cell. Biochem. 68: 3-10.
Sonnino, S., Bassi, R., Chignoro, V. and Tettamanti, G. (1990) J. Neurochem. 54: 1653-1660.
Sparrow, J.R. and Barnstable, C.J. (1988) J. Neurosci. 21: 398-409.
Suzuki, K. (1965) J. Neurochem. 12; 969-979.
Svennerholm, L. (1965) J. Lipid Res. 5: 145-155.
Svennerholm, L. (1977) IUPAC-IUP Commission on biochemical nomenclature. Eur. J. Biochem. 79, 11-21.
Svennerholm, L. (1984) In: Cellular and Pathological Aspects of Glycoconjugate Metabolism (H. Dreyfus, R. Massarelli, L. Freysz, G. Reel Eds) INSERM, Paris pp. 21-44.
Svennerholm, L., Boström, K., Fredman, P., Mansson, J.E., Rosengren, B. and Rynmark, B.M. (1989) Biochim. Biophys. Acta 1005: 109-117.
Takahashi, H., Hirokawa, K., Ando, S. and Obata, K. (1991) Acta Neuropathol. 81: 626-631.

Tettamanti, G., Ghidoni, R. and Trinchera, M. (1987) In: Gangliosides and Modulation of Neuronal Functions (H. Rahmann Ed) Springer NATO ASI Series H7 pp. 191-204.
Thangnipon, W. and Balazs, R. (1992) Neurochem. Res. 17: 45-59.
Thierfelder, S., Pini, S., Harrison, F. and Wiegandt, H. (1992) Differentiation 49: 7-15.
Van Echten, G. and Sandhoff, K. (1989) J. Neurochem. 52: 207-214.
Van Echten, G., Iber, H., Stotz, H., Takatsuki, A. and Sandhoff, K. (1990) Eur. J. Cell. Biol. 51: 135-139.
Vanier, M.T., Holm, M., Mansson, J.E. and Svennerholm, L. (1973) J. Neurochem. 21: 1375-1384.
Vanier, M.T., Holm, M., Ohman, R. and Svennerholm, L. (1971) J. Neurochem. 18: 591-592.
Wiegandt, H. (1982) Advances in Neurochemistry 4: 149-223.
Yavin, E. and Yavin, Z. (1979) Dev. Neurosci. 2: 25-37.
Yu, R.K. and Ando S. (1980) Adv. Exp. Med. Biol. 125: 35-45.
Yu, R.K. and Iqbal, K. (1979) J. Neurochenm. 32: 293-300.
Yu, R.K., Macula, L.J., Taki, T., Weinfeld, H.M. and Yu, F.S. (1988) J. Neurochem. 50: 1825-1829.
Yusuf, H.K., Merat, A. and Dickerson, J.W. (1977) J. Neurochem. 28: 1299-1304.

Polysialic Acid
J. Roth, U. Rutishauser and F. A. Troy II (eds.)

POLYSIALIC ACID IN ADULT BRAIN: ITS PRESENCE ON THE NEURAL CELL ADHESION MOLECULE AND NA^+-CHANNELS

Christian Zuber, Peter M. Lackie and Jürgen Roth

Division of Cell and Molecular Pathology, Department of Pathology, University of Zürich, Schmelzbergstrasse 12, CH-8091 Zürich, Switzerland

SUMMARY: We have studied the presence and distribution of α2,8-linked polysialic acid (poly Sia) and the neural cell adhesion molecule (N-CAM) in the adult rat brain by immunohistochemistry and Western blot analysis. Both molecules were widely distributed but not ubiquitous. Various brain regions showed colocalization of both poly Sia and N-CAM. However, strong immunoreactivity for poly Sia was seen in regions which were negative for N-CAM, such as parts of the main and accessory olfactory bulbs. Immunohistochemical evidence for the heterogeneity of poly Sia expression in different brain regions was confirmed by immunoblotting. We present evidence that N-CAM is not the only poly Sia bearing protein in adult rat brain. Specifically, immunoprecipitation using the poly Sia specific monoclonal antibody mAb 735 precipitated not only N-CAM isoforms carrying poly Sia, but also the sodium channel α subunit. Immunoblotting using sodium channel α subunit antibodies raised against the synthetic peptide SP20 revealed a smear from 250 kDa upwards. Poly Sia removal using an endoneuraminidase specific for homopolymers of α2,8-linked polysialic acid composed of 8 or more residues, reduced this smear to a single band at 250 kDa. Thus, both N-CAM and sodium channels carry homopolymers of α2,8-linked polysialic acid in adult rat brain.

INTRODUCTION

The ubiquitous presence of sialic acid residues at the non-reducing terminus of oligosaccharide side chains of membrane proteins modulates the biophysical and biological properties of the cell surface. This sialylation confers a net negative charge to the plasma membrane and is involved in a variety of specific receptor interactions. Homopolymers of α2,8-linked sialic acid (polysialic acid or poly Sia) occur more specifically. In *E. coli* K1 and group B meningococci the highly negatively charged poly Sia is found as the capsular polysaccharide (Troy and McCloskey, 1979; see also in this book). Recently, poly Sia has been demonstrated to exist in an insect, Drosophila melanogaster, and to be expressed in a highly regulated fashion during embryonic development (Roth et a., 1992). In lower vertebrates, protein-linked poly Sia is found in salmon egg alveoli (Iwasaki et al., 1990; see also the contribution of Inuoe in this book) and on the sodium channels of the electrical organ of the eel Electrophorus electricus (James and Agnew, 1987). In mammals, poly Sia has only been specifically demonstrated on the neural cell adhesion molecule (N-CAM) particularly in developing tissues such as brain (Finne, 1982; Hoffman et al. 1982), muscle (Rieger et al., 1985; Lackie et al., 1991) and kidney (Roth et al., 1987; Lackie et al., 1990) as well as a variety of tumors (Margolis and Margolis, 1983; Roth et al., 1988; Livingstone et al., 1988; Komminoth et al., 1991; Kibbelaar et al., 1989; Metzman et al. 1991; see also the contribution by Roth et al. in this book).

The poly Sia on N-CAM can directly modulate the homophilic binding properties of this cell adhesion molecule (Hoffman and Edelman, 1983; Sadoul et al., 1985; Rutishauser et al., 1985; Hall et al., 1990) and indirectly affects binding properties of other cell adhesion molecules such as L1 (Kadmon et al., 1990). In vitro experiments using mammalian cell cultures have shown that poly Sia expression modulates cell-cell interactions (Dickson et al., 1990; Acheson et al., 1991; Yang et al., 1992) and cell-substrate interactions (Acheson et al., 1991). During the embryonic development of the brain the ratio of sialic acid to N-CAM was observed to gradually decrease and in adult brain poly Sia could not be detected on N-CAM by chemical analysis (Finne, 1982; Hoffman et al., 1982). The recent immunohistochemical detection by antibody labeling of poly Sia in brain regions with known neural outgrowth, such as the hypothalamic region, led to the conclusion that expression of poly Sia is correlated with neural plasticity (Seki and Arai, 1991; Theodosis et al., 1991).

In the electroplax of the eel Electrophorus electricus the poly Sia bearing protein was identified as the voltage gated sodium channel (James and Agnew, 1987). The isolation of sodium channels from different sources and the resulting primary sequence data showed that a similar architecture existed for all the functional subunits of voltage dependent sodium channels yet identified (Gordon et al., 1988). Sodium channels from rat brain synaptosomes consist of one major α subunit of 260 kDa and two different smaller ß subunits of 36 and 33 kDa, respectively. For the α subunit three different mRNA species have been identified in rat brain (Noda et al., 1984). By immunohistochemistry applying peptide specific antibodies, two of the α subunits were shown to be not homogeneously distributed in the adult rat brain (Westenbroek et al., 1989). The functional sodium gating is associated in rat brain, muscle and heart with the 260 kDa α subunit which is the structural and functional homologue to the electroplax sodium channels. The ß1 subunit from rat brain has been shown to stabilize the α subunit structure during the opening period. The function of the ß2 subunit is not clear at present. In rat brain all three building blocks of the sodium channels are highly glycosylated. Approximately 30 % of the mass of the α subunit is due to carbohydrates. In the two smaller α subunits the carbohydrates contribute 25 % to protein masses

(Catterall, 1988 and therein). In the α subunit the major part of the carbohydrate is due to sialic acid. In rat brain sodium channel α subunit eight times more sialic acid is found than galactose (Elmer et al., 1985). Therefore, it was speculated that polymeric sialic acid chains exist to account for this unusual ratio between sialic acid and galactose residues (James and Agnew, 1987).

Among the reagents to detect and study poly Sia, two have been shown to be particularly useful for combined biochemical - morphological investigations. The IgG isotype 2a mouse monoclonal antibody 735 (mAb 735) used in this study has been shown to bind specifically to homopolymers of α2,8-linked poly Sia with a degree of polymerization of at least 8-10 (Frosch et al., 1985; Roth et al., 1987; Häyrinen et al., 1989; Husmann et al., 1990). The mAb 735 is equally useful for light and electron microscopic labeling techniques (Roth et al., 1987; Zuber and Roth, 1991; Husman et al., 1990) and immunochemical analysis of tissue homogenates by Western blotting (Roth et al., 1987; Zuber and Roth, 1991; Lackie et al., 1990, 1991). The endoneuraminidase N from the bacteriophage pK1F specifically hydrolyzes homopolymers of α2,8-linked poly Sia and has been shown to have a strict substrate requirement for poly Sia chains of at least 5-8 residues long (Hallenbeck et al., 1987). Therefore this enzyme abolishes the immunoreactivity of poly Sia containing molecules with the mAb 735 in a highly specific fashion (Roth et al., 1987; 1988; Lackie et al., 1990).

POLYSIALIC ACID IS HOMOGENEOUSLY DISTRIBUTED IN DEVELOPING RAT BRAIN

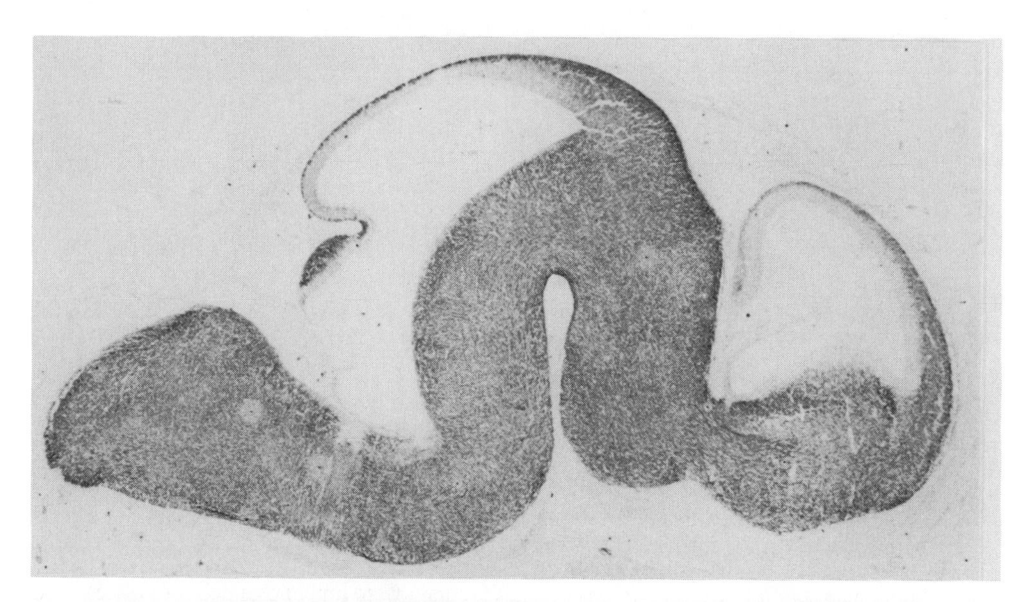

Figure 1. Parasaggital section from paraffin-embedded rat embryonic day 13 brain, immunogold silver staining with directly gold-labeled mAb 735. Immunostaining of similar intensity is observed in all regions of the developing brain.

Since its initial description in embryonic brain tissue (Finne, 1982), the existence and regulated expression of poly Sia in different cell types in embryonic brain has been ample confirmed (Nybroe et al., 1989). The immunohistochemical detection of poly Sia immunoreactivity with mAb 735 in the brain of an embryonic day 13 rat reveals at this stage of brain development a homogenous distribution of the polyglycan (Fig. 1).

POLYSIALIC ACID AND N-CAM ARE FOUND THROUGHOUT ADULT RAT BRAIN

In serial parasaggital sections of whole adult rat brain, immunoreactivity for poly Sia (Fig. 2 A) and N-CAM (Fig. 2 B) were widely distributed, but not ubiquitous.

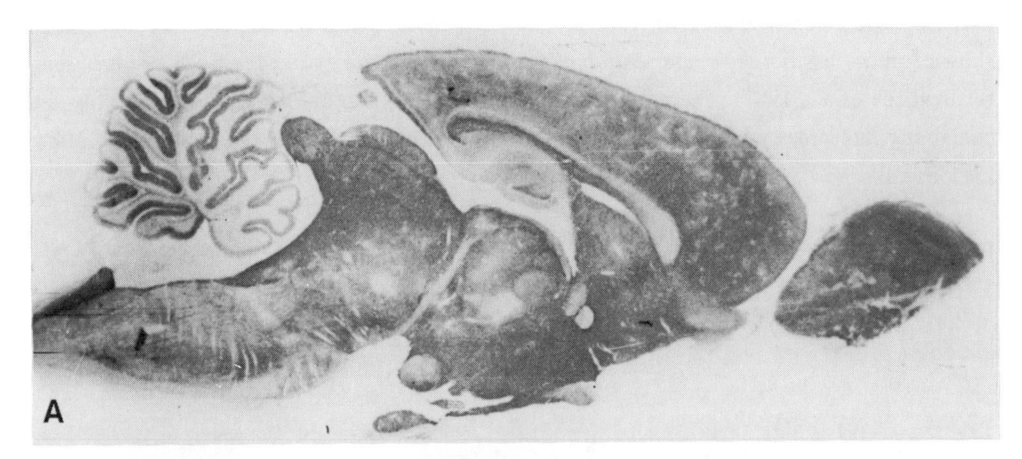

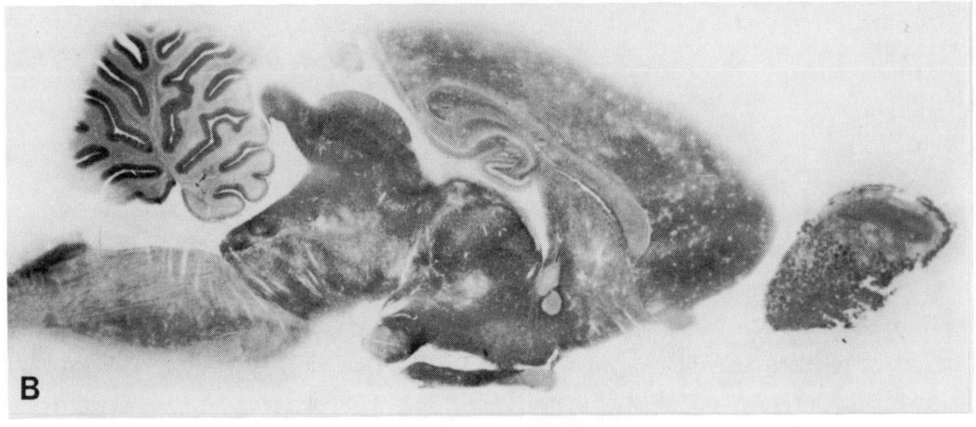

Fig. 2. Serial parasaggital paraffin sections of adult rat brain were immunostained for poly Sia (A) and for N-CAM (B). At this low magnification view both immunoreactivities can be localized in the cortex, the mesencephalon and the forebrain. Differential staining is observed in cerebellum and hippocampus.

Significant variations in the expression of these molecules were found in the cerebellum, hippocampus, corpus callosum, olfactory and accessory olfactory lobes. Brain subregions were found which were positive for both, only one, or neither immunoreactivity.

All regions of the adult rat brain which were examined showed poly Sia immunoreactivity on Western blots from homogenates resolved in 3-10% gradient SDS polyacrylamide gels in a smear extending from 150 kDa to greater than 300 kDa (Fig. 3).

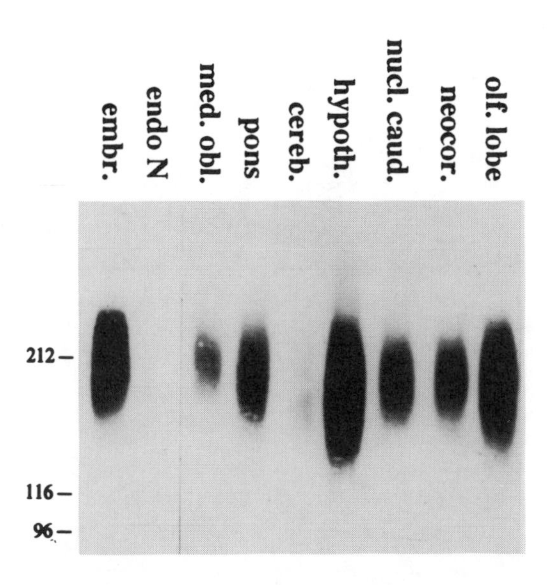

Fig. 3. Poly Sia immunoreactivity detected by western blot analysis of different regions of the rat brain. Protein samples of embryonic day 17 (embr.) rat brain and of different regions from adult rat brain were resolved using 3%-10% SDS polyacrylamide gradient gels. Blots on nitrocellulose were immunostained for poly Sia with directly gold-labeled mAb 735, which was detected using silver amplification. The amount of protein loaded was 20 µg for embryonic and 100 µg for adult samples. The weakest staining for poly Sia is detected in the cerebellum (cereb.). Stronger staining is detected in medulla oblongata (med. obl.), pons, nucleus caudatus (nucl. caud.) and neocortex (neocor.). Most prominent immunoreactivity is seen for the hypothalamus (hypoth.) and the olfactory lobe (olf. lob.). Control sample (endo N) was treated for 30 min with endoneuraminidase N at 37°C prior to SDS-PAGE.

The electrophoretic mobility of poly Sia bearing molecules in embryonic and adult brain samples were similar, indicative of a similar degree of polysialylation. However, the amount of poly Sia detected was different in embryonic and adult brain regions examined. To obtain staining of similar intensity to that seen for samples of olfactory bulb and hypothalamic region of adult brain ten times less protein from embryonic brain was needed. The amount of poly Sia detected by Western blot analysis of the different brain regions reflects well the poly Sia immunohistochemical staining pattern found in sections of rat brain (Fig. 2).

In the corpus callosum, poly Sia immunostaining was weak in a layer including the whole of the genu corpus callosum and extending to the dorsal half of the splenium corpus callosum, while the dorsal hippocampal commissure was also only weakly stained for poly Sia immunoreactivity (Fig. 4 A). The corpus callosum and the dorsal hippocampal commissure was however uniformly immunoreactive for N-CAM (Fig. 4 B).

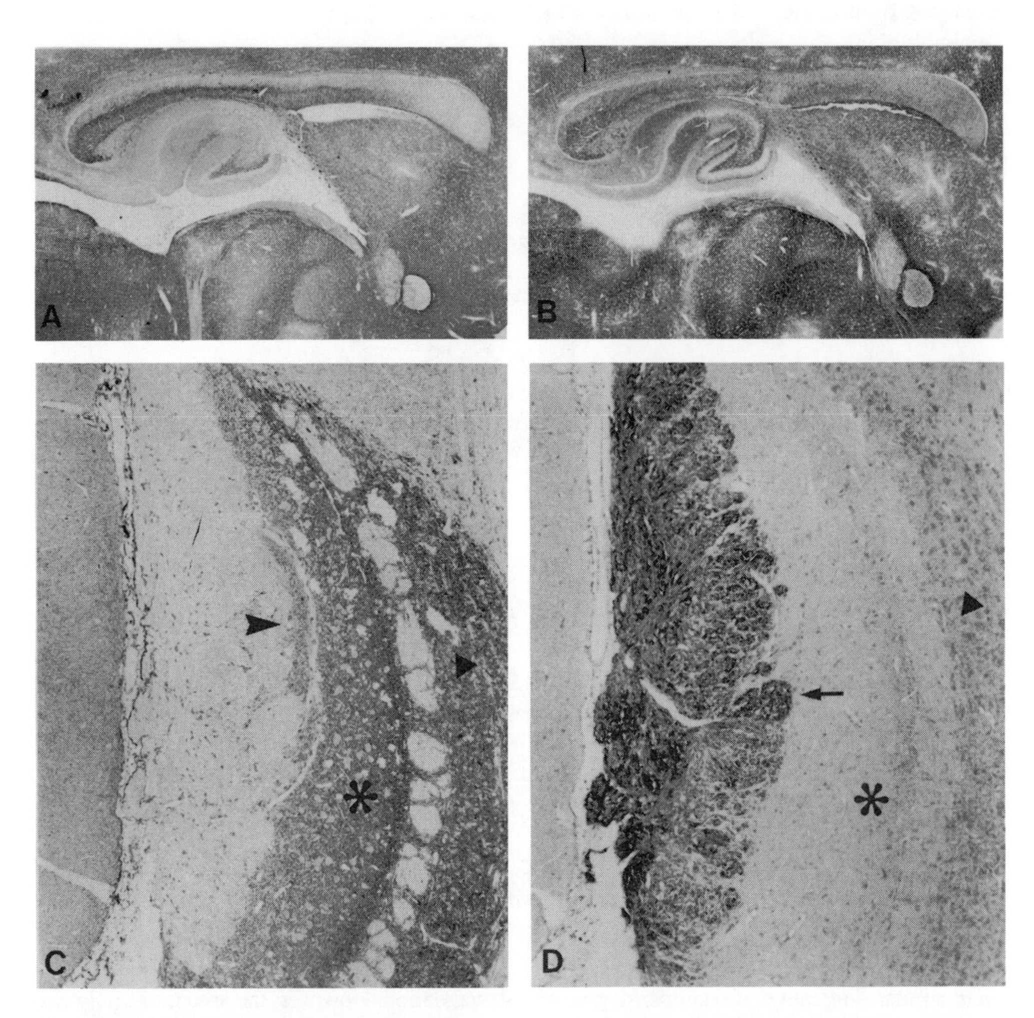

Fig. 4. Poly Sia and N-CAM immunoreactivities are not always colocalized. Poly Sia (A, C) and N-CAM (B, D) immunoreactivity in the hippocampus (A, B) and the accessory olfactory bulb (C, D) detected as indicated in Fig. 1. At higher magnification cell type specific differences in poly Sia (A, C) and N-CAM (B, D) immunolocalization are evident. Immunostaining for poly Sia (A) in the accessory olfactory bulb of adult rat brain reveals prominent labeling over the granular cell layer (triangle), parts of the external plexiform layer (asterisk) and over some periglomerular cells (arrowhead). N-CAM (B) immunoreactivity is strongly expressed in glomeruli (arrow) but only weakly in the granular cell layer (triangle) and absent in the external plexiform layer (asterisk). An adjacent region shows neither poly Sia nor N-CAM.

In the underlying Ammon´s horn only field CA4 and the hippocampal fissure were immunoreactive for poly Sia, although prominent staining for N-CAM was seen in addition in fields CA1 to CA3 and in the granular layer of the dentate gyrus . In the cerebellum (Fig. 2) the molecular cell layer of the cerebellar cortex was strongly stained down to the level of the Purkinje cell bodies, while very few structures in the underlying granule cell layer were immunoreactive for either poly Sia or N-CAM. The bundles of axons from Purkinje cells in the white matter of the cerebellar folia showed staining for both poly Sia and N-CAM (not shown).

In the accessory olfactory bulb a most striking differential staining pattern was observed. In this region the majority of cells within the external plexiform layer were immunoreactive for poly Sia (Fig. 4 C) but never for N-CAM (Fig. 4 D). The granular cell layer was strongly positive for poly Sia immunoreactivity, but only slightly for N-CAM. The glomerular layer was strongly positive for N-CAM immunoreactivity but negative for poly Sia, with the exception of a very few periglomerular cells. Between the external plexiform layer and the granular cell layer, a population of cells in a clearly defined band was immunoreactive for neither N-CAM nor poly Sia.

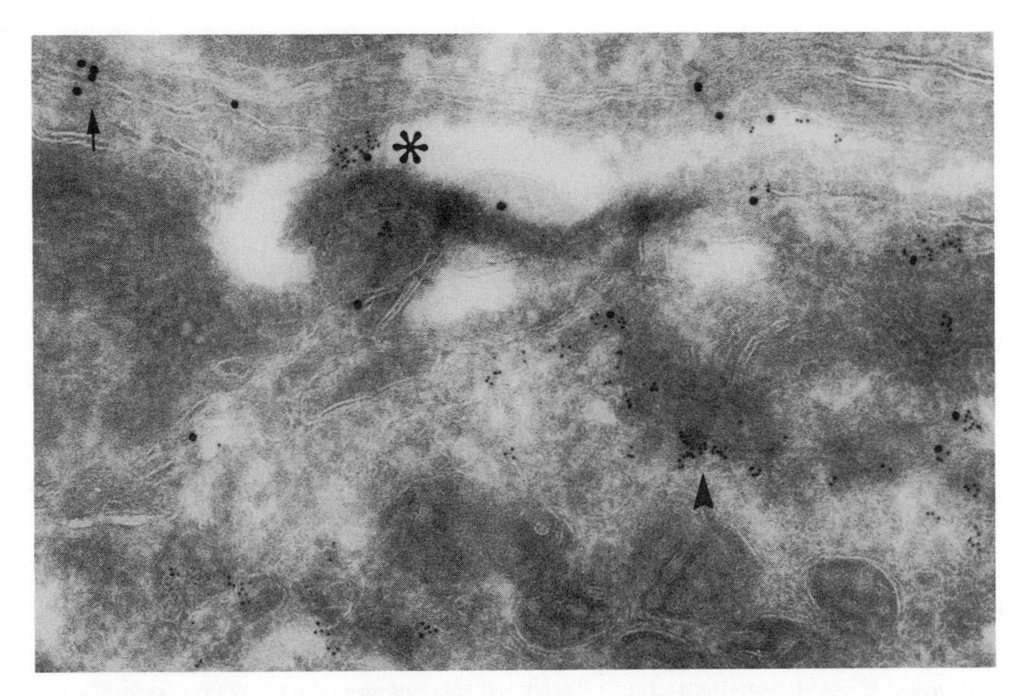

Fig. 5. Immunoelectron microscopic localization of poly Sia and N-CAM in the olfactory bulb. An ultrathin frozen section from the olfactory bulb was double immunostained for poly Sia (8 nm gold particles) and N-CAM (14 nm gold particles). In different regions of the section either N-CAM (arrow) or poly Sia (arrowhead) alone are detected. In other regions (asterisks) colocalization is observed.

In the olfactory bulb the whole of the external plexiform layer, including the middle tufted cells and the periglomerular cells of the glomerular cell layer were strongly immunoreactive for poly Sia but not for N-CAM. Conversely, the olfactory glomeruli were only slightly stained for poly Sia but strongly immunoreactive for N-CAM. Immunoreactivity for both poly Sia and N-CAM was seen in the cells of the outermost olfactory nerve layer, and in cells from the mitral layer until the ependymal layer which was itself negative for both poly Sia and N-CAM. In both olfactory bulbs therefore, cells of analogous function show the same immunostaining pattern for N-CAM. However, the glomerular cell layer of the main olfactory bulb showed strong poly Sia immunoreactivity which was lacking in the accessory olfactory bulb.

Double-immuno electron microscopic localization of poly Sia and N-CAM in the olfactory bulb showed membrane associated labeling for both molecules (Fig. 5). In some areas immunolabeling for poly Sia was associated with that for N-CAM, although regions labeled for either poly Sia or N-CAM alone were detected.

The diverse localization patterns for poly Sia and N-CAM immunoreactivity is indicative for the existence of other poly Sia bearing components in the adult rat brain.

IDENTIFICATION OF POLYSIALIC ACID BEARING PROTEINS IN ADULT RAT BRAIN

The molecular weight of poly Sia immunoreactivity ranges between 150 kDa and greater than 300 kDa in Western blots (3-10% gradient SDS-PAGE) of embryonic brain (Fig. 6). In fractions enriched for synaptosomal membranes (Fig. 6), however, a strongly immunoreactive band was seen around 250 kDa while a rather faint smear was ranging from 150 kDa to greater than 250 kDa. When the same synaptosomal membranes were probed for sodium channel immunoreactivity a strong band around 250 kDa was detected (Fig. 6). In addition a smear at higher and lower relative molecular weights with two minor bands at 170 and 140 kDa was positively stained. These two minor bands are probably due to degradation products of the sodium channel α subunit. Immunoblotting to detect N-CAM showed reactivity from approximately 110 kDa to 200 kDa with three major bands at 120, 140 and 180 kDa and two minor ones at 110 and 160 kDa. These bands were superimposed by a smear of polysialylated N-CAM isoforms expressed in the adult rat brain which extended to about 220 kDa. The polysialylated N-CAM isoforms could be further evaluated by immunoprecipitation analysis using the poly Sia specific mAb 735 for precipitation, endoneuraminidase N for the release of the bound material and N-CAM specific antibodies for the detection of the poly Sia bearing N-CAM isoforms (for technical details see Zuber et al., 1992). All three major N-CAM isoforms of 180, 140 and 120 kDa and the minor 160 kDa isoform were found to be polysialylated. The 110 kDa isoform, however, was not immunoprecipitated using the mAb 735 and therefore appears to be not polysialylated. The origin of this isoform is not clear and could be due to degradation during the preparation of the synaptosomal membrane fractions. For

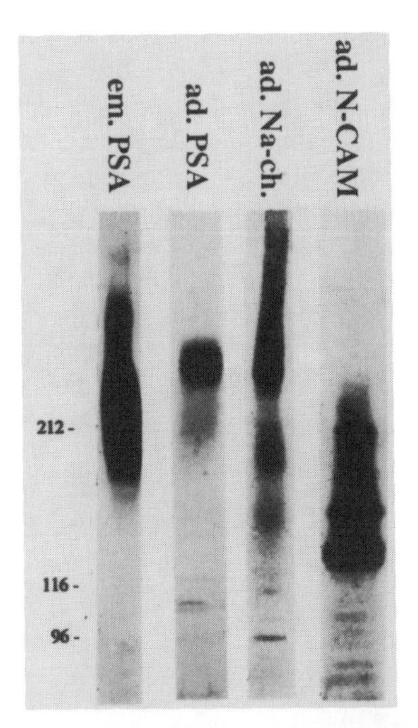

Fig. 6. Immunoblotting for poly Sia, sodium channel α subunit and N-CAM. Protein samples (50 μg) of embryonic rat brain (em. PSA) and synaptosomal membranes of adult rat brain (remaining lanes) were immunoblotted as indicated in Fig. 3. Synaptosomal membranes were immunostained for poly Sia (ad. PSA), sodium channel α subunit (ad. Na-ch.) and for N-CAM (ad. N-CAM).

the detection of minor soluble, polysialylated components we were using N-CAM depleted synaptosomal membrane-enriched fractions. The proteins which were removed by endoneuraminidase N treatment from the immunobeads were immunoblotted for sodium channel α subunit immunoreactivity (Fig. 7 B). The released immunoreactivity migrated at a distinct band of 250 kDa (Fig. 7 B). It corresponded to the lower molecular weight limit of the sodium channel α subunit immunoreactivity prior to immunoprecipitation and endoneuraminidase N treatment.

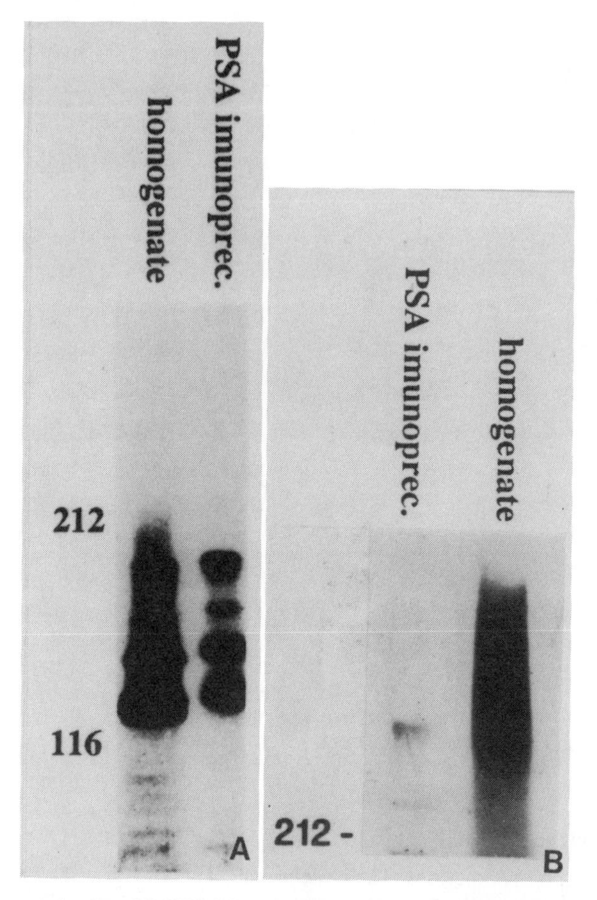

Fig. 7A. Immunoprecipitation N-CAM immunoreactivity with poly Sia specific mAb 735. Homogenate: N-CAM immunoreactivity from a sample of synaptosomal membrane fraction prior to immunoprecipitation and endoneuraminidase N treatment.PSA immunoprec.: Synaptosomal membranes of adult rat brain were immunoprecipitated with the poly Sia specific antibody mAb 735, treated with endoneuraminidase N and subjected to Western blot analysis for N-CAM.

Fig. 7B. Immunoprecipitation of sodium channel α subunit immunoreactivity with poly Sia specific mAb 735. PSA immunoprec.: N-CAM depleted synaptosomal membranes of adult rat brain were immunoprecipitated with the poly Sia specific antibody mAb 735, treated with endoneuraminidase N and subjected to Western blot analysis for sodium channel α subunit. Homogenate: sodium channel α subunit immunoreactivity for a sample of the N-CAM depleted synaptosomal membrane fraction prior to immunoprecipitation and endoneuraminidase treatment.

CONCLUSIONS

Our data show that poly Sia immunoreactivity is widely distributed in adult rat brain. N-CAM and the voltage-dependent sodium channel α subunit were identified as poly Sia-bearing proteins. The detection of poly Sia on N-CAM and the sodium channel α subunit in the adult rat brain raises various questions which are difficult to answer currently. It is not clear which relative amounts of poly Sia immunoreactivity in the adult brain are associated with N-CAM and the sodium channel α subunit. However, N-CAM polypeptides coexist as mAb 735 reactive and non-reactive forms.

Both molecules have completely different molecular structures. N-CAM contains a single membrane-spanning domain whereas the sodium channel α subunit is composed of 24 channel forming transmembrane domains. These structural differences result in a different solubilization behaviour of the two proteins. As already mentioned, three different mRNA species have been identified which code for the 260 kDa α subunit. Since the anti-peptide antibodies were raised against a conserved region of the sodium channel α subunit, it was not possible to distinguish amongst them.

Poly Sia immunoreactivity could be localized in the axonal bundles of the Purkinje cells where neither sodium channels could be detected (Westenbroek et al., 1989) nor neuronal plasticity can be expected. This poly Sia immunoreactivity could be due to an unknown polysialylated molecule, sodium channels not reactive with the used antibodies or polysialylated N-CAM expressed by these mature neurons. The latter possibility is most likely since N-CAM immunoreactivity was detected in the same structures by immunocytochemistry. Poly Sia immunoreactivity found in Western blots of cerebellum smeared to molecular weights above that for sodium channel α subunit. Further, the Western blot analysis of the different adult and embryonic brain regions gave qualitatively similar results. This indicates that, despite reduced amounts of polysialylated N-CAM polypeptides in adult brain, the degree of polysialylation of individual molecules in both embryonic and adult brain is similar. By Western blotting and immunolabeling, polysialylated molecules were found throughout the adult rat brain. Other investigators (Miragall et al., 1988; Seki and Arai, 1991a, b; Theodosis et al., 1991) by using different anti-poly Sia specific antibodies observed a more restricted distribution of poly Sia in adult rat brain. Specifically, the antibody applied by Theodosis et al. (1991) was found to require for binding a minimal binding length of 12 sialic acid residues (Hyärinen, personal communication). Therefore, homopolymers with a degree of polymerization < 12, which are alsorecognized by mAb 735 used by us, could not be detected in this study. It is unclear at present if the differential staining patterns observed with different antibodies are related to differences in the amount of poly Sia or in the degree of polymerization of poly Sia.

Possible roles of the carbohydrate moieties on the sodium channel α subunit have been tested. The inhibition of the synthesis of asparagine-linked oligosaccharides in the ER by tunicamycin treatment resulted in complete degradation of the sodium channel α subunit in the ER. However, inhibition of Golgi apparatus associated glycan processing did not interfere with the cell surface expression of the sodium channel α subunit (Schmidt and Catterall, 1987).

The sodium channel α subunit is a highly complex structure composed of four hexameric channel forming transmembrane domains. When an action potential reaches a closed, active sodium channel, charges are translocated inside the membrane. This translocation precedes the opening of the channel and the influx of sodium ions. Recio-Pinto et al. (1987) could show that the mid-point potential for this voltage-dependent activation process is strongly affected by desialylation of the α subunit. In the brain, activated sodium channels reach very quickly the typical 18 pSiemens conductivity. In desialylated rat brain sodium channels, however, additional conductances of 5, 8 and 14 pSiemens have been measured to occur (Scheuer et al., 1987). The poly Sia is therefore responsible for the stabilization of the activated and open channel to warrant the full conductance of 18± 2 pSiemens. The occurrence of aberrant conductance values in desialylated sodium channels would strongly interfere with the highly regulated process of the de- and repolarization essential for the proper propagation of the electrical signal along a nerve fibre. These data strongly suggested that the poly Sia is somehow stabilizing the channel structure in the membrane. These highly hydrophobic membrane proteins are therefore one exciting example how glycosylation seems to be needed for a functional stable membrane insertion.

ACKNOWLEDGEMENTS

The original work of the authors was supported by the Swiss National Science Foundation grant nr. 31-26273.89.

REFERENCES

Acheson, A., Sunshine, J.L. and Rutishauser, U. (1991) J. Cell Biol. 114: 143- 153
Catterall, W.A. (1988) Science 242: 50-61
Dickson, G., Peck, D., Moore, S.E., Barton, C.H. and Walsh, F.S. (1990) Nature 344: 348- 351
Elmer, l.W., O'Brien, B.J., Nutter, T.J. and Angelides, K.J. (1985) Biochem. 24: 8128-8137
Finne, J. (1982) J. Biol. Chem. 257: 11966- 11970
Frosch, M., Görgen, J., Bulnois, G.J., Timmis, K.M. and Bitter-Suermann, D. (1985) Proc. Natl. Acad. Sci. U.S.A. 82: 1194- 1198
Gordon, D., Merrick, D., Wollner, D.A. and Catterall, W.A. (1988) Biochem. 27: 7032- 7038
Häyrinen, J., Bitter-Suermann, D. and Finne, J. (1989) Mol. Immun. 26: 523- 529
Hall, A.K., Nelson, R. and Rutishauser, U. (1990) J. Cell Biol. 110: 817-824
Hallenbeck, P.C., Vimr, E.R., Yu, F., Bassler, B. and Troy, F.A. (1987) J. Biol. Chem. 262: 3553-3561
Hoffman, S. and Edelman, G.M. (1983) Proc. Natl. Acad. Sci. U.S.A. 80: 5762- 5766
Hoffman, S., Sorkin, B.C., Perrin, W.C., Brackenbury, R., Mailhammer, R., Rutishauser, U., Cunningham, B.A. and Edelman, G.M. (1982) J. Biol. Chem. 257: 7720- 7729
Husmann, M., Roth, J., Kabat, E.A., Weisgerber, Ch., Frosch, M. and Bitter-Suermann (1990) J. Histochem. Cytochem. 38: 209-215
Iwasaki, M., Inoue,S. and Troy, F.A. (1990) J. Biol. Chem. 265: 2596-2602
James, W.M. and Agnew, W.S. (1987) Biochem. and Biophys. Res. Com. 148: 817-826
Kadmon, G., Kowitz, A., Altevogt, P. and Schachner, M. (1990) J. Cell Biol. 110: 209-218
Kibbelaar, R.E., Moolenaar, C.E.C., Michalides, R.J.A.M., Bitter-Suermann, D., Addis, B.J.and Mooi, W.J. (1989) J. Pathol. 159: 23-28
Komminoth, P.,Lackie, P. M., Bitter-Suermann, D., Heitz, P. U. and Roth, J. (1991) Am. J. Pathol. 139: 297-304
Lackie, P.M., Zuber, C. and Roth J. (1991) Differentiation 47: 85-98
Lackie, P.M., Zuber, C. and Roth, J. (1990) Development 110: 933- 947
Livingstone, B. D., Jacobs, J. L., Glick, M. C. and Troy, F. A.(1988) J. Biol. Chem. 263: 9443-
Margolis, R. K. and Margolis, R. U. (1983) Biochem. Biophys. Res. Commun. 116, 889-
Metzman, R. A., Warhol, M. J., Gee, B. and Roth, J. (1991) Modern Path.41: 491-497
Miragall, F., Kadmon, G., Husmann, M. and Schachner, M. (1988) Dev. Biol. 129: 516-531
Noda, M., Shimizu, S., Tanabe, T., Takai, T., Kayano, T., Ikeda, T., Takahashi, H.,

Nakayama, H., Kanaoka, Y., Minamino, N., Kangawa, k., Matsuo, H., Raftery, M.A., Hirose, T., Inayama, S., Miyata, T., and Numa, S. (1984) Nature 312: 121-127
Nybroe, O., Linnemann, D. and Bock, E. (1988) Neurochem. Int. 3: 251-
Recio-Pinto, E., Thornhill, W.B., Duch, D.S., Levinson, S.R. and Urban, B. (1987) Soc. Neurosci. Abst. 17: 92
Rieger: F., Grumet, M. and Edelman, G.M. (1985) J. Cell Biol. 101: 285- 293
Roth, J., Taatjes, D.J., Bitter-Suermann, D. and Finne, J. (1987) Proc. Natl. Acad. Sci. U.S.A. 84: 1969-1973
Roth, J., Kempf, A., Reuter, G., Schauer, R. and Gehring, W. J. (1992) Science 256: 673-675, 1992
Sadoul, R.V., Hirn, M., Deagostini-Bazin, H., Rougon, G. and Goridis, C. (1985) Nature 304: 347-349
Scheuer, T., McHugh, L., Tejedor, F. and Catterall, W.A. (1988) Biophys. J. 53: 541a-
Seki, T. and Arai, Y. (1991) Anat. Embryol. 184: 395-401
Seki, T. and Arai, Y. (1991) Neurosci. Res. 12: 503-513
Theodosis, D.T., Rougon, G. and Poulain, D.A. (1991) Proc. Natl. Acad. Sci USA 88: 5494-5498
Troy, F.A. and McCloskey, M.A. (1979) J. Biol. Chem. 254: 7377- 7387
Westenbroek, R.E., Merrick, D.K. and Catterall, W.A. (1989) Neuron 3: 695- 704
Yang, P., Yin, X. and Rutishauser, U. (1992) J. Cell Biol. 116 1487-96
Zuber C, Roth J. (1990) Eur. J. Cell Biol. 51: 313-321
Zuber, C., Lackie, P.M., Catterall, W.A. and Roth, J. (1992) J. Biol. Chem. 267: 9965-9971

Polysialic Acid
J. Roth, U. Rutishauser and F. A. Troy II (eds.)

REGULATION OF NCAM POLYSIALYLATION STATE DURING INFORMATION PROCESSING IN THE ADULT RODENT

Ciaran M. Regan

Department of Pharmacology, University College, Belfield, Dublin 4, Ireland.

SUMMARY: The neural cell adhesion molecule (NCAM) is shown to increase polysialylation of the hippocampal synapse-specific isoform in a region-specific manner during consolidation of a passive avoidance response. Antibody interventive studies demonstrated NCAM to be integral to the memory consolidation process at a time which proceeded the increase in sialylation. These observations are interpreted to reflect a role for NCAM in synapse structuring and for polysialylation in signaling the extent of synaptic growth which is believed to underlie information processing.

NCAM STRUCTURE AND FUNCTION DURING DEVELOPMENT

The neural cell adhesion molecule (NCAM) is a complex of cell surface glycoproteins which are derived from the alternate splicing of a single gene product (for review see Walsh and Doherty, 1991). Three major isoforms, with molecular weights of 180 (NCAM180), 140 (NCAM140) and 120 (NCAM120) kDa, are apparent in the central nervous system and these differ in their mode of membrane attachment through selective exon use. In human neural tissue a secreted NCAM isoform has been demonstrated to arise through the alternate use of a stop codon-containing exon. NCAM180 contains an extensive cytoplasmic domain which interacts with the cytoskeleton via brain spectrin (Pollerberg et al., 1987). This isoform is only present on differentiated neurons and is located to the postsynaptic density (Persohn et al., 1989). NCAM140 contains a smaller cytoplasmic domain and NCAM120 is membrane attached by a glyco-phosphatidyl-inositol moiety. The extracellular domains of these NCAM isoforms contain 5 disulphide loops which are representative of the C2-type immunoglobulin structural motif. This region is coded by exons 1-10 and is strongly conserved between all NCAM isoforms. A single splicing exception occurs in the fourth immunoglobulin domain by the alternative incorporation of the VASE (variable alternatively spliced exon) exon which encodes a 10 amino acid insert the use of which increases with differentiation (Small and Akeson, 1990).

NCAM is believed to mediate calcium-independent homophilic (*trans*) binding through regions which are located in the third Ig domain on molecules of the apposing cells (Hoffman and Edelman, 1983; Frelinger et al., 1986). In addition a *cis* binding mechanism with the L1 immunoglobulin superfamily member and to the extracellular matrix, through a heparin binding site, have also been reported (Cole and Glaser, 1986; Kadmon et al., 1990a,b). The extent of NCAM homophilic binding has been demonstrated to be dependent on the degree to which its glycan moiety is sialylated. The NCAM glycan is believed to be a complex carbohydrate containing three to four antennae capped with sialic acid in an α2-3 linkage. The glycan structure is unique in that one or more antennae bear α2-8 linked polysialosyl units the length of which is varied during discrete periods of neural structuring (for reviews see Regan, 1991; Troy, 1992).

In early development NCAM exists in a sialic acid-poor form (A-form) when structural integrity of cell collectives is necessary during force driven events such as neurulation. Later NCAM increases in amount and extent of sialylation (E-form) during the periods of growth which involve cell migration and neurite extension. NCAM returns to the sialic acid-poor phenotype during periods of synapse formation when the neural structure is finally elaborated (Choung and Edelman, 1984; Sunshine et al., 1987). This cycle of events, which is illustrated in Fig. 1, also occurs at temporally distinct times in discrete brain regions.

In the postnatal development of the cerebellum NCAM increases sialylation during periods of cell acquisition, this is retained during axon extension and granule cell migration and only decreases during periods of synapse formation (Meier et al., 1984). In the postnatal period NCAM does not extend the homopolymers of sialic acid to the same length observed during embryonic development as the high molecular weight bands associated with polysialylation are discrete entities of approximately 210kDa rather than the polydisperse bands which appear as a smear (Breen and Regan, 1988).

The mechanisms which regulate NCAM sialylation state remain to be elaborated. An α2-8 polysialyltransferase has been demonstrated to be markedly active in the early embryo and to be developmentally downregulated in the postnatal period (Regan, 1991). However this enzyme complex may not directly sialylate NCAM as the α2-8 sialic acid homopolymers appear to be transferred *en bloc* to the polypeptide. In the postnatal period a developmentally regulated endogenous Golgi glycoprotein sialyltransferase complex has been demonstrated to be comprised of at least four isoforms, two of which show maximal activity at times of increased NCAM sialylation. Not all NCAM isoforms are amenable to sialylation by this complex during the individual phases of neural structuring. NCAM140 is only sialylated up to and including postnatal day 10 whereas NCAM180 only serves as a substrate from postnatal day 12 onwards. In *in vitro* systems neurite extension is dependent on the expression and polysialylation of NCAM 140 and

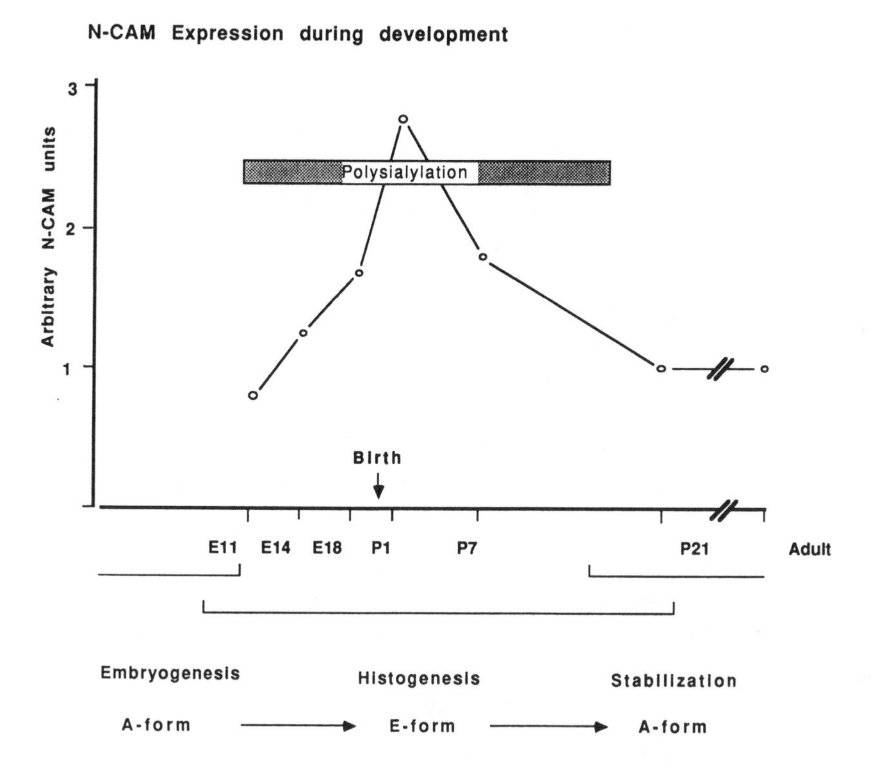

Figure 1. Temporal change in average NCAM concentration and expression of polysialic acid (stippled bar) during the development of individual brain regions. The olfactory bulb is an exception as polysialylation and maximal NCAM concentration is retained into adulthood. Adapted from Choung and Edelman, 1984.

NCAM120 whereas NCAM180 is inhibitory (Doherty et al., 1990a,b; 1992). Thus differentiation-dependent polysialylation may facilitate the individual recognition functions of NCAM isoforms.

These prevalence and chemical modulations have been suggested to reflect a morphological role for NCAM during neural structuring (Edelman, 1984). In this model increased NCAM expression and sialylation generate a refractory neuronal phenotype which is amenable to structural change whereas reduced expression and sialylation produce the permissive phenotype associated with adult

dependent on the expression and polysialylation of NCAM 140 and NCAM120 whereas NCAM180 is inhibitory (Doherty et al., 1990a,b; 1992). Thus differentiation-dependent polysialylation may facilitate the individual recognition functions of NCAM isoforms.

These prevalence and chemical modulations have been suggested to reflect a morphological role for NCAM during neural structuring (Edelman, 1984). In this model increased NCAM expression and sialylation generate a refractory neuronal phenotype which is amenable to structural change whereas reduced expression and sialylation produce the permissive phenotype associated with adult structures. Thus in the refractory phenotype NCAM and its posttranslational modifications provide the signals which are necessary for growth and structural change. Regulation of NCAM signaling would be a prerequisite for adult neuroplasticity and the predicted modulations of NCAM function are now demonstrated to occur during memory formation in the adult animal.

MEMORY AS A MODEL OF INDUCED SYNAPTIC PLASTICITY

Change in synapse connectivity pattern is the most consistent structural feature associated with information storage. In the majority of studies which have manipulated adult experience synapse number is affected in a manner which has led to the suggestion that processes such as memory and learning may modulate synapse turnover similar to the synapse overproduction observed during development (for review see Greenough and Chang, 1985). Induction of longterm potentiation (LTP), which is accepted widely as a model of memory and learning (Brown et al., 1988), results in rapid synapse formation both *in vitro* and *in vivo* (Lee et al., 1980; Chang and Greenough, 1984). Increased synapse formation is also observed following acquisition of conditioned avoidance tasks. Chicks trained to avoid pecking at a foul tasting bead show synaptic change in brain regions which increase their metabolic activity during learning. The intermediate hyperstriatum ventrale (IMHV) exhibits change in synapse length and presynapse volume and vesicle content whereas the lobus parolfactorius (LPO) shows an increase in synapse number (Stewart et al., 1984; 1987). Similarly, in the rodent hippocampal formation synapse number is dramatically increased within 70 min following acquisition of an avoidance task. In the latter model synapse number returned to that observed in the control animal within a two week period (Wenzel et al., 1980). The increase in synapse number may be associated with memory consolidation as lesion studies in the chick have demonstrated the IMHV to be required for task acquisition whereas the LPO is required at later times for task consolidation (Rose, 1991) .

Antibody interventive studies also support a role for *de novo* synapse formation in learning. Monoclonal antibodies to hippocampal neuronal surface glycoproteins or peptide antagonists of integrin function prevent the maintenance of LTP *in vitro* in hippocampal slices (Stanton et al., 1987; Staubli et al., 1990). Intraventricular infusions of antibodies specific for a variety of cell surface glycoconjugates also potently attenuate information storage (Karpiak et al., 1987; Nolan et al., 1987; Doyle et al., 1990a; Jork et al., 1991). The importance of their associated glycans is also highlighted by the ability of the deoxy-galactose antimetabolite to impair glycan structuring and, concomitantly, memory formation (Jork et al., 1989; Krug et al., 1991). Thus information processing provides an ideal model to study region-specific change in NCAM expression.

CHANGE IN NCAM FUNCTIONAL STATE DURING LEARNING

Method of Investigation: The passive avoidance response is an ideal paradigm to study NCAM-mediated events in memory formation. In this task animals are required to remain on a platform, for a criterion time of 5 min, in order to avoid receiving a footshock. The paradigm is rapidly acquired (5-8 min; 1-2 footshocks), thereby synchronizing subsequent biochemical events, and recall, as assessed by latency to step-down within a 3 min criterion at 24 and 48h after training, is stable over the experimental period. The confounding variable of footshock is accommodated by using yoked animals which receive the same number of footshocks as the trained animals with whom they were paired. Thus an effect is considered to be significant only when statistically different from both the yoked and passive animal. Change in NCAM sialylation state was monitored following intraventricular infusion of the sialic acid precursor - N-acetyl-D-mannosamine (ManNAc) delivered 4h prior to the time of sacrifice. Incorporation of the ManNAc precursor into glycoconjugates was evaluated in acid-insoluble fractions of whole homogenates and P2 pellet fractions derived from each hand-dissected brain region and was expressed as a ratio of the precursor remaining in the acid-soluble fraction in order to account for regional variations in the diffusion and uptake of the sugar precursor. Incorporation of labeled ManNAc into the NCAM glycan was established using NCAM immunoprecipitates obtained from nonionic detergent solubilised samples and expressing NCAM cpm as a percentage of those obtained in 1 mg of the solubilised P2 pellet fraction. These procedures have been described in greater detail (Doyle et al., 1992a).

Paradigm-specific Change in NCAM sialylation: Precursor incorporation studies showed no significant change in the overall sialylation of glycoconjugates associated with whole homogenates or P2 pellet fractions in any brain region. In contrast NCAM sialylation was significantly increased in the hippocampus in the 12-24h posttraining period when compared to both the yoked and passive controls. Surprisingly, the yoked control showed a significant increase in NCAM sialylation, when compared to that observed in the trained and passive animals, at the 48h recall time.

NCAM Isoform-specific Sialylation during Learning: Western blotting demonstrated the NCAM immunoprecipitates to contain all expected isoforms however estimations on the relative incorporation of labelled precursor into the excised bands indicated NCAM180 to be preferentially sialylated (Fig. 2). The paradigm-associated sialylation was further confirmed by its specific immunoprecipitation with a monoclonal antibody which recognises the α2-8 linked polysialosyl units of the capsular polysaccharides associated with the coat of meningococcus group B bacteria (Rougon et al., 1986). This demonstrated NCAM180 to be polysialylated. In approximately 50% of the animals examined an additional polysialylated version of NCAM was observed as a discrete band at 210kDa. The infrequent occurrence of this isoform may reflect the thermolability of polysialic acid during the solubilisation procedure required for denaturing gel electrophoresis.

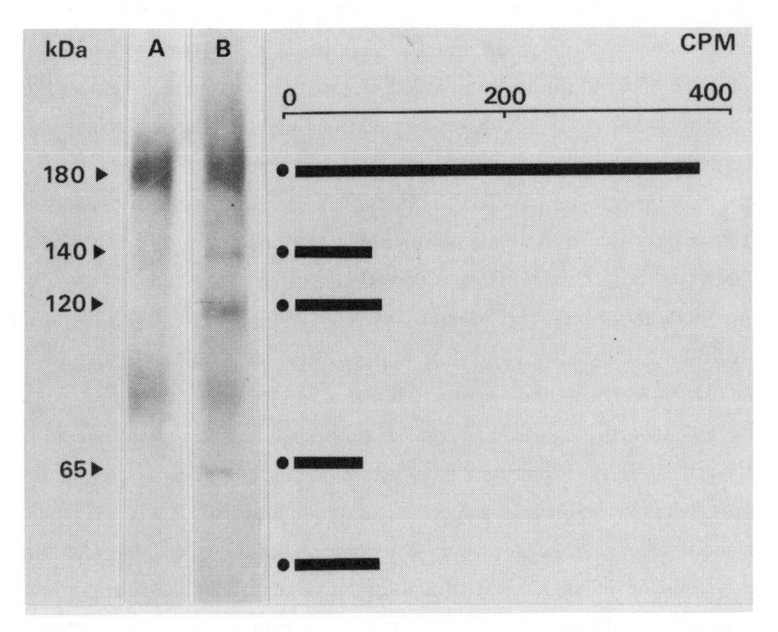

Figure 2. Hippocampal immunoprecipitates obtained from animals at 24hr after training, separated by denaturing gel electrophoresis, transferred to nitrocellulose and immunoblotted with anti-PSA (lane A) or anti-NCAM (lane B). The relative amounts incorporated into each NCAM polypeptide, and a blank region, in lane B is indicated by the histogram to the right of the gel panels. Modified from Doyle et al., 1992a.

Anti-NCAM Interventive Studies: To determine if NCAM was integral to memory formation an antibody interventive approach was adopted. Intraventricular infusions of anti-NCAM during training, or in the immediate posttraining period had no effect on task recall. In contrast, administration of anti-NCAM in the discrete 6-8h posttraining period resulted in an amnesia which became apparent at the 48h recall time and this was not obvious when tested at 24h. This effect could not be obtained at any other time within a 16h posttraining period or with an antiserum directed to an intracellular epitope (anti-neurofilament protein) and could not be attributed to the surgical procedure or antibody-induced immobility. Studies with ^{125}I-labelled anti-NCAM IgG demonstrated the antibody to penetrate all brain regions albeit in an erratic manner as would be expected from diffusion across the ependymal layer which lines the ventricles. The infused antibodies appeared to be intact and their levels tended to peak within 1-2h of infusion and had returned to basal levels within 6h. The above results have already been published (Doyle et al., 1992a,b).

CONCLUSIONS

These studies demonstrate NCAM to play a role in memory consolidation processes and prompt a number of observations.

1) Change in NCAM180 sialylation state suggests a role involved in synapse structure and/or function as this isoform is located to the postsynaptic density. The specific sialylation of NCAM180 may restrict change to local synapse remodelling as *in vitro* studies suggest it to inhibit extensive axonal growth (Doherty et al., 1992). Further, a precedent exists for the specific sialylation of NCAM180 as this also occurs during postnatal periods of synapse formation (Breen and Regan, 1988).

2) The magnitude of the response suggests that it is unlikely to be associated *de novo* synapse formation alone but that all synapses in the region of the hippocampus are undergoing change. The size of the response would be further accentuated by the synchronization of biochemical events imposed by the rapid training procedure.

3) The time factor must also be reconciled as the period of paradigm-specific change in NCAM sialylation state extends beyond that usually accepted as necessary for memory consolidation. However the time period associated with increased NCAM sialylation is well within that required for increased natural synapse turnover which occurs in response to deafferentation following a short lag phase (for review see Cotman et al., 1987). Thus if synapse remodeling is associated with memory processes then a lengthy consolidation period may be expected. In this regard it is interesting to note that the NCAM sialylation response to an unconditioned stimulus, as in the yoked control, is delayed.

4) The regional specificity of increased NCAM sialylation also requires comment. The hippocampus is generally accepted as being integral to memory processing (Teyler and DiScenna, 1985), however the striatum appears to be involved in the acquisition of avoidance responding (Doyle et al., 1990b). The lack of change in striatal, but not hippocampal, NCAM sialylation state

may parallel the situation observed in the chick where increase in synapse number is only observed in the LPO and not in the IMHV which is involved in the early phases of memory storage (Rose, 1991). A further role for NCAM in the mechanisms of information acquisition may yet be defined. Recent *in vitro* studies suggest that NCAM may be involved in the early destabilisation of synaptic contacts during information processing as serotonin-induced synaptic remodelling is associated with a rapid internalisation of the *Aplysia* homologue of NCAM (Bailey et al., 1992; Mayford et al., 1992).

5) The amnestic action of anti-NCAM in the early posttraining period appears to represent a perturbation of NCAM function which is separate to the increase in NCAM sialylation state but contributary to the overall process of memory formation. The lack of amnesia observed at the 24h recall time following an anti-NCAM infusion suggests the information is stored, but extremely labile, during the period of NCAM sialylation. This may be consistent with the antibody blocking a process, such as sequestration of NCAM into the new synapse, which is necessary for the proper maturation of a fully functional synapse.

Irrespective of the precise role of NCAM in information storage processes these studies contribute to understanding the mechanisms by which broadly distributed CAMs may regulate defined neural functions. These studies illustrate the ability of NCAM to regulate its sialylation state in a region- and isoform-specific manner in response to signals, which remain to be identified, which direct information storage processes. Region-specific change in NCAM sialylation state is likely to influence specific transduction mechanisms which may result in altered synaptic plasticity (Schuch et al., 1989; Doherty et al., 1991). This may not necessarily involve a *trans* interaction as glycosylation-dependent NCAM *cis* interactions with the L1 immunoglobulin superfamily member have been reported (Kadmon et al., 1990a,b). Thus selective isoform use/activation may underlie the specificity of NCAM function in the regulation of neural plasticity.

ACKNOWLEDGEMENTS

These studies were supported by EOLAS - the Irish Science and Technology Agency and BioResearch Ireland.

REFERENCES

Bailey C.H., Chen, M., Keller, F., Kandel, E.R. (1992) Science 256: 645-649.
Breen, K.B. and Regan, C.M. (1988) J. Neurochem. 50: 712-716.
Brown, T.H., Chapman, P.F., Kariss, E.W.and Keenan, C.L. (1988) Science 242: 724-728.
Chang, F.-L.F. and Greenough, W.T.(1984) Brain Res 309: 35-46.
Choung, C.-M., Edelman, G.M. (1984) J. Neurosci. 4: 2354-2368.
Cole, G.J. and Glaser, L. (1986) J. Cell Biol. 102: 403-412.

Cotman, C.W., Gibbs, R.G. and Nieto-Sampedro, M. (1987) In: The Neural and Molecular Bases of Learning (J.-P. Changeux and M. Konishi, Eds.) Wiley, London, pp. 375-396.
Doherty, P., Rimon, G., Mann, D.A. and Walsh, F.S. (1992) J.Neurochem. 58: 2338-2341.
Doherty, P., Ashton, S.V., Moore, S.E. and Walsh, F.S. (1991) Cell 67: 21-33.
Doherty, P., Fruns, M., Seaton, P., Dickson, G., Barton, C.H., Sears, T. and Walsh, F.S. (1990a) Nature 343: 464-466.
Doherty, P., Cohen, J. and Walsh, F.S. (1990b) Neuron 5: 209-219.
Doyle, E., Nolan, P., Bell, R., and Regan, C.M. (1992a) J. Neurosci. Res. 31: 513-523.
Doyle, E., Nolan, P.M., Bell, R. and Regan, C.M. (1992b) J. Neurochem. In Press.
Doyle, E., Bruce, M.T., Breen, K.C., Smith, D.C., Anderton, B. and Regan, C.M. (1990a) Neurosci. Lett. 115: 97-102.
Doyle, E., Nolan, P. and Regan, C.M. (1990b) Neurochem. Res. 15: 551-558.
Edelman, G.M. (1984) Proc. Natl. Acad. Sci. USA 81: 1460-1464.
Frelinger, A.L. and Rutishauser, U. (1986) J. Cell Biol. 103: 1729-1737.
Greenough, W.T. and Chang, F.-L.F. (1985) In: Synaptic Plasticity (C.W. Cotman Ed.) The Guilford Press, New York and London, pp. 335-372.
Hoffman, S. and Edelman, G. (1983) Proc. Natl. Acad. Sci. USA 80: 5762-5766.
Jork, R., Potter, J., Bullock, S., Grecksch, G., Matthies, H. and Rose, S.P.R. (1989) Neurosci. Res. Comm. 5: 105-110.
Jork, R., Smalla, K.-H., Karsten, U., Grecksch, G., Ruthrich, H.-L. and Matthies, H. (1991) Neurosci. Res. Comm. 8: 21-27.
Kadmon, G., Kowitz, A., Altevogt, P. and Schachner, M. (1990a) J. Cell Biol. 110: 193-208.
Kadmon, G., Kowitz, A., Altevogt, P. and Schachner, M. (1990b) J. Cell Biol. 110: 209-218.
Karpiak, S.E., Graf, L.and Rapport, M.M. (1978) Brain Res 151: 637-640.
Krug, M., Jork, R., Reymann, K., Wagner, M. and Matthies, H. (1991) Brain Res. 540: 237-242.
Lee, K.S., Schottler, F., Oliver, M. and Lynch, G. (1980) J. Neurophysiol. 44: 247-258.
Mayford, M., Barzilai, A., Keller, F., Schacher, S. and Kandel, E.R. (1992) Science 256: 638-644.
Meier, E., Regan, C.M. and Balazs, R. (1984) J. Neurochem. 43: 1328-1335.
Nolan, P.M., Bell, R. and Regan, C.M. (1987) Neurosci. Lett. 79: 346-350.
Persohn, E., Pollerberg, G.E. and Schachner, M. (1989) J. Comp. Neurol. 288: 92-100.
Pollerberg, E.G., Burridge, K., Krebs, K.E., Goodman, S.R. and Shachner, M. (1987) Cell Tiss. Res. 250: 227-236.
Regan, C.M. (1991) Int. J. Biochem. 23: 513-523.
Rose, S.P.R. (1991) Trends Neurosci. 14: 390-397.
Rougon, G., Dubois, C., Buckley, N., Magnani, J.L. and Zollinger W. (1986) J. Cell Biol. 103: 2429-2437.
Schuch, U., Lohse, M.J. and Schachner, M. (1989) Neuron 3: 13-20.
Small, S.J. and Akeson, R.A. (1990) J. Cell Biol. 111: 2089-2096.
Stanton, P.K., Sarvey, J.M. and Moskal, J.R. (1987) Proc. Natl. Acad. Sci. USA 84: 1684-1688.
Staubli, U., Vanderklish, P. and Lynch, G. (1990) Behav. Neural Biol. 53: 1-5.
Stewart, M.G., Rose, S.P.R., King, T.S., Gabbott, P.L.A. and Bourne, R. (1984) Dev. Brain Res. 12: 261-269.
Stewart, M.G., Csillag, A. and Rose, S.P.R. (1987) Brain Res. 426: 69-81.
Sunshine, J., Balak, K., Rutishauser, U., Jacobson, M. (1987) Proc. Natl. Acad. Sci. USA 84: 5986-5990.
Teyler, T.J. and DiScenna, P. (1985) Neurosci. Biobehav. Rev. 9: 377-389.
Troy, F.A. (1992) Glycobiol. 2: 5-23.
Walsh, F.S. and Doherty, P. (1991) Semin. Neurosci. 3: 271-284.
Wenzel, J., Kammerer, E., Kirsche, W., Matthies, H. and Wenzel, M. (1980) J Hirnforsch. 21: 647-654.

Polysialic Acid
J. Roth, U. Rutishauser and F. A. Troy II (eds.)

IS PSA-NCAM A MARKER FOR CELL PLASTICITY ?

Geneviève Rougon, Sylviane Olive and Dominique Figarella-Branger.

Biologie de la Différenciation Cellulaire; CNRS URA 179. Case 901. Parc Scientifique et Technologique de Luminy. 13289 Marseille Cedex 9; France.

SUMMARY

The spatio-temporal sequence of expression of PSA-NCAM in comparison with total NCAM has been investigated in several developing and adult mammalian tissues.
In development processes, such as the differentiation of spinal cord and of neural crest derivatives or in muscle regeneration, PSA-NCAM expression is transient and restricted to cells already committed to a neuronal or a myoblastic phenotype. The common denominator of differentiating PSA-NCAM positive cells is that they are undergoing a change in their shape.
In normal adult tissues, PSA-NCAM expression occurs in several discrete areas of the brain, spinal cord and muscle. Some of them, such as the hypothalamo-hypophyseal system, are known to undergo cytoarchitectural changes under various physiological stimuli.
Finally, some cells express PSA-NCAM under pathological conditions. This is the case of some neuroectodermal tumors, regenerative muscle fibers, reactive gliosis and of sprouting neurons observed in kainic acid treated animals with status epilepticus.
These data let us propose that PSA-NCAM can be considered as a marker for plasticity. During development, a PSA-NCAM positive cell is able to move, change its shape, size and molecular phenotype according to the maturation process; in adult tissues PSA expressing cells are able to modulate their morphology in response to various physiological events or to re-initiate mechanisms occurring normally only during development.

INTRODUCTION.

Informative interactions between cell surfaces are thought to be essential to the processes of cell recognition, sorting and migration during development and for subsequent stabilization or remodelling of the tissues in the adult. A variety of cell surface proteins involved in cell contact has been identified, the best characterized are a group of high molecular weight glycoproteins known collectively as the neural cell adhesion molecules (NCAMs).

They have been implicated in a wide range of morphogenetic events including cell segregation, axonal bundling, migration of axons along glial pathways and in several aspects of myogenesis including myoblastic fusion and muscle innervation. During these processes, the adhesive properties and functions of NCAMs are apparently modulated by the differential expression of several NCAM isoforms whose diversity results both from alternative splicing of a single RNA primary transcript and from various post-translational modifications (Santoni et al. 1989; Rougon et al. 1990; Doherthy et al. 1990).

One of the most striking features of NCAM is its degree of polysialylation (Finne et al. 1983). The presence of large amounts of polysialic acid (PSA) on NCAM is known to decrease adhesiveness. In addition PSA expression appears to affect not only NCAM function itself but also a number of other types of cell-cell interaction. Some of these effects appear to be independent of NCAM's intrinsic properties as an adhesion promoting molecule, suggesting that PSA may serve as an overall regulator of contact-dependent physiological events (Rutishauser et al. 1988; Acheson et al. 1991). We used a monoclonal antibody specific for high PSA-NCAM (Rougon et al. 1986) together with a site-directed polyclonal antibody recognizing all NCAM isoforms (Rougon and Marshak 1986) to study the spatio-temporal expression of PSA-NCAM isoforms in comparison with "total" NCAM in several normal, developing or adult, and pathological tissues. Here we report data obtained in both human and rodents using various models. Our aim is to correlate PSA -NCAM expression with the different cellular events observed during development and with physiological changes occurring in adults or during various repair processes.

PSA-NCAM IN DEVELOPING SPINAL CORD AND NEURAL CREST DERIVATIVES.

The temporal and spatial expression of PSA-NCAM was studied in developing spinal cord and neural crest derivatives of mouse truncal region. Temporal expression was analyzed on immunoblots of extracts of spinal cord and dorsal root ganglia (DRGs) microdissected at different developmental stages (from embryonic day 9 to birth). Data indicated that sialylation was regulated independently of the expression of NCAM polypeptide chains. In developing spinal cord the occurrence of PSA on NCAM appears to be correlated with the acquisition of the neuronal phenotype by neuroepithelial cells, since PSA expression is concomittant with that of the 180 KD isoform of NCAM and with that of neurofilaments.

When studied using immunohistochemistry, developing motoneurons and commissural neurons presented a strong expression of PSA-NCAM both on their cell bodies and their neurites. In particular PSA-NCAM was heavily expressed on their growing and fasciculating axons (Fig. 1). Although PSA-NCAM might be expressed on neurons after their aggregation in peripheral ganglia (for example DRG ganglia at day E12), the expression was only weak and variable according to the ganglion studied. From this study we concluded that PSA-NCAM is associated with differentiating neurons and that sialylation is lost on most neurons once the neuronal networks are established (Boisseau et al. 1991).

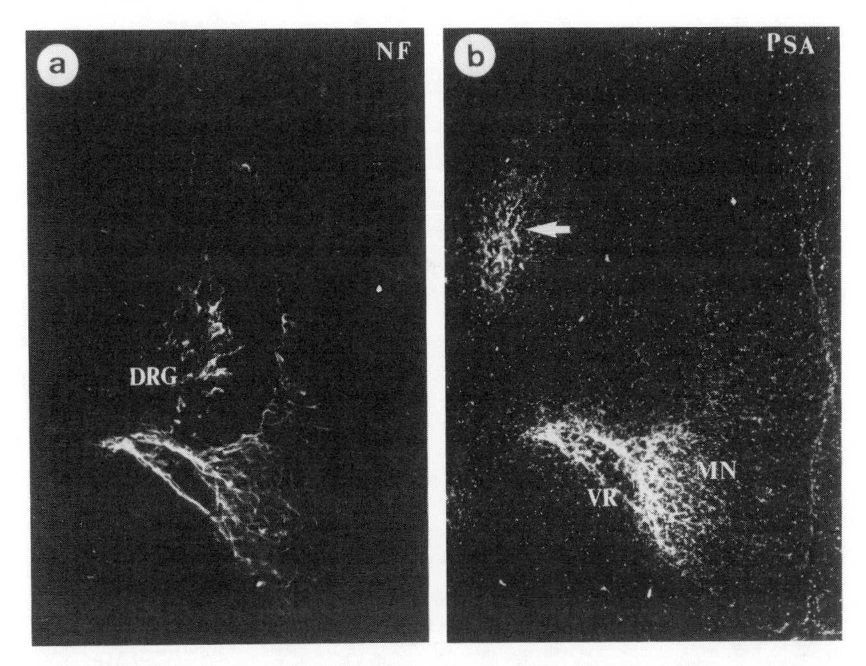

Fig 1 : Neurofilament (A) and PSA-NCAM (B) expression in a 10.5 day mouse embryo. Motoneurons somata (MN) and ventral roots (VR) are expressing PSA-NCAM and neurofilament whereas the DRG primordium is negative. Somites also express PSA- NCAM (arrow).

PSA-NCAM ON DEVELOPING HUMAN FOETAL MUSCLES.

Immunohistochemical analysis of NCAM expression in developing human psoas and quadriceps muscles from 15 weeks of gestation to birth showed that 1) developing myofibers (15 weeks of gestation to 28 weeks in psoas) express PSA-NCAM on their surface membrane 2) NCAM remains expressed on myotubes until innervation then disappears from the fibers; at birth most of the fibers are NCAM negative. After birth, extrafusal myofibers do not express NCAM. NCAM is concentrated at the neuromuscular junction, in satellite cells, and on some muscle spindle cells. Interestingly, satellite cells never express PSA-NCAM, but part of neuromuscular junction and some muscle spindle cells do (Figarella-Branger et al. 1990 a and 1992 b).

PSA-NCAM IN DEVELOPMENT :

From this series of sudies we concluded that resting precursor cells, such as the neuroepithelial cells in the neural tube or satellite cells in the muscle, do not express the PSA isoforms of NCAM although they constantly express NCAM and very likely the 140 KD isoform (Gennarini et al. 1984).

It is only when cells are activated and differentiate toward their respective phenotype (neuron or myoblast) that they start to express PSA often concomitantly with the occurrence of several other NCAM isoforms. This expression appears to be correlated with movement of the cells and more generally with changes in the cell shape. Fig. 2 schematizes the postulated sequence of expression of NCAM isoforms for both of neurons and muscles during development.

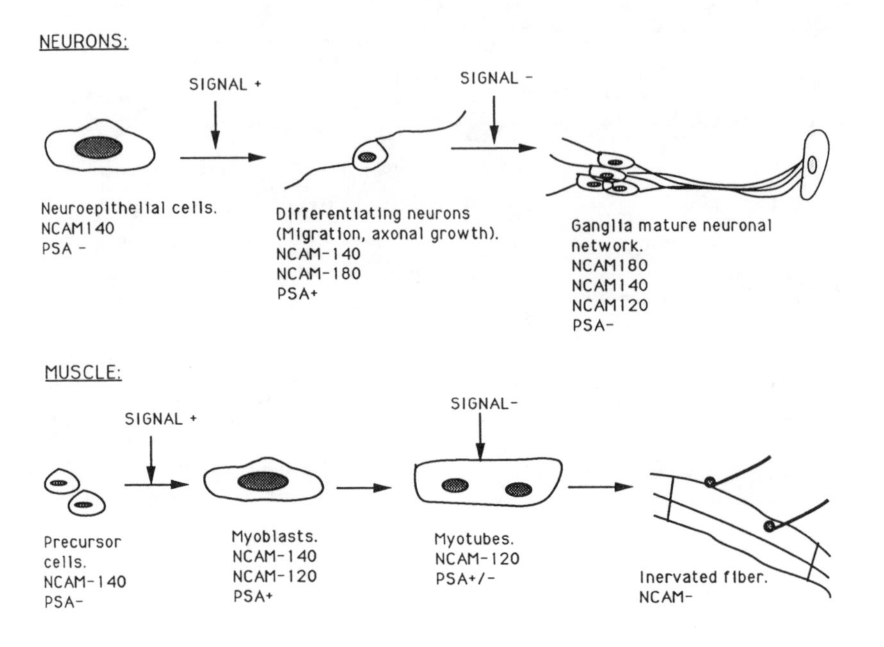

Fig 2 : Schematic representation of PSA- NCAM and other NCAM isoform expression in developing neural and muscle tissues.

PSA-NCAM IN THE NORMAL ADULT NERVOUS SYSTEM.

The examination of PSA-NCAM expression in the normal adult nervous system led to the conclusion that PSA could be constantly visualized in several discrete areas of the brain and spinal cord. An intercellular punctuate immunolabelling characterized the staining in certain hypothalamic and thalamic nuclei, superficial laminae of the dorsal horn of the spinal cord, ventral portion of the dentate gyrus of the hippocampus and lateral geniculate, parabranchial and habenular nuclei, bed nucleus of the stria terminalis, mesencephalic central grey and olfactory bulb. In other areas such as pyriform cortex, dorsal aspect of the dentate gyrus and fimbria and lamina X of the spinal cord, isolated neuronal-like cells were completely filled with immunolabel or showed a surface reaction on their cell bodies and processes. Highly immunoreactive isolated glial cells were also noticed within the ependymal layer of the central canal and lateral ventricules (Bonfanti et al. 1992). Among these areas many such cells are known to undergo structural reorganisation under particular physiological and experimental conditions.

We believe that PSA may be a crucial factor in allowing certain neurons and glial cells to manifest their capacity for structural plasticity in adulthood. For example activation of certain neurosecretory systems of the mammalian hypothalamus induces remodelling of their neurons and glial cells. During physiological stimulation of this system (parturition, lactation), astrocytic coverage of oxytocinergic somata and dendrites diminishes and their surfaces become extensively juxtaposed. In the neurohypophysis and median eminence, stimulation evokes a retraction of glial processes and an increase in the contact area between neurosecretory terminals and the perivascular space. These changes are reversible and glial coverage returns to normal upon cessation of stimulation. Our understanding of the significance of such modifications is speculative. In hypothalamic nuclei, they may permit synaptic remodelling that takes place concurrently; in neurohaemal structures, they may facilitate neuropeptide release. The expression of PSA in the hypothalamo-hypophyseal system was studied at the electron microscop level (Theodosis et al. 1991 and Kiss et al. submitted for publication). PSA-NCAM immunoreactivity is present in the supraoptic and paraventricular nuclei of the hypothalamus and in the neurohypophysis. The immunoreactivity was seen in dendrites, axons, and terminals and in associated astrocytes but not at the level of neuronal somata (Fig. 3).

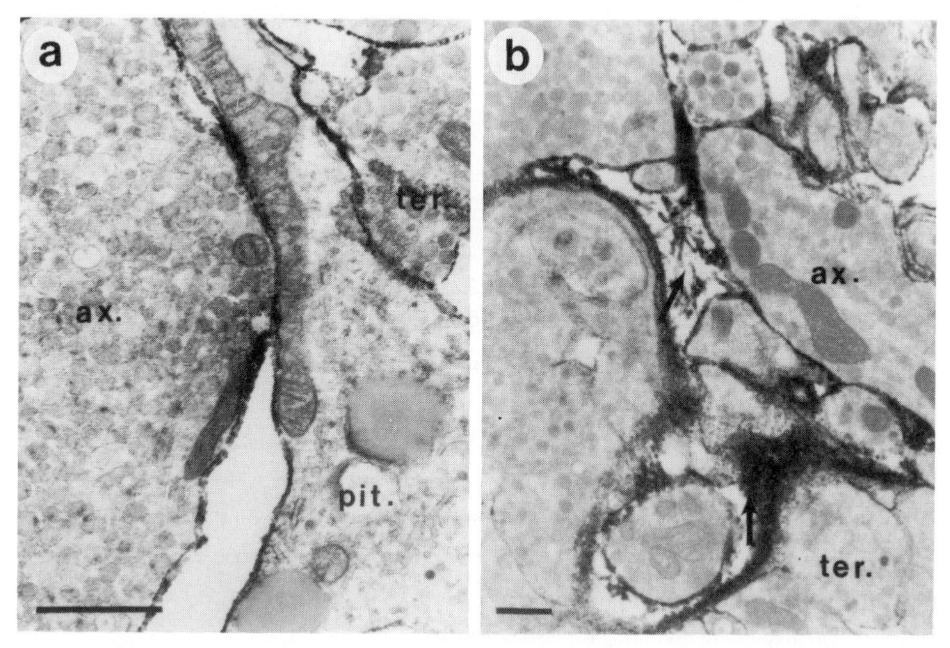

Fig 3 : Electron microscopy of PSA-NCAM localization in the adult rat neurohypophyi.s (A) Immunoreactivity is seen on the surface of neurosecretory axons (ax.) and terminals (ter.) and of astrocytic-like pituicytes (pit). PSA-NCAM could also be seen in the extracellular matrix (B) showed by arrows. Calibration bar: 1µm.

We propose that the continued expression of PSA-NCAM allows magnocellular neurons and their astrocytes to reversibly change their morphology in adulthood.

The surgical transection of the hypophyseal stalk, which eliminates descending neurosecretory axons from the neurohypophysis, results in a complete loss of PSA immunoreactivity in the neurohypophysis. Unlike that of PSA the immunoreactivity of NCAM on pituicytes was unaltered. These results suggest that the presence of the neurosecretory axons in the neurohypophysis is necessary to retain pituicytes ability to express PSA (Kiss et al. submitted). Although it is known that astrocytes are able to synthesize PSA-NCAM further studies are necessary to determine whether pituicytes synthesize PSA or bear them on their surface as a consequence of fixation of PSA liberated from neighbouring axons.

PSA-NCAM IN NERVOUS SYSTEM PATHOLOGIES.

Epilepsy: In order to acquire some insight into potential histogenetically plastic functions of PSA-NCAM, adult rats were treated with kainic acid, a powerful exitotoxic and convulsant glutamate analogue. When injected into amygdala it induces short and long-term effects, including status epilipticus, local and specific remote degenerative lesions, reactive gliosis, sprouting in different subsets of structures, particularly the hippocampus, and late spontaneous recurrent seizures. The treated animals were examined for PSA-NCAM expression, and an intense labelling found to be associated with glial-like cells, particularly those in the hippocampal formation, corresponds to reactive gliosis. This is confirmed by staining with anti-glial fribrillary acidic protein antibodies (Fig. 4).

This expression was detectable from about 3 days after kainic acid injection, persisted for at least 12 weeks and developed according to an observable spatio-temporal distribution pattern. In animals submitted to amygdala kindling, a non-lesional model of epilpsia, no such re-expression of PSA-NCAM was observed. Therefore the occurence of paroxymal activity alone does not seem sufficient *per se* to induce the re-expression of PSA-NCAM. This re-expression is very likely to be associated with remote brain damage induced by long-lasting status epilepticus.

In these kainic acid treated animals, sprouting was observed from neurons either directly or in the immediate vicinity of cells expressing PSA-NCAM in the adult animals (Le Gal La Salle et al. 1992). This tends to support the hypothesis that retention of PSA-NCAM expression by neurons may be related to their sprouting capabilities. From the conceptual point of view, these results would tend to indicate that continous expression of PSA-NCAM on neurons in adulthood or its re-expression on cells such as astrocytes may be part of a process that reactivates developmental mechanisms involved in remodelling of tissue structures.

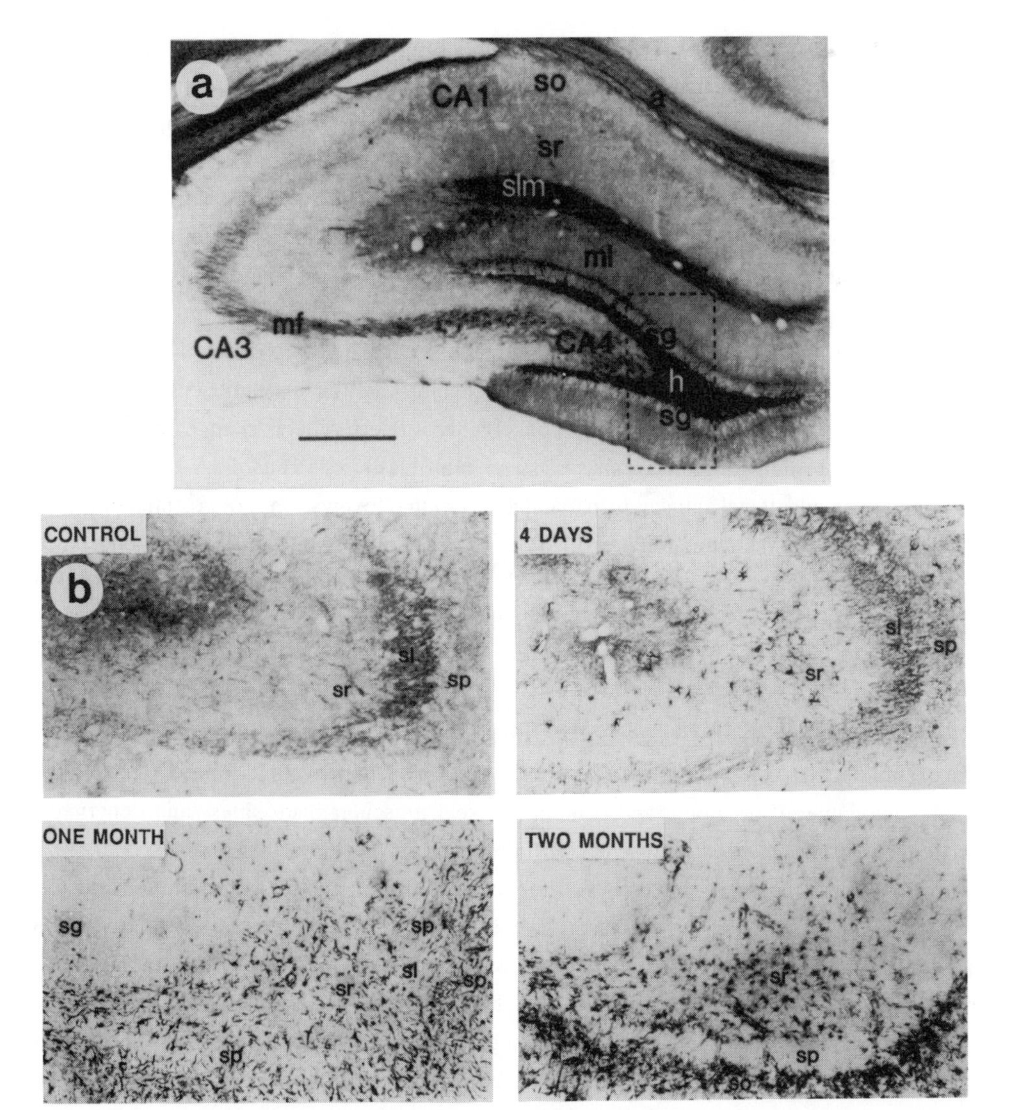

Fig 4 : (A) Immunolocalization of PSA-NCAM in a frontal section of normal adult rat hippocampus.
The immunoreactivity is detectable in a zone lying just beneath the granule cell layer. The *hilus* of the *dentate gyrus*, the *stratum lacunosum moleculare* and the mossy fibers display also moderate immunoreactivity (h=*hilus*, mf= mossy fibers, ml= molecular layer, slm= *stratum lacunosum moleculare*, so= *stratum oriens* sr= *stratum radiatum* , scale bar= 300 μm).
(B) Temporal pattern of PSA-NCAM immunopositive cell from 4 days to 2 months following kainic acid injection in the dorsal hippocampus. Control (left upper panel) and treated animals. Emergence of PSA-NCAM positive cells occured 4 days after the injection. One month post-injection numerous cells are labeled in the *stratus radiatum* and *stratus oriens* in the CA3 region (scale bar= 100 μm).

Neuroectodermal tumour: Adhesiveness and migratory properties are clearly central properties for tumour behavior and in particular for the metastasis process. Another characteristic feature of tumours is the expression of molecules and in particular carbohydrates resembling those found during the embryonic development of the tissue from which they originate (Roth et al. 1988). We investigated neuroectodermal tumours for their expression of PSA-NCAM and other NCAM isoforms (Figarella-Branger et al. 1990 b; 1992 a). All neuroectodermal tumours we analyzed (neuroblastomas, medulloblastomas, gliomas, ependymomas) were found to express NCAM. However, a large diversity was found among the isoforms expressed. All medulloblastomas studied, as well as a majority of neuroblastomas, express PSA-NCAM. Negative cases corresponded to tumours that had been subjected to chemotherapy or to ganglioneuroma. The PSA toward "mature" NCAM interconversion conceivably reflects cellular changes critical in the conversion into benign ganglioneuromas of some neuroblastomas. Thus, PSA-NCAM may be considered as an oncofoetal antigen. Its particular spectrum of expression is of potential interest to pathologists. In addition, tumour expressing PSA-NCAM are known to show a high tendency to metastatize. The high expression of PSA-NCAM in these tumours might be correlated to their migratory properties if we consider that in developing tissues migratory neuronal cells also express PSA-NCAM.

PSA-NCAM IN DISEASED HUMAN MUSCLES.

We examined NCAM expression in patients with inflamatory, dystrophic (Duchenne's, facioscapulohumeral, myotonic and oculopharyngeal muscular distrophies) and denervating (amyotrophic lateral sclerosis, chronic neuropathies) diseases (Figarella-Branger et al. 1990 a).

The main observations were that the regenerative fibers of inflamatory and dystrophic muscles, characterized on the basis of their hematoxylin eosin staining, strikingly expressed NCAM and its sialylated isoforms on their membranes and in some instances in their cytoplasm. All PSA-NCAM positive cells were NCAM positive. However, the number of NCAM positive cells was always higher than that of PSA positive cells. This is in concordance with the fact that PSA is borne by NCAM. Also, at early stages of regeneration, both types of antibody staining were detected intracellularly with a punctuate distribution (Fig. 5).

Contrastingly, denervated muscles re-expressed NCAM but never their PSA isoforms. In all denervated muscles, most of the fibers in atrophic groups were positive for NCAM (Fig. 5).

The distribution of staining differed from that observed in regenerative fibers in that it was more concentrated on the membrane. As already reported by several other groups, when reinnervation successfully occured NCAM expression was lost from the fibers as confirmed by observation of type grouping.

We propose that during regeneration expression of PSA is transient and parallel to that observed in normal muscle development. Therefore, it is likely that in adult muscle, PSA could only be expressed by "activated" satellite cells in developing regenerative fibers, committed to the myoblastic phenotype. A mature fiber, if still able to re-express NCAM, whose expression is under the control of innervation, appears to have lost the ability to synthesize PSA.

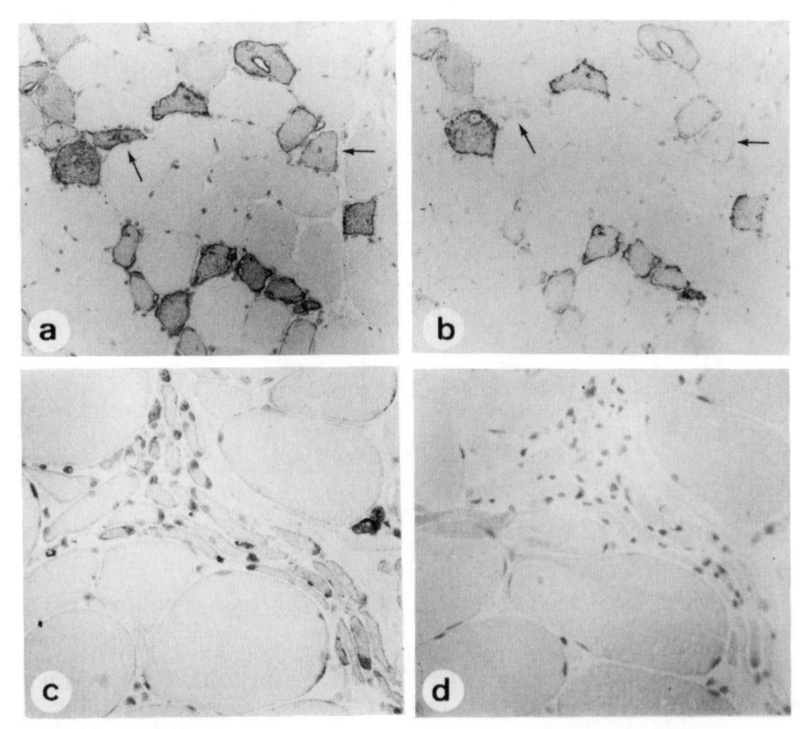

Fig 5 : Expression of PSA-NCAM and "total" NCAM in diseased human muscles : Regenerative fibers strongly express NCAM (A) and PSA-NCAM (B) both in their cytoplasm and on their membrane. Some fibers (arrow) do not express PSA while they still express NCAM. Denervated fibers forming a group of atrophic fibers express NCAM on their membrane (C) but not PSA (D).

CONCLUSION.

In the light of the different models studied it is clear that PSA is always associated with cells already committed toward a phenotype and never on resting or pluripotential precursor cells. Moreover, a PSA positive cell exhibits plasticity according to its ability to move, and, for adult neuron its ability to re-initiate mechanisms only normally occurring during development (Fig. 6). This leads us to proprose that PSA can be considered as a marker for plasticity.

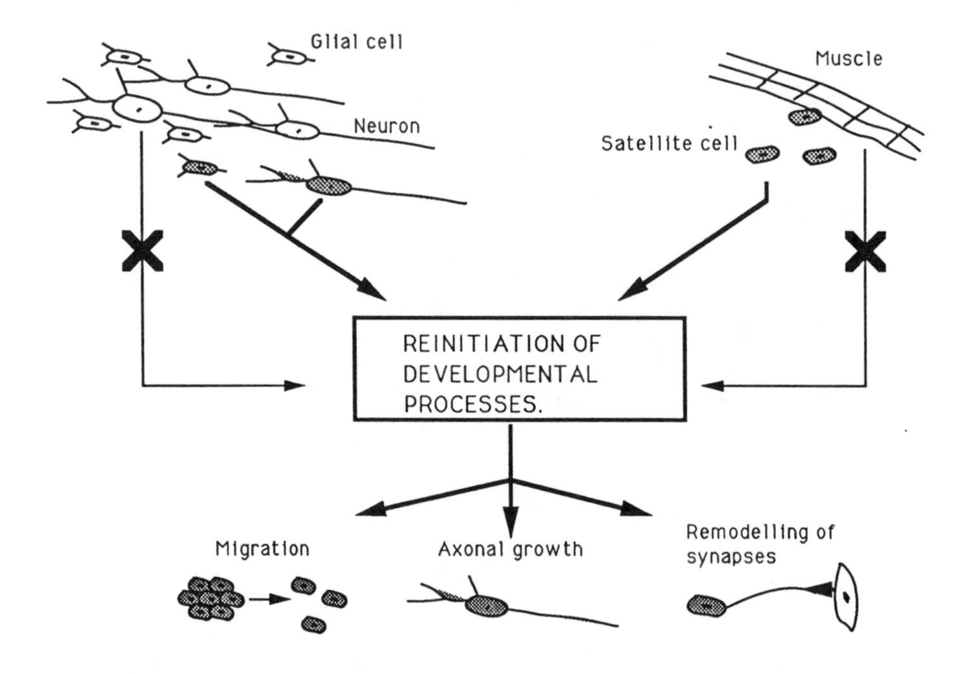

Fig 6 : Schematic representation of the potentialities of PSA-NCAM positive cells in the adult central nervous system.
Neurons exhibiting potential for plasticity remain PSA-NCAM positive all over the adult life. By contrast, in muscle regeneration and may be also in reactive gliosis, the NCAM positive cells derived from the activation of a resting precursor.

All of these properties are compatible with a decrease in the adhesive properties of the cells and with the fact that PSA modulates the binding of NCAM itself (Sadoul et al 1983) and of other adhesion molecules (Rutishauser et al. 1990, Acheson et al. 1991). The question yet not answered is whether the role of PSA occurs mainly at the cell surface by mere mechanical mechanisms, steric hindance and charge repulsion. An alternative mode of action could be that PSA is recognized by a ligand and participates in the transmission of intracellular signals which would in turn modulate the cell's surface properties (see for example Saffell et al. 1992).

Another question yet to be answered is whether or not an adult cell that has lost the ability to synthesize PSA units is ever able to recover this ability. So far all the models we studied provided evidence against this idea. In any case, the identification and cloning of enzymes and factors that participate in PSA synthesis and in the control of their expression remain a priority. This will open the door to the manipulation of their expression and the understanding of their precise role in cell differentiation and plasticity.

ACKNOWLEDGEMENTS.
This work was supported by grants from Association Française contre les Myopathies, Association de Recherche contre le Cancer, Fédération Nationale des Centres de Lutte contre le Cancer and Groupement des Entreprises Françaises dans la Lutte contre le Cancer to G. R. The authors also wish to thank all the collaborators who participated to this work and in particular Drs. D. Théodosis, G. Le Gal La Salle, J. Kiss and M. Simonneau.

REFERENCES.

Acheson, A., Sunshine, J. and Rutishauser, U. (1991) J. Cell Biol. 114 : 143-153.
Boisseau, S., Nedelec, J., Poirier, V., Rougon, G. and Simonneau, M. (1991) Development 112 : 69-82.
Bonfanti, L., Olive, S., Poulain, D. and Theodosis, D. (1992) Neurosci.49 : 419-436.
Doherty, P., Cohen, J. and Walsh, F.S. (1990) Neuron 5 : 209-219.
Figarella-Branger, D., Nedelec, J., Pellissier, J.F., Boucraut, J., Bianco, N. and Rougon, G.(1990 a) J. Neurol. Sci. 98 : 21-36.
Figarella-Branger, D., Durbec, P. and Rougon, G. (1990 b) Cancer. Res. 50 : 6364-6370.
Figarella-Branger, D., Pellissier, J. F., Daumas-Duport, C., Delisle, M., Pasquier, B., Parent, M., Gambarelli, D. and Rougon, G. (1992 a) Am. J. Surg. Path. 16 : 97-109.
Figarella-Branger, D., Pellissier, J.F., Bianco, N., Pons, F., Leger, J.J. and Rougon, G. (1992 b) J. Neuropath. Exp. Neurol. 51 : 12-23.
Finne, J., Finne, U., Deagostini-Bazin, H. and Goridis, C. (1983) Biochem. Biophys. Res. Commun. 112 : 482-487.
Gennarini, G., Rougon, G., Deagostini-Bazin, H., Hirn, M. and Goridis, C. (1984) Eur. J. Biochem. 142 : 57-64.
Kiss, J., Wang, C. and Rougon, G. Neurosci. (submitted).
Le Gal La Salle, G., Rougon, G. and Valin, A. (1992) J. Neurosc. 12 : 872-882.
Rougon, G., Dubois, C., Buckley, N., Magnani, J.L. and Zollinger, W. (1986) J. Cell Biol. 103 : 2429-2437.
Rougon, G. and Marshak, D. (1986) J. Biol. Chem. 261 : 3396-3401.
Rougon, G., Nédelec, J., Goridis, C. and Chesselet, M.F. (1990) Acta Histologica 38 : 51-57.
Roth, J., Züber, C., Wagner, P., Taatjes, D.,, Weisgerber, C., Heitz, P., Goridis, C and Bitter-Suermann, D. (1988) Proc.. Natl. Acad. Sci. USA. 85 : 2999-30003.
Rutishauser, U., Acheson, A., Hall, A., Mann, D. and Sunshine, J. (1988) Science (Washington,D.C) 240 : 53-57.
Sadoul, R., Hirn, M., Deagostini-Bazin, H., Rougon, G. and Goridis, C. (1983) Nature 304 : 347-349.
Saffell, J., Walsh, F. and Doherthy, P. (1992) J. Cell Biol. 118 : 663-670.
Santoni, M. J., Barthels, D., Vapper, G., Baud, A., Goridis, C. and Wille, W. (1989) EMBO J. 8 : 385-392.
Theodosis, D., Rougon, G., Poulain, D. Proc. Natl. Acad. Sci. USA. 88 : 5494-5498.
Walsh, F.S. (1988) Neurochem. Int. 12: 262-267.

Polysialic Acid
J. Roth, U. Rutishauser and F. A. Troy II (eds.)

EXPRESSION OF POLYSIALIC ACID IN HUMAN TUMORS AND ITS SIGNIFICANCE FOR TUMOR GROWTH

Jürgen Roth, Christian Zuber, Paul Komminoth, E. Paul Scheidegger, Michael J. Warhol*, Dieter Bitter Suermann** and Philipp U. Heitz

Division of Cell and Molecular Pathology, Department of Pathology, University of Zürich, Schmelzbergstr. 12, CH-8091 Zürich, Switzerland, * Ayer Clinical Laboratory, Department of Pathology, Pennsylvania Hospital, Philadelphia, PA 19107, USA, Institut für Medizinische Mikrobiologie, Medizinische Hochschule Hannover, Konstanty-Gutschow-Str. 8, 3000 Hannover 61, Federal Republic of Germany

SUMMARY: Polysialic acid has been shown to modulate the adhesive properties of the calcium-independent neural cell adhesion molecule. Polysialylated NCAM is widely expressed during organogenesis and can be found in certain tumors. This review summarizes findings obtained in various human tumors such as Wilms tumor, teratoma and various neuroendocrine tumors. The results demonstrate the usefulness of poly Sia as a marker in tumor diagnosis and suggest that it may be of importance in certain aspects of tumor growth.

INTRODUCTION

Cell surface carbohydrates have been suggested to play an important role during histo- and organogenesis and in tumors by mediating intercellular recognition and adhesion. The onco-developmental expression is one of the highly interesting aspects of these glycoconjugates since it is related to the control of differentiation programs.

Polysialic acid (poly Sia) consisting of homopolymers of sialic acid with varying chain lengths and joined by α 2,8-, α 2,9-, or alternating α 2,8- and α 2,9 - ketosidic linkages has a limited and specific distribution. It is found as the capsular polysaccharide of certain pathogenic bacteria (Bhattacherja et al., 1976; Egan et al., 1977; Troy, 1979; Roth and Troy, 1980), on major glycoproteins of cortical alveoli of a number of Salmonidae fish eggs (Inoue and Iwasaki, 1978; Kitajima et al., 1988; see chapter of Inuoe in this book) and in the voltage-sensitive Na+ channels of the electrical organ from Electrophorus electricus (James and Agnew, 1988) and adult rat brain

(Zuber et al., 1992, see also this book). Poly Sia also occurs on the neural cell adhesion molecule (N-CAM) (Nybroe et al., 1988; Rutishauser and Jesselt, 1988).

NCAM is a membrane glycoprotein that can promote cell - cell adhesion through a homophilic binding mechanism (Doherty et al.1990; Hall et al.1990; Rutishauser et al. 1983). The kinetics of this adhesion correlate inversely with the degree of glycosylation of NCAM (Doherty et al.1990; Hoffmann et al.1983; Rutishauser et al. 1985), which specifically arises from differences in the length of poly Sia (Cunningham et al.1983; Finne, 1982). Changes of the length of poly Sia on NCAM have been shown to modulate the adhesive properties of this molecule during embryonic development (Landmesser et al.1990; Rothbard et al. 1982; Rutishauser et al. 1988; Rutishauser et al. 1985; Sadoul et al. 1985). Through the use of bacteriophage endoneuraminidase (endo N), which specifically hydrolyzes poly Sia composed of 8 and more residues to leave 4 or 5 a2,8 linked sialic acid residues (Hallenbeck et al. 1987; Rutishauser et al. 1985; Vimr et al. 1984), it has been demonstrated in vitro that poly Sia can influence the function not only of NCAM but also of other cell surface molecules (Acheson and Rutishauser, 1988; Acheson et al. 1991; Pinfen Yang et al. 1992). Rutishauser and coworkers (Pinfen Yang et al. 1992) have proposed that one potential underlying mechanism for these effects is due to the unusual physicochemical properties of this very large and abundant linear cell surface polyglycan. The abundance of membrane-associated sialic acid represents a major component of cell surface charge and constitutes a hydrated matrix of appreciable size and structural complexity (Murray and Jenssen, 1992).

The poly Sia of NCAM was shown to be an oncodevelopmental antigen when it was found to be present in embryonic human kidney and Wilms tumor (Roth et al., 1987, 1988). It was speculated that the PSA of NCAM may be related to the invasive and metastatic growth potential of human tumor cells (Livingstone et al., 1989; Roth et al. 1988a, b 1990). Investigations of a wide range of tumors have indicated that many neuroendocrine tumors express polysialylated NCAM. These include pheochromocytoma, small cell carcinoma of the lung, neuroblastoma, rhabdomyosarcoma, pituitary tumors, teratomas and medullary carcinoma of the thyroid (Roth et al., 1989; Metzman et al., 1991).This paper will summarize the data on poly Sia expression in human tumors. The results were obtained using the monoclonal antibody mAb 735 labeled with colloidal gold for immunocytochemistry and Western blotting. This antibody reacts exclusively with homopolymers of α 2,8-linked sialic acid of 8 or more residues long. Furthermore, bacteriophage encoded endoneuraminidase N was applied which reduces the length of poly Sia to less than 8 residues and abolishes binding of mAb 735.

POLYSIALIC ACID IS AN ONCO-DEVELOPMENTAL ANTIGEN IN HUMAN KIDNEY

Kidney organogenesis in mammals is characterized by the sequential appearance of three distinct organs (for review see Saxén, 1987). First, the transient pronephros is formed, which initiates the caudal growth of the nephric duct. The mesonephros is, in turn, induced by the extending nephric duct which subsequently branches to form the uretic bud. Finally the metanephros, which becomes the permanent kidney, is formed by the inductive interaction of branches of the uretic bud with undifferentiated (metanephrogenic) mesenchyme. Both embryonic anlagen are of mesodermal origin.

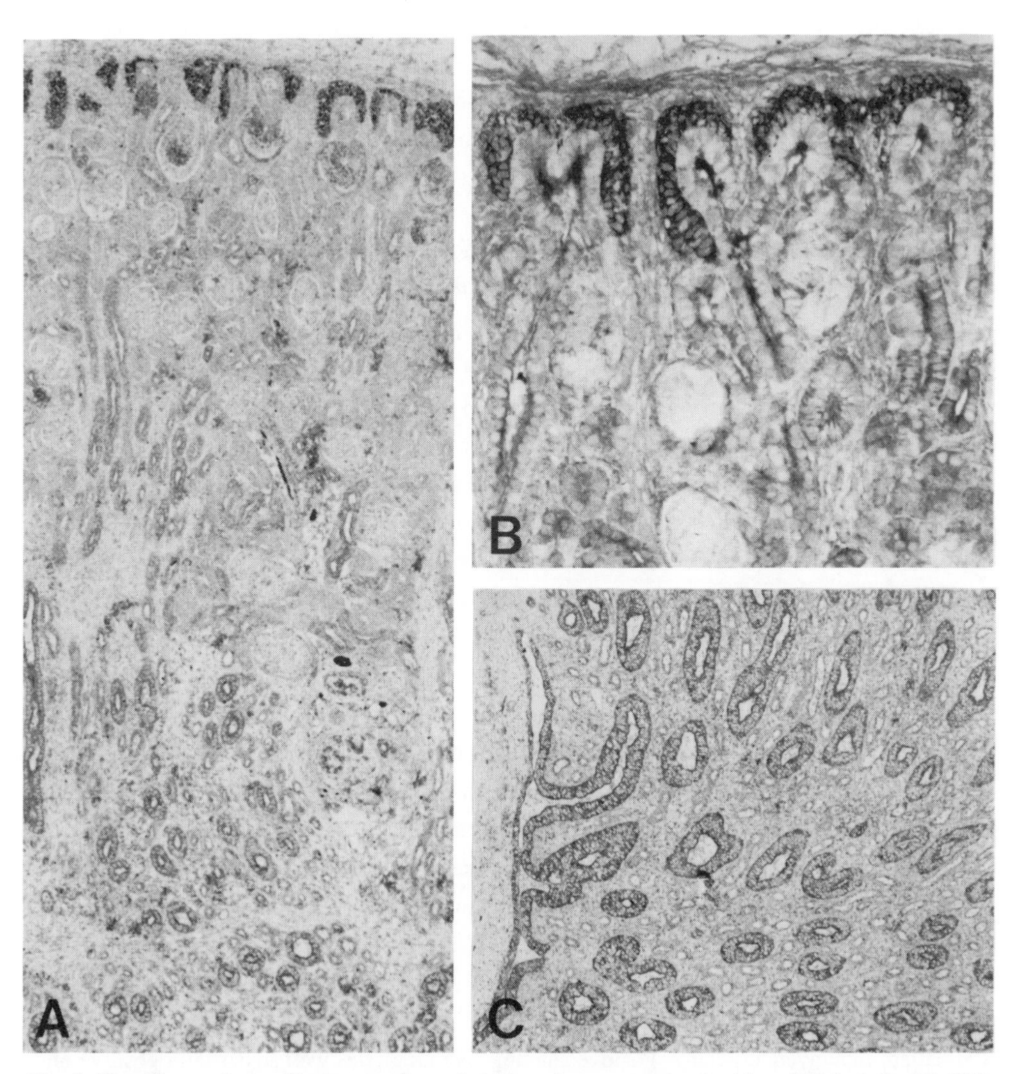

Fig. 1. Paraffin sections of human embryonic kidney immunostained with gold-labeled mAb 735. (A) Immunostaining is detectable both i the cortical and medullary regions of the developing kidney. (B) Detail of the nephrogenic zone with poly Sia positive metanephrogenic mesenchymal condensations and terminal ampulla of collecting ducts. (C) Poly Sia positive collecting ducts in the medulla.

The results of our studies on the occurrence and distribution of poly Sia and NCAM in developing kidney (Roth et al., 1987; Lackie et al., 1990) are detailed in the chapter by Lackie et al. in this book. They can be summarized as follows. It was observed that (i) PSA and N-CAM are present in both embryonic kidney anlagen (Figs. 1 and 4) during, and subsequent to, organ formation and (ii) that the predominant N-CAM polypeptide isoform present during late embryonic development has a Mr of 140 kD and is polysialylated. Both PSA and N-CAM become undetectable in parenchymal elements of fully differentiated rat and human kidneys, clearly demonstrating the strict developmental regulation of this molecule in kidney. This contrasts with the continued expression of N-CAM (for review see Nybroe et al., 1988) in the brain of adult rats.

The most prevalent malignant kidney tumor of childhood is the Wilms' tumor (nephroblastoma). It is a complex mixed embryonal neoplasm and in its classic form is composed of three basic elements: (i) nonorganoid compact masses formed by blastemal cells, (ii) tubular structures formed by cuboidal to columnar cells contained in or adjacent to the former, and (iii) a so-called fibrous stroma; as well as minor components such as glomeruloid bodies, muscle fibres, adipose tissue, cartilage or bone (Schmidt et al., 1982; Gonzalez-Crussi, 1984). The most widely accepted concept of histogenesis of Wilms' tumor assumes its origin from primitive metanephric cells that, although malignant, retained a differentiative potential which probably accounts for the complexity in the histological appearance of this neoplasm (Mierau et al., 1987). This concept is strongly supported by similarities in structure, epithelial antigens, cytoskeletal elements, extracellular matrix and lectin binding sites with normal developing and adult human kidney. As indicated above, a characteristic feature of Wilms´ tumor is the presence of structural elements resembling those found during embryonic development of human kidney. These similarities prompted us to investigate Wilms´ tumor for the possible presence of polysialic acid and N-CAM by immunohistochemistry, immunoblotting, and in situ hybridization.

All histological types of Wilms' tumor, i.e., the mixed "triphasic" type, the monomorphous tubular type, the biphasic blastemal-epithelial type, the undifferentiated blastemal type, and the focal anaplastic type exhibited immunostaining for polysialic acid (Fig. 2). The most constant finding was related to the blastemal cells which displayed intense cell surface staining in both the mixed and the undifferentiated blastemal tumor type. Peripheral cells of blastemal cell masses exhibiting certain signs of epithelial differentiation such as polarity and palisade formation showed intense immunostaining. Another constant observation was the nonreactivity of the monocional anti-polysialic acid antibody with the stroma (Fig. 2). The transition from positively stained blastemal cells to nonstained stromal cells was always sharp and abrupt. Immunostaining for polysialic acid of the other tumor components often displayed considerable variability in intensity and pattern in a given structure. Positive and negative tubules as well as partially positive ones were frequently observed close to each other (Fig. 2). Glomeruloid bodies (Fig. 2) and elements resembling S-shaped bodies were positive, the latter often presented stained and non-stained

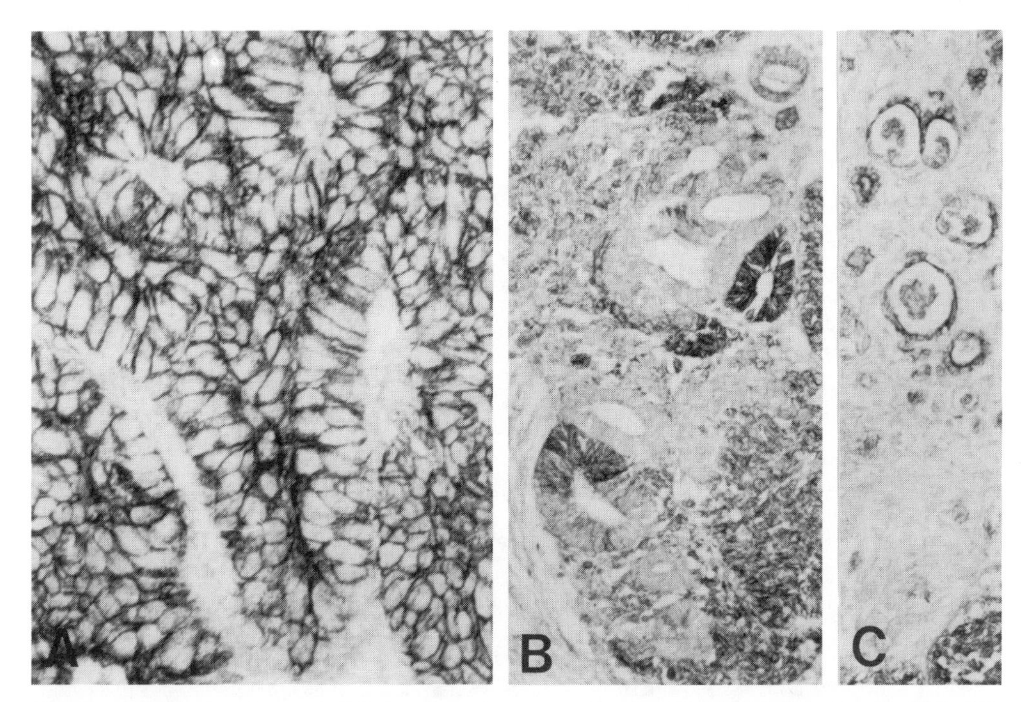

Fig. 2. Paraffin sections of Wilms tumor immunostained with gold-labeled mAb 735. (A) Blastemal component is positive for poly Sia and the stroma is not stained. (B) Epithelial differentiations in the form of tubules can exhibit poly Sia immunoreactivity (C) as can so called glomeruloid bodies.

portions. Muscle fibers, if encountered, exhibited cell surface staining.The immunostaining was completely abolished when sections were pretreated with endosialidase E and under the various other control conditions.

Immunohistochemical studies have revealed the presence of laminin, fibronectin and a basement membrane antigen in Wilms tumor (Kumar et al., 1986; Sariola et al., 1985). Furthermore, ultrastructural studies have revealed the presence of an amorphous electron dense material at the tumor cell surfaces and in the extracellular space which due to its similarity with a lamina densa has been referred to as basement membrane, basement membrane-like or extracellular matrix (Garvin et al., 1987; Ito and Johnson, 1969; Schmidt et al., 1982). In the electron microscopic investigations of the blastemal component of Wilms tumor a layer of an amorphous electron dense material along the plasma membrane were clearly recognizable (Fig. 3). Some peculiarities in its distribution were noted. Along a single cell, plasma membrane regions which were covered only by a very thin layer could be distinguished from others which exhibited an enormously developed coat. When the mAb 735 was applied to ultrathin frozen thawed sections, gold particle staining of varying degree was present along the surface of the blastemal cells (Fig.

3). At regions of close membrane-to-membrane contact characterized by the presence of a thin surface coat only sparse, often cluster-like immunolabeling existed. In contrast, plasma membrane regions covered by a thick surface coat exhibited a dense immunolabeling throughout the entire thickness of the amorphous material.

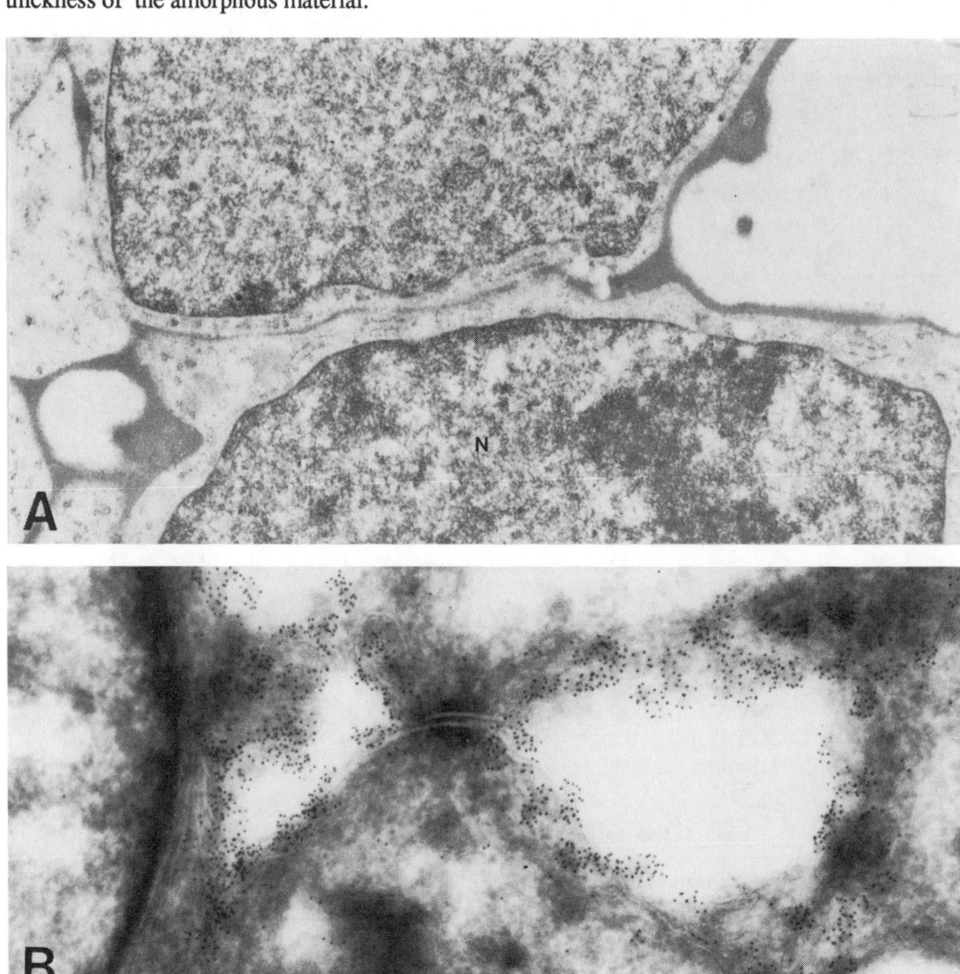

Fig. 3. Electron microscopic demonstration of the surface coat of Wilms tumor cells (A). This surface coat is intensely labeled with the mAb 735 as indicated by the presence of numerous gold particles (B)

These results clearly demonstrates that the cell surface material often referred to as basement membrane, basement membrane-like or extracellular matrix is composed of poly Sia. Therefore, the characteristic cell surface coat of blastemal cells in Wilms tumor can be considered chemically and morphologically the counterpart of material forming the cell wall in certain highly pathogenic bacteria. Polysialic acid has been shown to modulate the adhesive properties of N-CAM. It is also noteworthy, that no 180 kDa isoform of N-CAM is detectable in Wilms tumor (see below) which is reported to be concentrated at sites of cell-cell contact, is restricted in its mobility in the plane of the plasma membrane and can bind spectrin-like cytoskeletal elements. These findings may bear some importance for the growth behaviour of the Wilms tumor cells since they directly demonstrate that the presence of poly Sia hinders membrane-membrane contact and often results in the formation of small intercellular lumina. The presence of polysialic acid along the cell surface may contribute to the invasive and metastatic potential of the tumor cells. Furthermore, by forming a dominant cell surface component, the poly Sia may not only prevent a variety of ligands from interacting with their cell surface receptors but also influence specific immunological host-tumor interactions by its extreme weakly immunogenic properties.

POLYSIALIC ACID IN WILMS TUMOR IS ON NCAM

What is the evidence that poly Sia detected in Wilms tumor (and embryonic kidney) is on NCAM? Indirect evidence consists in the immunoelectron microscopic detection of NCAM polypeptide immunoreactivity (Zuber and Roth, 1990) and NCAM mRNA by in situ hybridization (Roth et al., 1988). Immunoblot analysis of whole homogenates from embryonic brain and kidney with the

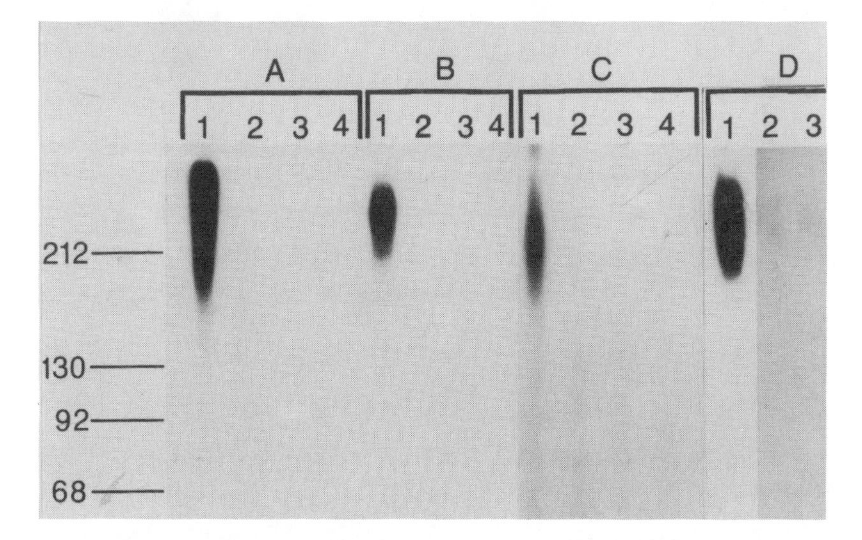

Fig. 4. Immunoblot analysis with mAb 735 of human embryonic brain (A 1), human embryonic kidney (B 1, D 1), Wilms tumor (C 1), normal adult kidney adjacent to Wilms tumor (D 2) and embryonic human liver (D 3).A, B, C 2: Pretreatment with endoneuraminidase N; A, B, C 3: Colominic acid preabsorbed mAb 735; Omission of mAb 735.

gold-labeled mAb 735 revealed a single broad band with similar electrophoretic mobility (Fig. 4, panel A 1 and B 1) which was undetectable in whole homogenates from embryonic liver (Figure 7, panel D 3). Homogenates from Wilms' tumor (Fig. 4, panel C 1 and D1) showed an immunoreactive band with the same mobility as observed for embryonic kidney and brain, whereas no band was detectable in homogenates of the adjacent normal kidney (Fig. 4, panel D 2). No band was observed under the control conditions (Fig. 4 , A-C 2-4). In order to characterize the polysialic acid carrying polypeptide(s), tissue homogenates from human and rat brain as well as Wilms tumor were subjected to combined immunoprecipitation and immunoblot analysis. Immunoblot analysis with mAb 735 of the immunoprecipitate obtained with the anti-NCAM antibodies revealed a single broad band for both brain and Wilms tumor (Fig. 5, lane A 1, 2). When the antibodies were used in reversed order, the results were the same although the bands were broader. In order to compare the N-CAM polypeptide pattern of Wilms tumor to that of embryonic brain, homogenates were desialylated with endoneuraminidase prior to immunoblot analysis with anti-N-CAM antibodies. For rat (Fig. 5, lane B 2) and human (Fig. 5, lane B 4) brain the typical 180 kD, 140 kD, and 120 kD polypeptides were observed. However, in Wilms tumor none of the 180 kD polypeptide was detectable but polypeptides of approximately 120 kD

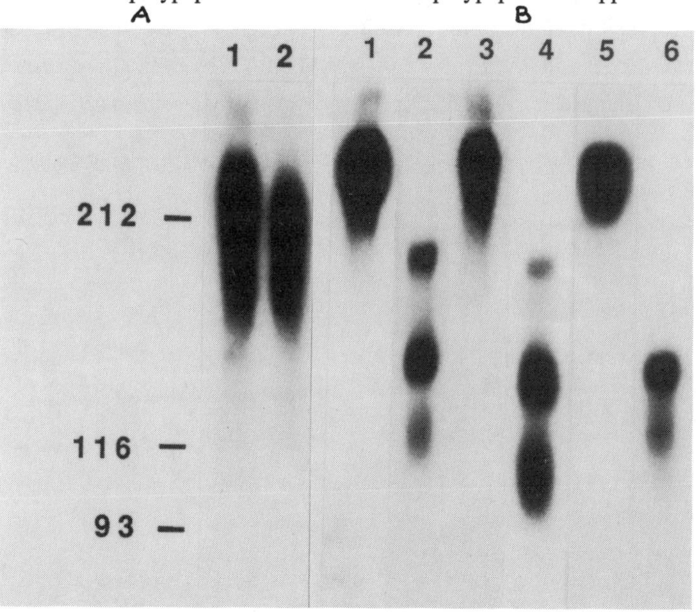

Fig. 5. Combined immunoprecipitation - immunoblot analysis with mAb 735 and anti-NCAM antibody of human embryonic brain (A 1) and Wilms tumor (A 2) reveals structural relationship between poly Sia and NCAM.
Immunoblot analysis for polysialic acid of postnatal day 3 rat brain (B 1), embryonic human brain (B 3), and Wilms tumor (B 5) . Desialylation by endoneuraminidase N of an aliquot from the same homogenate samples followed by immunoblot analysis for N-CAM reveals the three typical isoforms in rat (B 2) and human (B 4) brain but only two isoforms of approximately 120 kD and 140 kD in Wilms tumor (B 6).

and 140 kD were observed (Fig. 5, lane B 6). The 18 Wilms tumors revealed a certain variability in the relative amounts of the two N-CAM polypeptides which was not related to a particular histological type. These data are direct evidence for the presence of two highly sialylated isoforms of N-CAM of approximately 120 kD and 140 kD in Wilms tumor. This polypeptide pattern which shows an inherent variation is expressed by Wilms tumors of different histological types. This indicates that cellular components of both blastema and epithelial differentiations may express common N-CAM isoforms.

PRESENCE OF POLYSIALIC ACID IN TERATOMAS AND NEUROENDOCRINE TUMORS

TERATOMAS

Teratomas are tumors comprise of tissues derived from all three germ layers: ectoderm, mesoderm and endoderm. These tumors can contain only mature tissue elements and are completely benign. However, the presence of immature elements establishes a malignant potential that has both prognostic and therapeutic implications. The immature elements most commonly seen are neural in origin and primarily consist of immature glia, ependyma, neurotubules and neuroepithelial rosettes. A histologic grading system (Thurlbeck and Scully, 1960) for the grading of teratomas is based on: (a) the degree of immaturity; (b) the presence of neuroepithelium, and (c) the quantity of neuroepithelium. Thus, the identification of immature neural tissue elements is essential for the accurate and precise grading of teratomas. Commonly used markers of neural tissues such as S-100, glial fibrillar acidic protein (GFAP), nerve growth factor receptor and neurofilament protein (NFL) have not been useful for identifying neuroblasts and other immature neural elements in teratomas. The application of mAb 735 to detect poly Sia in sections of human teratomas proved that it was the only reagent that consistently marked all types of neural tissue, both mature and immature (Table 1).

Table 1. Immunocytochemical reactivity of teratoma neural components

	POLY SIA	S 100	NFL	GFAP	NGF
Neuroblasts	+	0	0	0	0
Immature Glia	+	+/-	+/-	+/-	0
Primtive Neuro-epithelium	+	+/-	+/-	+/-	0
Peripheral Nerves	+	+	+	+/-	+/-
Mature Glia	+	+	+/-	+	+/-
Mature Neurons / Ganglia	+	+	+	0	0

+ : positive; +/- : variable positive; 0 : negative. GFAP: glial fibrillar acidic protein; NGFR: nerve growth factor receptor; NFL: neurofilament protein.

These results indicate the poly Sia is a sensitive, reliable, consistent and specific marker of immature neural elements in solid teratomas.

TUMORS OF THE ADRENAL GLAND

The adrenal gland consists of the medulla and the cortex. Pheochromocytomas are tumors of the medulla and stain positively with mAb 735. This property is retained in cell cultures of pheochromocytomas (Margolis and Margolis 1983).In contrast, adrenal carcinomas derived from the cortex exhibit no immunostaining with mAb 735.

ENDOCRINE TUMORS OF THE PANCREAS

Insulinoma is a commonly observed tumor among the endocrine pancreatic tumors and derives from the insulin-producing ß cells. During development of the pancreas both the exocrine and endocrine elements exhibit immunostaining for both NCAM and poly Sia (unpublished data, see also Lackie et al., this book). The endocrine islets in the pancreas of adults are positive for NCAM and show no reactivity for poly Sia by immunohistochemistry and immunoblotting (Zuber et al., 1991). The study of both benign and malignant human insulinoma revealed presence of NCAM but no poly Sia by immunohistochemistry and immunoblotting (Zuber et al., 1991).

THYROID TUMORS

The carcinomas of the thyroid can be divided histologically into: (a) papillary, (b) follicular, (c) anaplastic, and (d) medullary ones. Medullary carcinoma of the thyroid is usually diagnosed by the immunohistochemical detection of calcitonin which is not detectable in the other carcinomas. Poly Sia can be detected in the medullary carcinoma of the thyroid but not in the other carcinomas and therefore is another specific marker for the distinction among these diverse tumors arising in a single endocrine gland.

SMALL CELL LUNG CARCINOMA

Bronchopulmonary neuroendocrine neoplasms such as small cell lung carcinoma and bronchial carcinoid are assumed to derive from normal neuroendocrine cells of the lung. Currently, it is not clear if there is a common (neuroendocrine) cell of origin for the spectrum of neuroendocrine neoplasms of the lung.

A number of monoclonal antibodies have been raised against small cell lung carcinoma specimens or cell lines in various laboratories in the search for discriminating antibodies in lung tumor classification. Many of these monoclonal antibodies exhibited also reactivity towards neural tissues and were grouped as SCLC-cluster 1 antibodies (Souhami et al., 1987). Subsequent immunofluorescence and immunochemical studies on 3T3 cells transfected with a full-length clone of human NCAM revealed that they were reactive with NCAM (Patel et al., 1989). Monoclonal antibody 123C3, an NCAM-reactive SCLC-cluster 1 antibody, has been reported to stain in

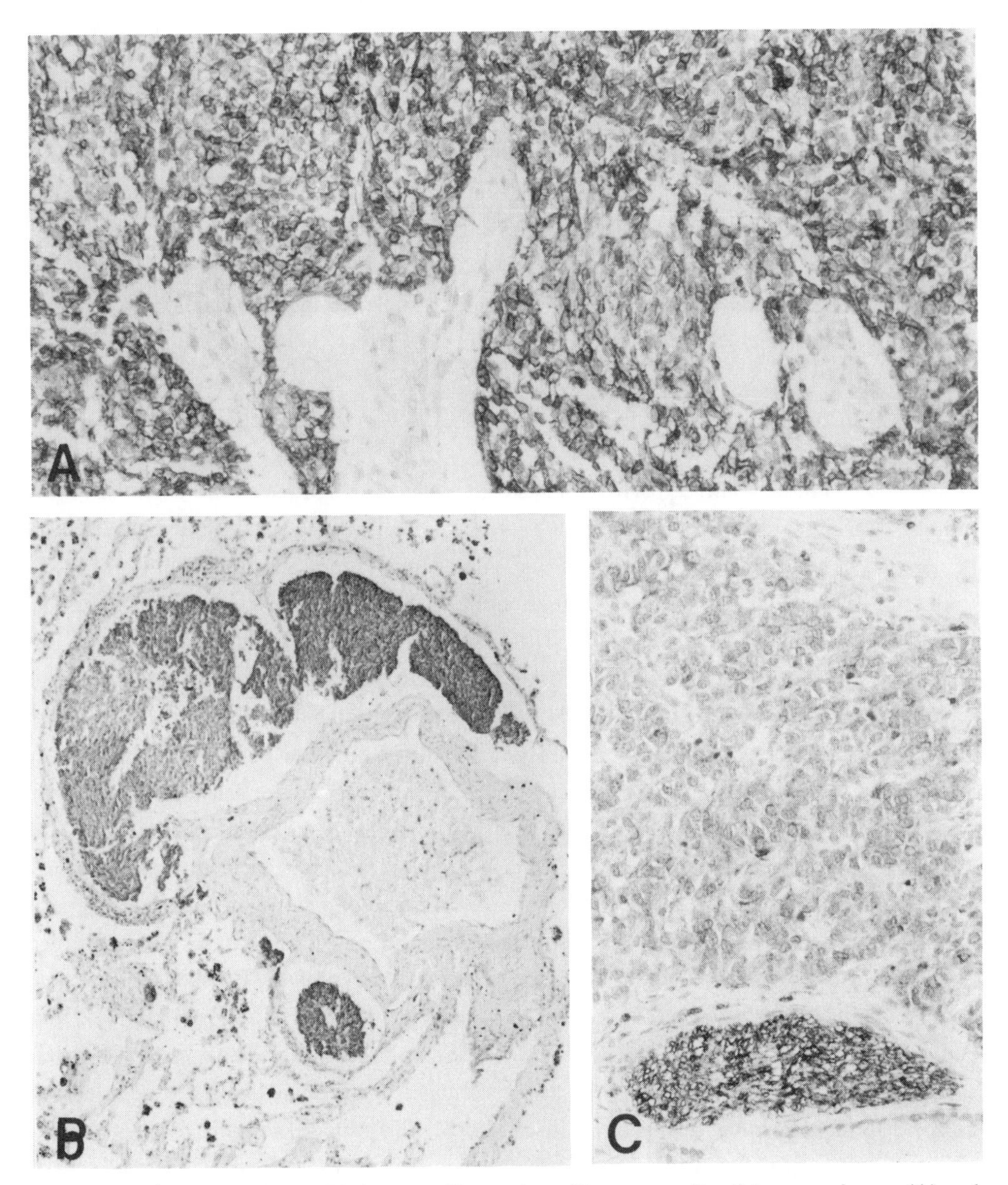

Fig. 6. Demonstration of poly Sia in a paraffin section of human small cell lung carcinoma (A) and tumor masses in a lymphatic vessel (B) adjacent to a blood vessel. A bronchial carcinoid is not stained by mAb 735 but a peripheral nerve show immunostaining.

addition to small cell lung carcinomas also bronchial carcinoids and a whole spectrum of neuroendocrine and non-neuroendocrine cell types (Schol et al., 1988). With the use of the gold-labeled mAb 735, poly Sia expression was studied in neuroendocrine and non-neuroendocrine tumors of the lung. We found all cases of small cell lung carcinoma immunohistochemically positive for poly SiA (Fig. 6) whereas the bronchial carcinoids (Fig. 6) as well as squamous cell carcinoma and adenocarcinoma were unreactive. Therefore, the monoclonal antibody against poly Sia can be used to distinguish small cell carcinoma from other types of lung tumors. Both, small cell lung carcinoma and bronchial carcinoids exhibited positive immunostaining for NCAM. However, they differed in the extent of polysialylation of NCAM.

POLYSIALIC ACID AND ITS SIGNIFICANCE FOR TUMOR GROWTH

Is the presence of cell surface poly Sia related to the invasive and metastatic potential of the tumors? At the first glance one could favour such a hypothesis. Poly Sia as reported above could be detected in highly invasive tumors such as Wilms tumor (Bitter-Suermann and Roth, 1987; Roth et al., 1988 a, b), neuroblastoma (Livingstone et al., 1989), and immature teratoma Metzman et al., 1991) and small cell lung carcinoma (Kibbelaar et al., 1989; Komminoth et al., 1991) and other tumors with malignant potential such as pheochromocytoma and medullary carcinoma of the thyroid. No poly Sia despite detectable NCAM was found in insulinoma and carcinoids. However, normal peripheral nerves and benign adenomas of the anterior pituitary were found to be positive for poly Sia. Obviously, there are many highly invasive human tumors which exhibit no cell surface poly Sia and, on the other hand, certain normal adult tissues which contain poly Sia. Therefore, a general role of poly Sia of NCAM in invasiviness and metastatic potential in tumors is difficult to assume. However, in specific situations the presence of poly Sia together with a variety of other factors may contribute to the invasive and metastatic growth potential of tumors. Recent results obtained with the small cell lung carcinoma speak in favour of this assumption (Scheidegger et al., submitted). We have established and characterized subclones of the NCI-H69 cell line derived from a human small cell lung carcinoma. These subclones expressed different levels of cell surface PSA, from zero to 95 % positive cells as determined by immunostaining with gold-labeled monoclonal antibody mAb 735, but all exhibited NCAM. This provided us with a model system to look at the effect of poly Sia expression on the behaviour of tumor cells from a common origin. Methods were applied to study the importance of poly Sia in modulating cell-cell and cell-matrix interactions. Under our experimental conditions cells with poly Sia on their surface formed significantly less aggregates than those which have no cell surface poly Sia. Further, poly Sia positive cells which had their cell surface poly Sia removed by endo N treatment had significantly more aggregates than the controls. This could allow the poly Sia positive cells to detach more easily from the primary tumor and provide a molecular explanation for how they complete the first step in the metastatic cascade. The established subclones of cell lines of small cell carcinoma of the lung with and without poly Sia should provide a valuable model systems to study the importance of this molecule in tumor cell biology.

ACKNOWLEDGEMENTS

The original research of the authors reported in this summary was supported by the Krebsliga beider Basel, Krebsliga des Kantons Zürich and the Swiss National Science Foundation.

REFERENCES

Acheson, A. and Rutishauser, U. (1988) J. Cell Biol. 106: 479-486.

Acheson, A., Sunshine, L.J. and Rutishauser, U. (1991) J. Cell Biol. 114: 143-153.

Bhattacherjee, A. K., H. J. Jennings, C. P. Kenny, A. Martin, and I. C. P. Smith, I. C. P. (1976) Can. J. Biochem. 54: 1-8.

Bitter-Suermann, D. and Roth, J. (1987) Immunol. Res. 6: 225-237.

Cunningham, K.L., Hoffmann, S., Rutishauser, U., Hemperley, J.J. and Edelman, G.M. (1983) Proc Natl Acad Sci USA 80: 3116-3120.

Doherty, P., Fruns, M., Seaton, P., Dickson, G., Barton, C.H., Sears, T.A. and Walsh, F.S. (1990) Nature 343: 464-466.

Egan, W., T.-Y. Liu, D. Dorow, J. S. Cohen, D. J. Robbins, E. C. Gotschlich, and J. B. Robbins, J. B. (1977) Biochemistry 16: 3687-3692.

Finne, J. (1982) J. Biol. Chem. 257: 11966-11970.

Garvin, A. J., Sullivan, J. L., Bennett, D. D., Stanley, W. S. and Inabett T, Sens, D. A. (1987) Am. J. Pathol. 129: 353.

Gonzalez-Crussi F: The pathology of Wilms' tumor, Wilms' Tumor (Nephroblastoma) and Related Renal Neoplasms of Childhood. Edited by F. Gonzalez-Crussi. Boca Raton, FL, CRC Press Inc., 1984, 177-206.

Hall, A.K., Nelson, R. and Rutishauser, U. (1990) J Biol Chem 110: 817-824.

Hallenbeck, P.C., Vimr, E.R., Yu, B., Bassler, B. and Troy, F.A. (1987) J. Biol. Chem. 262: 3553-3561.

Hoffmann, S.B. and Edelman, G.M. (1983) Proc Natl Acad Sci USA 80: 5762-5766.

Inoue, S. and Iwasaki, M. (1978) Biochem. Biophys. Res. Commun. 83: 1018-1023.

Ito, J. and Johnson, W. W. (1969) J. Natl. Cancer. Inst. 42: 77.

James, W. M. and Agnew, W. S. (1988) Biochem. Biophys. Res. Commun. 148: 817-826.

Kibbelaar, R.E., Moolenaar, C.E.C., Michalides, R.J.A.M., Bitter-Suermann, D., Addis, B.J. and Mooi, W.J. (1989) J. Pathol. 159: 23-28.

Kitajima, K., S. Inoue, Y. Inoue and Troy, F. A. (1988) J. Biol. Chem. 263: 18269-18276.

Komminoth, P., Lackie, P.M., Bitter.Suermann, D., Heitz, P.U. and Roth, J. (1991) Am. J. Pathol. 3: 297-304.

Kumar, S., Carr, T., Mardsen, H. B., and Calabuig-Crespo, M. C. (1986) J. Clin. Pathol. 39: 51.

Lackie P. M., Zuber C., Roth J. (1990) Development 110: 933-947.

Landmesser, L., Dahm, L., Tang, J. and Rutishauser, U. (1990) Neuron 4: 655-667.

Livingstone B., Jacobs J.L., Glick M.C. and Troy F. A. (1989) J. Biol. Chem. 263: 2999-3003.

Margolis, R. K. and Margolis, R. U. (1983) Biochem. Biophys. Res. Commun. 116: 889-894.

Metzman, R. A., Warhol, M. J. and Roth, J. (1991) Modern Pathol. 4: 491-197.

Mierau, G.W., Beckwith, J.B. and Weeks, D.A. (1987) Ultrastruct. Path. 11: 313-333.

Murray, B.A. and Jensen, J.J. (1992) J. Cell Biol. 117: 1311-1320.

Nybroe, O., Linnemann, D. and Bock, E. (1988) NCAM biosynthesis in brain. Neurochem Int 1988; 12: 251-262.

Patel, K., Moore, S.E., Dickson, G., Rossel, R.J., Beverley, P.C., Kemshead, J.T. and Walsh, F.S. (1989) Int. J. Cancer 44: 573-578.
Pinfen Yang, , Xinghua Yin, and Rutishauser, U. (1992) J. Cell Biol. 116: 1487-1496.
Rohr, T. E. and Troy, F. A. (1980) J. Biol. Chem. 255: 2332-234.
Roth, J. and Zuber, C. (1990) Lab Invest 62, 55-60.
Roth, J., D. J. Taatjes, D. Bitter-Suermann and Finne, J. (1987) Proc. Natl. Acad. Sci. U.S.A. 84: 1969-1973.
Roth J, Wagner P, Zuber C, Weisgerber C, Heitz PU, Goridis C, and Bitter-Suermann D. (1988a) Proc. Natl. Acad. Sci. USA 85: 2999 (1988 a).
Roth, J., Zuber, C., Wagner, P., Blaha, I., Bitter-Suermann, D. and Heitz, P. U. (1988b). Am. J. Path. 133: 227.
Rothbard, J.B., Brackenbury, R., Cunningham, B.A. and Edelman, G.M. (1982) J. Biol. Chem. 157: 11064-11068.
Rutishauser, U., Grumet, M. and Edelman, G.M. (1983) J Cell Biol 97: 145-152.
Rutishauser, U., Watanabe, J., Silver, J., Troy, F.A. and Vimr, E.R. (1985) J Cell Biol 101: 1842-1849.
Sadoul, R.V., Hirn, M., Deagostini-Bazin, H., Rougon, G. and Goridis, C. (1985) Nature 304: 347-349.
Sariola H, Eckblom P, Rapola J, Vaheri A, Timpl R: Am J Pathol 1985, 118: 96-107.
Saxén L: Organogenesis if kidney. In Development and Cell Biology Series, edited by Barlow P W,Green P B, Wylie C C, vol 19, pp 1-165. Cambridge University Press, 1987.
Schmidt, D., Dickersin, G.R., Vawter, G.F., Mackay, B. and Harms, D. (1982) Pathobiol. Ann. 1982, 12:281-300.
Schol D.J., Mooi, W.J., Van der Gugten, A.A., Wagenaar, S.S. and Hilgers, J. (1988) Int. J. Cancer Suppl. 2: 34-40.
Souhami, R.L., Beveley, P.C.L. and Bobrow L. (1987) Lancet 2:325-326.
Thurlbeck, W.M. and Scully (1960) Cancer 13: 804
Troy, F. A. (1979) Ann. Rev. Microbiol. 33: 519-560.
Vimr, E.R., McCoy, R.D., Vollger, H.F., Wilkison, N.C. and Troy, F.A. (1984) Proc. Natl. Acad. Sci . USA 81: 1971-1975.
Zuber, Ch., and J. Roth, Eur. J. Cell Biol. 51, 313-321, 1990.
Zuber, C., Lackie, P. M., Klöppel, G., Heitz, P.U. and Roth, J. (1991) Verh. Dt. Ges. Path. 75: 279.
Zuber, C., Lackie, P. M., Catterall, W. A. and Roth, J. (1992) J. Biol. Chem. 267: 9965-9971.

Index